Introduction to Space Science

William M. Kaula
AN INTRODUCTION TO PLANETARY PHYSICS

Thomas L. Swihart
ASTROPHYSICS AND STELLAR ASTRONOMY

Frank D. Stacey
PHYSICS OF THE EARTH

R. C. Whitten and I. G. Poppoff
FUNDAMENTALS OF AERONOMY

Robert C. Haymes
INTRODUCTION TO SPACE SCIENCE

A. J. Dessler and F. C. Michel
PARTICLES AND FIELDS IN SPACE

S. Matsushita and W. H. Campbell
INTRODUCTION TO THE MAGNETISM OF THE EARTH IN SPACE

Introduction to Space Science

Robert C. Haymes

Rice University
Houston, Texas

John Wiley & Sons, Inc.

New York   London   Sydney   Toronto

Library of Congress Catalog Card Number: 78-140550 ISBN 0-471-36500-9

Printed in the United States of America
10 9 8 7 6 5 4 3 2

To my wife Jamie.
Without her constant encouragement,
this book would never
have been completed.

Preface

Space science may be defined as the scientific study of cosmic physical systems. It is among the oldest of intellectual pursuits and yet is one of the most challenging endeavors of the modern era.

The use of the word cosmic is intended to emphasize the importance of large-scale, collective phenomena; for example, the properties of collisionless plasmas of cosmic dimensions that have magnetic fields embedded in them cannot be studied in the laboratory because of the difficulty in scaling, yet shocks formed in such plasmas appear to be a very general and important feature of the behavior of matter. They may be responsible for the apparent acceleration of charged particles in the vicinity of the earth and also for the heating of electrons to high temperatures in the neighborhood of the huge discrete radio sources found throughout the known universe.

Space science is an interdisciplinary field; an understanding of many sciences, such as physics, chemistry, and geology, is required in order to understand our total environment. Space science draws on all of these, but the whole is greater than the sum of its parts.

This book is an outgrowth of a one-year course of the same name that my colleagues and I developed in the Department of Space Science at Rice University. The course is open to seniors and to first-year graduate students. A familiarity with college physics, such as the usual first-year course, is assumed, but no previous acquaintance with any of the areas of space science is necessary for the course. Another assumption is that the student is familiar with ordinary differential equations.

The planetary system and the interplanetary medium are discussed in the first 10 chapters. The earth and the near-earth space are used as models whenever possible. Most of the material contained within the first 10 chapters is usually covered during the first semester; the remainder, which emphasizes the sun and the contents of the space beyond the solar system, occupies the second term.

It will be seen that the length of the various chapters is uneven. The fundamental cause is that some areas of space science are better defined and more knowledge is available than in others. I am most grateful to my colleagues within the Department at Rice, most notably H. R. Anderson, P. A. Cloutier, A. J. Dessler, J. W. Freeman, Jr., D. Heymann, R. P. Kovar, N. Soga, and W. H. Tucker, who have made many valuable contributions to various sections of the book. The responsibility for any errors or for obscurity of presentation, however, is mine alone.

We are not concerned, in space science, with technology. Very little will be found in this volume concerning observatories or vehicles such as balloons, rockets, satellites, or deep-space probes; the emphasis throughout is on the basic principles of the physical phenomena, together with a little speculation concerning future possibilities. We should note, however, that science and technology go hand in hand. One cannot proceed without advances in the other.

Space science may well result in some practical applications. Indeed, it has already resulted in many useful things, ranging from navigation and the calendar to the energy released by nuclear fusion and to improved global communication and weather forecasting.

Applications, however desirable, are not the main goal of science. Space science, like any of the other physical sciences, attempts to increase our understanding of nature and also to describe it as accurately and concisely as possible.

Applications of knowledge to the betterment of man's life cannot result unless there is a body of knowledge to draw upon. Space science represents the attempt by man to extend his knowledge of the physical universe. In my view, it requires no further justification than this for its continued vigorous existence.

Space science has raised more questions than it has answered; this is the excitement of the field. Many problems will undoubtedly remain unsolved for a very long period, but the attempt to resolve them is both fascinating and rewarding. It is the purpose of this book to introduce the reader to our quest for knowledge of the universe, and hopefully this introduction will lead to a sympathetic understanding of that quest.

R. C. HAYMES

Houston, Texas
March, 1971

Contents

Chapter 1

CELESTIAL COORDINATES AND TIME

1.1 The Celestial Sphere

In order to discuss the various celestial objects, we must first locate them as precisely as possible. The sky is most easily thought of in terms of a spherical surface, called the celestial sphere, when we attempt to describe the angular positions of celestial objects such as the stars. The celestial sphere is defined as a great sphere of infinite radius, with the observer at the center.

On the earth we may describe the positions of places in terms of latitude and longitude on a globe. A similar description may be made for celestial objects in terms of their coordinates on the celestial sphere. This description does not yield information on the distances to the stars, however, for all the stars are assumed to be infinitely far away, "on" this rigid surface.

1.2 Coordinate Systems

The direction of a star may be considered as the coordinates of the point on the celestial sphere. The coordinate system employed, if it is to be generally useful, should be independent of the various positions of different observers on the earth. It is also desirable that the system be independent (or nearly so) of the motions of the earth itself. In other words, the coordinates must be permanently fixed into the celestial sphere and not move with respect to it.

There are three coordinate systems now in use in astronomy. They include the equatorial system, the ecliptic system, and the galactic system of coordinates. The first two will give the same results for all observers on the earth; the third is valid for all observers within the solar system.

1.3 Equatorial Coordinate System

The equatorial system is the most popular coordinate system. It is most easily thought of in terms of the geographical coordinate system used to locate points on the earth. In the geographical system the *latitude* of a point is its distance in degrees north or south from the equator, so that its value ranges from 0° (at the equator) to 90° at either pole. The poles are distinguished by the letters N and S (north and south). The *longitude* of the point is measured in degrees east or west, along the equator, to the meridian containing the

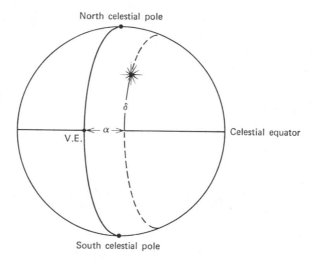

Figure 1.1   The right ascension α and declination δ of a star in the northern hemisphere of the celestial sphere. The celestial equator is in the plane of the earth's equator.

point (from some reference meridian). A meridian is half a great circle joining the poles and passing through the point. A meridian is perpendicular to the equator; the reference, or *prime* meridian, is chosen as that which passes through the original site of the Royal Greenwich Observatory in England.

Now in the equatorial system the angular coordinate corresponding to geographical longitude is called the *right ascension,* α. The coordinate corresponding to latitude is the *declination,* δ. Thus the declination of a star is its angular distance (in degrees) measured north (+) or south (−) of the celestial equator. The celestial equator, by definition, is a great circle 90° away from either celestial pole, and it is in the plane of the earth's equator (Figure 1.1).

1.4   Celestial Poles

The north and south celestial poles are found by extending the rotational axis of the earth to the celestial sphere. Since the stars "rise" in the east and "set" in the west, the celestial sphere appears to rotate about the earth. The celestial poles, however, do not appear to rotate; the stars describe circles, called diurnal circles, in time about them (Figure 1.2).

The north celestial pole is easily found from the position of the "north star" Polaris, which is now almost exactly at that pole. Polaris has a declination of +89°; it lies only 1° from the north celestial pole.

Figure 1.2   Photograph of polar star trails (time exposure). The stars seem to circle about the (north) pole. Polaris is only 1° from that pole and barely seems to move at all.

1.5 Right Ascension

The reference ("zero") meridian of right ascension is an arc of a great circle on the sphere that passes through the celestial poles and through a point called the vernal equinox (V.E.). The V.E. is a *point on the celestial sphere* (just as a star is); it is the direction or place where the sun crosses the celestial equator northward sometime on or about 21 March. (The variation in dates is caused by the plan of leap years.)

The right ascension of a star is thus the angular distance measured from the V.E. along the celestial equator to the great circle passing through the celestial poles and the star (see Figure 1.1). By convention, α is measured eastward from the V.E. to this great circle, which is known as the "hour circle" of the star.

Right ascension is not customarily measured in degrees. Instead "sidereal time units" are employed. The value of α ranges from $0^h0^m0^s$ at the V.E., to $24^h0^m0^s$. In this system of units one hour (written as 1^h) corresponds to $15°$. To give an example, $\alpha = 12^h0^m0^s$ at the autumnal equinox. The autumnal equinox is the point where the sun crosses the celestial equator southward on or about 22 September. The superscripts are a convention of astronomers to denote hours, minutes, and seconds of time. We shall postpone a discussion of the measurement of time until Section 1.16.

The stars, to a very good approximation, do not change their equatorial coordinates in time. On the other hand, the members of the solar system ("planets" means wanderers) do sensibly move with respect to the celestial sphere. Thus, to give another example, we, may cite some of the ever-changing coordinates of the sun:

Position on 21 March (V.E.): $\alpha = 0^h$, $\delta = 0°$.
Position on 23 June (summer solstice): $\alpha = 6^h$, $\delta = +23.5°$.
Position on 22 September (autumnal equinox): $\alpha = 12^h$, $\delta = 0°$.
Position on 22 December (winter solstice): $\alpha = 18^h$, $\delta = -23.5°$.

The summer and winter solstices are those points where the sun is most northerly and most southerly, respectively.

1.6 Units of Angular Measurement

Since total solar eclipses occur in the manner that is observed, we may conclude that the sun and the moon subtend equal angles at the earth. That is, the two objects must subtend just about the same angle, since the moon just masks the sun during such an eclipse. This angle turns out to be about 30 minutes of arc, so that both the sun and the moon appear to $\frac{1}{2}°$ "wide." Thus, for example, Polaris is "two moons" (or suns) away from the north celestial pole. Another useful scale of angular measure is provided by the

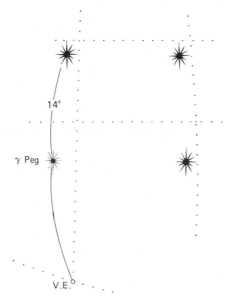

Figure 1.3   A star chart that shows the relationship of the vernal equinox to the Square
of Pegasus.

clenched fist of an observer. If you hold your arm straight out in front of you,
your clenched fist will be approximately 10° across; your fist subtends an arc
of roughly 10° at your eyes.

The common observation that both the sun and moon appear larger when
viewed near the horizon is evidently an optical illusion caused by the availa-
bility of reference objects near the horizon. Photographs do not show any
change in size with angle from the vertical.

On a star map the vernal equinox is located about 14° south of a configuration of stars called the Square of Pegasus (Figure 1.3). This angle is about
equal to one of the sides of the Square, which helps in locating the V.E. in
the absence of the sun. Once the V.E. is found, the right ascension of any star
may be measured in terms of the angle between the hour circle of the star
and that of the V.E. This angle, because of the units of α, is called the "hour
angle" of the star.

It is interesting that the smallest angular distance easily perceived by the
unaided eye is about 0.1 "sun," or about 3 arc-minutes. Modern earth-based
telescopes are reliable to 0.01 arc-second, a factor of 18,000 smaller than the
naked-eye value.

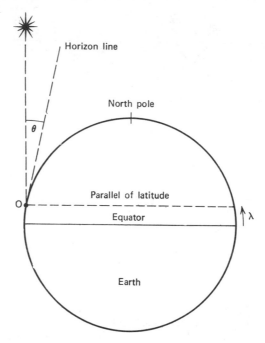

Figure 1.4   An observer at O sees a star at an angle θ above the horizon (θ = altitude angle) that is equal to his latitude λ.

1.7   Measurement of Latitude

An observer near the north pole of the earth sees Polaris to be directly overhead. An observer on an idealized smooth spherical earth at the equator would see Polaris just above the horizon (Figure 1.4). A generalization of this results in the fact that the *altitude angle of the north celestial pole is equal to the observer's latitude*. This fact is extremely important for navigation. For many practical purposes the altitude angle (the angle measured from the horizon to the star) of Polaris may be substituted instead. Unfortunately there is no bright star in the southern hemisphere as useful for this purpose as Polaris is in the northern hemisphere. The configuration of stars called the Southern Cross is centered at a declination of about $-59°$, much too far away from the southern celestial pole to serve.

1.8   Measurement of Declination

Another useful relation is that the difference between the observer's latitude and the declination of a star is equal to the "zenith distance" at meridian transit of the star. The zenith distance or angle is measured from the zenith; it is the complement of the altitude angle we met above. A star is said to "transit the meridian" when it crosses the observer's meridian, the great circle that passes through the two poles and the observer's zenith. The star *culminates* then, for it reaches its maximum altitude angle at that time.

To cite some examples, an observer at a latitude of 40° N sees the bowl of the Big Dipper ($\delta = +58°$) culminate at a zenith distance of $40° - 58° = -18°$, or 18° north of the zenith. The Crab nebula ($\delta = +22°$), on the other hand, achieves a zenith distance of $40° - 22° = +18°$, so that the Crab culminates 18° south of zenith at a latitude of 40° N. It is obvious that measurement of the minimum zenith distance of a celestial object permits measurement of the declination, once the observer's latitude is known. Thus both α and δ are "directly observable" quantities.

1.9   Wandering of the Celestial Poles

It is an unfortunate fact that the equatorial system is not exactly fixed in the celestial sphere, because of the precessional motion of the earth's axis. That is, the axis of rotation moves relative to the stars.

Any rotating asymmetric distribution of matter that is acted on by external torques will experience precession and nutation of its rotational axis. A spinning top on the earth is a good example.

In the case of the earth the rotation of the planet causes an "equatorial bulge," owing to the centrifugal forces that arise from the rotation. The equatorial diameter is 26.70 statute miles greater than the polar diameter of 7899.98 statute miles. In addition to the oblateness of the earth caused by its rotation, observations of the changing planes of artificial satellite orbits reveal that the earth is somewhat "pear-shaped"; more mass is present south of the equator than north of it. We shall return to this interpretation of the observations in Chapter 2.

The combined torques of the sun and the moon cause the terrestrial axis to precess. The moon is most important because it is closest, but both bodies exert torques on the bulge that attempt to align the earth's equator with the ecliptic. The ecliptic is the apparent annual path of the sun's center on the celestial sphere; the moon and the planets also all follow the ecliptic quite closely. Precession results from the torques because of conservation of angular momentum $\mathbf{L}$; $\mathbf{L}$ is parallel to the angular velocity ω and is therefore along the rotational axis of the earth (Figure 1.5). The radius of the $\mathbf{L}$-executed circle is just the obliquity of the ecliptic. In other words, the torques continuously attempt to align the earth's spin axis so that it is perpendicular to the ecliptic, but "conservation of $\mathbf{L}$" prevents the alignment.

This precession has a period of about 26,000 years, so that the equinoxes, the two points of intersection of the equator and the ecliptic, slide westward

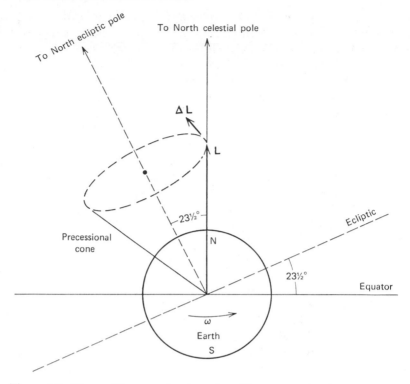

Figure 1.5 The earth's precessional motion. The *magnitude* of the angular momentum **L** is constant, but the direction of **L** constantly changes; Δ**L** is always perpendicular to **L** and this produces a conical motion of the celestial pole, against the background of the fixed stars. The time required for the celestial pole to move 360° (in an arc of radius $23\frac{1}{2}$° about the ecliptic pole) is approximately 26,000 years.

along the ecliptic by about 50.26 arc-seconds per year. Thus, in this approximation, the celestial poles would execute circles about the ecliptic poles once every 26,000 years with a radius of 23.5°, the inclination or "obliquity" of the ecliptic (Figure 1.6).

Nutation also occurs; the angular momentum vector appears to "nod up and down" relatively rapidly as it circles the ecliptic poles. The inclination of the lunar orbit to the ecliptic is chiefly responsible for the nutation, the period of which is 18.6 years. The maximum nutational departure from the mean circle executed by the poles because of lunar and solar effects is only 9".23.

The additional effects caused by the other planets of the solar system increase the nutation and make the precessional motion irregular. Therefore the motion of the poles is not quite closed on the celestial sphere. In the year

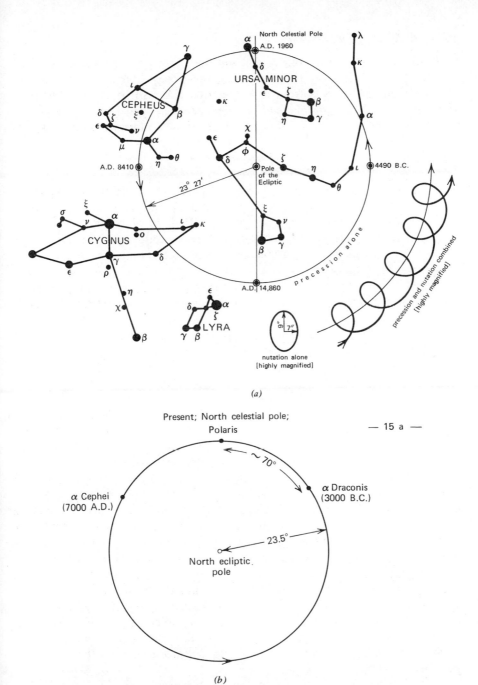

Figure 1.6 The precession of the north celestial pole about the north ecliptic pole; one "revolution" takes about 26,000 years. The picture is idealized, since the precessional path is not quite closed. Reprinted from D. H. Menzel, *A Field Guide to the Stars and Planets* (H. A. Rey, illustrator), Houghton Mifflin Co., 1964.

2100 Polaris will be only "one sun" from the north celestial pole, half the present angular distance. From then on the pole will move away from Polaris. Some 5000 years from now the pole star will be Alpha Cephei, which is about 28° from Polaris. The pole star was Alpha Draconis around 3000 B.C., 5000 years ago. The celestial poles have moved roughly 70° in recorded history.

1.10   Exact Stellar Coordinates

Whenever the exact position of a celestial object is required, the *date* of the equatorial coordinates must be specified in addition to α and δ. In most of the older star catalogs we find $\alpha(1900.0)$ and $\delta(1900.0)$, which means that the data refer to the values on 1 January 1900. Newer catalogs list values for the epoch 1950, on which we make small corrections to find the exact α and δ at the time of observation. The correction factor for α is $T \times 3^s.07$, where T is measured in years from 1 January 1950 and the numerical factor represents the annual displacement of the V.E. in right ascension. The corresponding correction for declination is $T \times 20.05$ arc-seconds. These figures apply only to the coordinates of the vernal equinox. Stellar coordinates will be changed by different amounts, depending on α and δ as well as on the motion of the star.

1.11   Ecliptic Coordinate System

This system has now been largely replaced by the equatorial system described above, but it is still useful for discussion of solar system phenomena. The ecliptic system used coordinates called *ecliptic longitude* and *ecliptic latitude* and it was originally introduced for planetary studies.

The sun, moon, and planets (the brightest objects in the skies) are all in the *zodiac*, the belt of sky 16° wide centered on the ecliptic (Figure 1.7). If one looks upward at dusk, the (fading) sun, the moon, and the bright planets all seem to lie on almost a "straight line," or common arc segment. It is this fact that led to the introduction of the ecliptic coordinate system; all these objects may be approximately located in terms of one coordinate only, the ecliptic longitude. This coordinate is the angular distance from the vernal equinox, measured in degrees eastward along the ecliptic.

The other coordinate, ecliptic latitude, is the angular distance of the object from the ecliptic (Figure 1.8). The north and south ecliptic poles are two points 90° distant from the ecliptic. Since the ecliptic is inclined 23° 27′ 8″.26 to the equator, these poles are about 23.5° distant from the celestial poles (see Section 1.9).

The ecliptic system is based on the earth's *revolution* about the sun, whereas the equatorial system is based on the *rotation* of our planet about its axis. That the ecliptic system is based on revolution may be seen from the

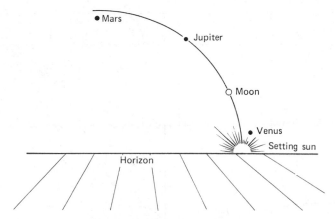

Figure 1.7 Schematic illustration of the zodiac; all of the brighter celestial object's (the solar system members) appear to be nearly on a common segment of arc.

fact that the ecliptic may be defined as the great circle resulting from the intersection of the earth's *orbital* plane with the celestial sphere. The ecliptic system is still used in planetary astronomy and occasionally in describing the motion of interplanetary spacecraft.

1.12 Galactic Coordinate System

When one studies galactic phenomena, such as the galaxy's brightness in the radio, optical, or gamma wavelengths, it is more convenient to use a galactic system of coordinates rather than a system based on the earth's rotation or revolution.

The various coordinate systems may be distinguished from each other in terms of *reference circles*. In the equatorial system the reference circle is the earth's equator; the reference circle for the ecliptic system is the ecliptic. The reference circle for the galactic system is the galactic equator.

1.13 Galactic Equator

The galactic equator is a great circle that most nearly follows the center line of the visible Milky Way. The Milky Way is a belt of sky that seemingly glows; it is the result of the concentration in the galactic "plane" of a vast number of unresolved stars (Figure 1.9).

Since intervening bands of dust obscure large portions of Milky Way from our optical view, the most precise location of the galactic equator has been at radio wavelengths that are not seriously attenuated by dust. The best results have come from locating the circle on the celestial sphere along which radiation at a wavelength of 21 cm has the strongest intensity. As we shall

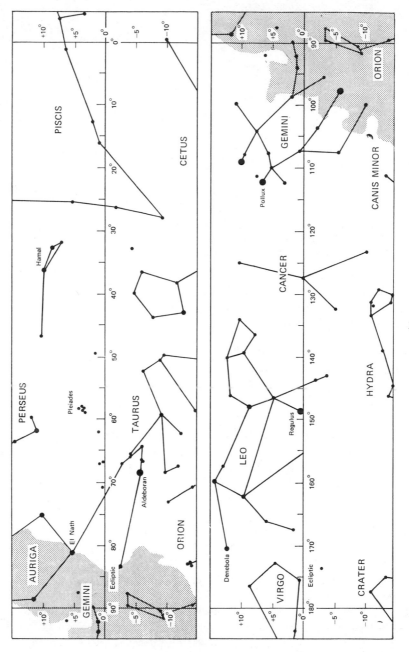

(a)

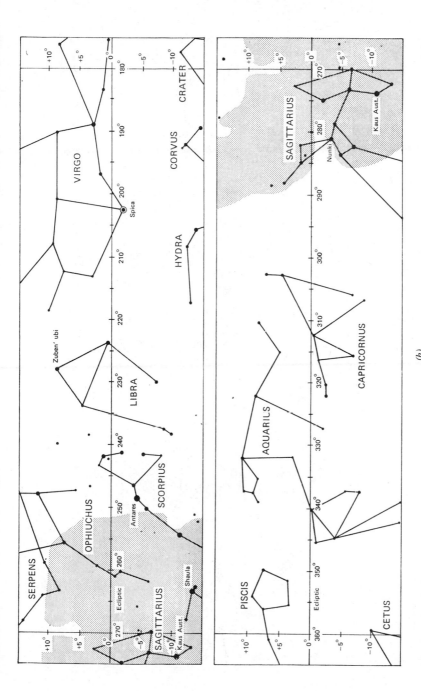

Figure 1.8 Ecliptic-coordinates star map. Ecliptic longitude and latitude in the zodiac are shown. From D. H. Menzel *Field Guide to the Stars and The Planets* (H. A. Rey, illustrator), Houghton Mifflin Co., 1964.

(b)

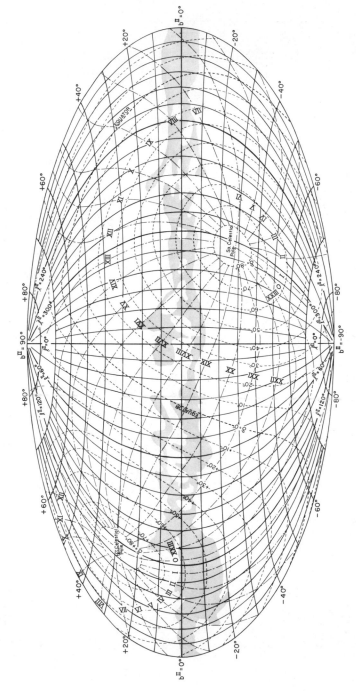

Figure 1.9 Galactic coordinates map. The general outline of the Milky Way is shown in the shaded area. Also shown are celestial equatorial coordinates.

see later, radio waves that are 21 cm in length may be emitted by cold neutral hydrogen atoms.

The galactic equator is sharply inclined to the celestial equator. The angle of inclination is about 62°. The two intersection points are located in the stellar configurations or constellations of Aquilla and Monoceros, respectively.

1.14   Galactic Latitude

By definition, the north galactic pole (N.G.P.) lies in that hemisphere that includes the north celestial pole, although the two poles are 62° apart. The N.G.P. lies at $\alpha = 12^h49^m$, $\delta = +27.4°$ (1950.0), while the south galactic pole is found at $\alpha = 0^h49^m$, $\delta = -27.4°$ (1950.0). *Galactic latitude*, written nowadays as b^{II}, is measured from the equator (0°) to either galactic pole ($\pm90°$).

1.15   Galactic Longitude

The values of *galactic longitudes*, l, were revised in 1961. In the new system the coordinate l^{II} is measured eastward along the galactic equator from a prime meridian that passes through the galactic poles and through a strong discrete radio source believed to be in the direction of the galactic center. The strong radio source is in the constellation Sagittarius and is called Sagittarius A. The equatorial coordinates of Sagittarius A are $\alpha = 17^h42.4^m$ and $\delta = -28° 55'$ (1950.0). (This declination tells us that the galactic center is best studied from the southern hemisphere of the earth, for the center culminates highest there.) Galactic longitude is measured in degrees, and ranges in value from 0 to 360°. If one adds 32° to the old longitudes, known as l^I, the new values are obtained.

1.16   Time and its Measurement

The measurement of time is based on the rotation of the earth and therefore on the apparent rotation of the celestial sphere. The interval between successive meridian transits of a selected reference object is defined as a *day*. The reference object, such as a fixed star or the sun, that is chosen will determine the "kind" (i.e., sidereal or solar) of day we are dealing with. Each kind of day is divided into 24 equal parts, called hours. We may now define time itself (in geometrical terms) as the hour angle of the reference object (Figure 1.10).

However, the rotational speed of the earth varies irregularly, although slightly; for example, the melting of the Greenland ice caps changes the moment of inertia of the planet and hence the angular speed of rotation. The lunar (and, to a lesser degree, the solar) tides are also slowing the earth through friction, mainly in the shallow Bering sea. It has therefore become

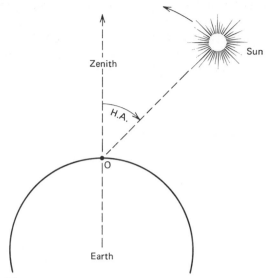

Figure 1.10   Geometrical definition of solar time; if
the hour angle (H.A.) of the sun is zero, it is local
noon for an observer at O.

clear that some more stable time reference will be required, and it appears
that the periods of the electromagnetic waves emitted by atoms and molecules
will serve this purpose. Lasers or masers are devices that presently appear
to be promising in this regard. Even more promising is the Mossbauer effect
of recoilless radiation, for energy differences corresponding to one part in
10^{12} can be detected. Since the energy of a photon is linearly related to its
frequency, hence to the period of the radiation, the Mossbauer effect holds
promise for clocks that are stable to 1 second in 30,000 years. This is about a
factor of 15,000 more stable than time measurements based on the rotation
of the earth.

1.17   Sidereal and Solar Time

If the reference object or point is the vernal equinox, the time is called
sidereal time. If, on the other hand, the reference object is the sun, we have
solar time.

Stars "obey" sidereal time because the V.E. is fixed, to a first approximation, in the celestial sphere, as are the stars. We define *sidereal noon* to occur
when the V.E. undergoes meridian transit, and we further define sidereal noon
to be the beginning of a sidereal day. If we had a sidereal clock, it would
therefore read $0^h00^m00^s$ at sidereal noon. A sidereal clock is a 24-hour
clock; it reads up to $23^h59^m59^s.9$. Since the stars obey sidereal time, sidereal
time is very important for navigation. The precession of the equinoxes

affects accurate work; the length of the mean sidereal day is about 8 milli-seconds shorter than the interval between two consecutive transits of a given star across the meridian of the observer.

Whereas stars are "regulated" by sidereal time, human activities are not, since sidereal noon occurs at various times in the day or night. Humans therefore follow solar time. The solar day is the 24 hours between suc-cessive meridian transits of the sun. Unlike the sidereal day, the solar day is defined to start at midnight ($0^h00^m00^s$). Meridian transit of the sun, called "local noon," is labeled $12^h00^m00^s$ in this system and it is therefore different from the sidereal labeling.

1.18   Relationship between Sidereal and Solar Days

It is evident that the sidereal day is a little shorter than the solar day. On 21 March, for example, when the sun crosses the equatorial plane north-ward, the V.E. is on the meridian at local noon. We see immediately from this that the sidereal and solar clocks disagree by 12 hours on 21 March.

On the following day, 22 March, however, the V.E. transits the meridian before the sun does. This is because the earth has moved in its orbit about the sun in the interim, as well as having executed rotation about its axis. Hence the earth must rotate a little more before the sun is on the meridian. In other

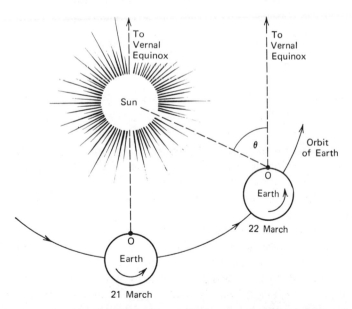

Figure 1.11   The sidereal day is shorter than the solar day because of the earth's motion about the sun; the Vernal Equinox transits the observer's (O) meridian on 22 March before the sun does. The difference is given by the angle θ, which is exaggerated in the diagram; θ is about $1°$ per day.

words, by the time local noon arrives, the new sidereal day is already a few solar minutes old (see Figure 1.11).

We may estimate how many minutes old it is. There are (approximately) 365 days in a solar year. Thus the earth moves $360°/365 \approx 1°$ per day. Since there are 15° to each hour of time, 1° corresponds to about 4 minutes. More precisely, it turns out that one sidereal day is $23^h56^m4^s.091$ of mean solar time.

This 4 minutes per day may seem small at first glance. However, another way of saying this is that a given star rises about 2 hours earlier each month. This gives rise to the concept of "seasonal constellations"; some stars will be overhead at night only during the winter, while others can be optically observed only during the summer.

Now our two clocks (sidereal and solar) agree only once a year, at the autumnal equinox, approximately 22 September. On that date the V.E. transits the meridian at local midnight; both clocks read $0^h00^m00^s$ at that instant. The maximum difference between the readings of the ordinary solar and 12-hour sidereal clocks occurs on 21 March, when, as noted before, the difference is 12 hours.

1.19   Right Ascension and Sidereal Time

In Section 1.5 we noted that the zero meridian of right ascension passes through the celestial poles and through the V.E. Hence we see that as the V.E. transits the meridian, the right ascension of this point is $0^h00^m00^s$. By definition, we also have that the sidereal time is $0^h00^m00^s$ at this instant. More generally, the *right ascension of a star is equal to the sidereal time at the instant of meridian transit.*

This principle may be extended to cover stars that are not on the observer's meridian. The relationship is

$$\text{S.T.} = \alpha + \text{H.A.}, \tag{1.1}$$

where H.A. is the hour angle of the star and S.T. represents sidereal time. Consequently α may be determined by the use of a sidereal clock. Conversely—and this is a more correct ordering—our sidereal clock may be set if we know the right ascension of a point on the celestial sphere that is at the point of culmination.

The fundamental first step is an accurate measurement of the position of the vernal equinox. This may be obtained from very careful observations of the apparent coordinates of the sun throughout the year, so that the ecliptic may be accurately plotted on the celestial sphere; the place where it crosses the celestial equator (northward bound) is the location of the vernal equinox. Once the position of the V.E. is known, we know where the coordinate $\alpha = 0^h00^m00^s$ lies, and our sidereal clock may consequently be set, enabling measurements of right ascension to be made for any point on the celestial sphere. The results of the ecliptic observations, along with other similarly useful sightings, are tabulated in nautical almanacs such as the *American Ephemeris* for every day in the year.

1.20 Examples of Time System Conversion

The right ascension of the Crab nebula is shown on star maps to be approximately 5^h30^m. We ask, At what local time of day will the Crab transit the meridian on (a) 21 May and (b) 21 January?

(a) Local time is the hour angle of the sun plus 12 hours. We defer a more precise definition of this "time" until Section 1.22, where we shall see that the "time" in question is known as mean solar time. Now we note that 21 May occurs 62 days after 21 March. Therefore the difference between the sidereal and solar clocks on 21 May is about 62 days $\times$ 3^m56^s/day = 244^m = 4^h04^m, which confirms our 2 hours per month estimate.

Now we may use (1.1) to solve this problem. We see that S.T. = α at meridian transit, so that the local time of culmination on 21 May will be $5^h30^m + 12^h00^m - 4^h04^m = 13^h26^m$. Since the local time at sidereal noon on 21 March is 12^h00^m, our result may be re-expressed as the more familiar 1:26 P.M.

(b) We notice that 21 January occurs 121 days after 22 September. The local time at sidereal noon is 0^h00^m at the autumnal equinox. Now the difference between the sidereal and solar clocks on 21 January = 121 days $\times$ 3^m56^s/day = 7^h54^m, so that the local time = $5^h30^m + 0^h00^m - 7^h54^m$ = 9:28 P.M. on 20 January, the preceding evening. On 21 January, therefore, the local time of meridian transit for the Crab nebula is 9:24 P.M. It is instructive to recall from Section 1.8 that the nebula will be observed 18° south of the zenith for an observer located at a latitude of 40° N on these two (or any other) meridian transits.

The reader will observe that we used the two different equinoxes as reference dates for the two computations. This was done merely for convenience in counting days from the reference date. We could, of course, have used either equinox to solve both problems. As we shall see in Section 1.29, use of the *American Ephemeris* listing of Julian date numbers would have simplified this even further.

1.21 Navigation

One can see that if an accurate, reliable clock is available, navigation on the earth's surface is possible through observations of stellar transits. The measurement of the observer's latitude was discussed in Section 1.7. Our calculation (a) in Section 1.20 indicated that meridian transit of the Crab nebula should occur at 1:26 P.M., in our somewhat simplified case. If our chronometer reads something else, the difference is due to a *longitude* difference from the place at which the clock was originally set. The difference is such that it amounts to 15° per hour. Thus, if the clock reads 3:26 P.M. at meridian transit, we know that we are 30° east of the longitude for which the clock was set.

1.22   Mean Solar Time

We have so far discussed apparent solar time as though it were quite constant. However, the solar day is not precisely constant in duration throughout the year. There are several causes for this, apart from the irregular variations in the earth's rate of rotation.

One such cause is that the hour angle of the sun is measured along the celestial equator. Consequently it is determined by the projection of the ecliptic on the equator. The fact that the obliquity of the ecliptic is not zero but 23.5°, however, means that the value of the projection changes slightly throughout the year in a sinusoidal fashion.

In addition, it happens that the earth's orbit around the sun is not perfectly circular, but is instead a slightly eccentric ellipse. The laws of planetary motion cause the orbital speed of the planet to vary throughout the year as a consequence of the eccentricity. These two effects (ellipticity and speed variation), when superimposed, cause the length of the apparent solar day to vary rather irregularly throughout the year. The apparent day is a little longer in the winter months, around December.

It is desirable to correct for this behavior, and a concept called the *mean solar day* has accordingly been developed. This "day" is the 24 hours between meridian transits of an imaginary point called the *mean sun*. The mean sun moves eastward along the celestial equator, completing its cycle as does the apparent sun along the ecliptic. *Mean solar time* is reckoned with this fictitious sun; the mean solar second is 1/86,400 of the mean solar day, or of the average apparent solar day. Mean solar time is announced by various radio stations in different countries around the globe. In the United States, for example, stations WWV and WWVH are available; from transmitters located in Colorado and Hawaii, respectively, they announce the time that we set our watches by.

A useful special case, called universal time (U.T.), is the mean solar time at the longitude of Greenwich, England. U.T. is useful in correlating studies of world-wide phenomena on a common time basis. Universal time, based on the earth's rotation, is therefore somewhat irregular and is corrected for irregularities in that rotation by frequent sights on the stars. For high-accuracy prediction of the universal times of celestial events, a sort of "mathematical fiction" time called *ephemeris time* has been invented. This latter "time" runs on uniformly; its contant arbitrary unit equals the length of the tropical year at the beginning of the year 1900 divided by 31,556, 925.97474, which was the number of seconds in that year. Corrections for converting ephemeris time to U.T. are published in the *Nautical Almanac*; they are determined frequently by observing the universal times when the moon arrives at positions in the sky predicted on ephemeris time. In the present century ephemeris time has been gaining on U.T.; it was ahead of U.T. by 35 seconds in 1970. In other words, stars that culminated exactly "on schedule" in 1900 are now about $\frac{1}{2}$ minute late (U.T.).

1.23 The Equation of Time

The difference between apparent and mean time is called the equation of time E. Symbolically we may write E as

$$E = \text{L.A.T.} - \text{M.S.T.}, \qquad (1.2)$$

where the sign convention is in accordance with the practice of the *American Ephemeris*. In equation (1.2), local apparent time is denoted by L.A.T. and mean solar time is represented by M.S.T. E is never very large; the largest values are about $+16^m$ on 1 November and approximately -14^m on 1 February; that is, the sundial reads 16 minutes ahead (or faster) than local mean time on 1 November (Figure 1.12).

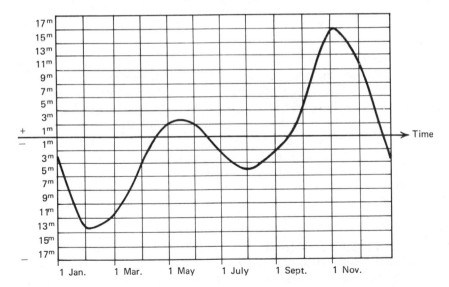

Figure 1.12   The equation of time E and its variations throughout the year.

1.24 Civil Time and Time Zones

Mean solar time, while advantageous, is not practical for everyday use; "observations of the hour angle of the mean sun" imply that observers, even at slightly different longitudes, observe different times. It would be awkward to go elsewhere in a city and have to reset our watches.

Therefore "time zones" have been established such that all clocks in the same zone read by edict the same solar time, called "civil time." The United States has four such zones (Eastern, Central, Mountain, and Pacific), each of which differs from its neighbors by 1^h00^m (Figure 1.13). The boundaries of these zones, which are nominally 15° (of longitude) wide, have been drawn to create the minimum confusion. There are 24 such zones around the world; the time observed by each is the mean time of the meridian that passes through the center of the zone. It should be noted, however, that zone time should not be used in investigations, say, of the world-wide simultaneity of cosmic ray events; mean solar or sidereal time will be more precise.

Other edicts have occasionally been promulgated. Consequently systems like "daylight saving time" and "double time" have been employed, but we shall not discuss them further.

Another edict—a useful one—has resulted in the *international date line*. It was created in order that time shall be single-valued. It has been drawn along the 180° meridian, but there are departures from this meridian for the sake of convenience. The rule is that in crossing the line from west to east, one decreases his date by one full day; a crossing in the reverse direction produces a one-day advancement.

1.25 The Year and the Calendar

The *sidereal year* is the "true" period of the earth's revolution, for it is the length of time that the sun takes to return to the same point on the celestial sphere. Humans, however, base their calendar on the *tropical year*, the period that the sun takes to return to the vernal equinox. This calendar keeps in step with the seasons. The tropical year is a little shorter than the sidereal year because of the precession of the equinoxes; the difference amounts to about (50 arc-seconds/360°) times the length of the sidereal year (which is 365.2564 mean solar days). Thus the difference is only about 20 minutes per year.

The history of the calendar is interesting, but we shall confine our attention to a few of the more prominent developments.

1.26 Development of the Calendar

The first mention of a calendar concerns the execution of two astronomers. Two Chinese, named Hi and Ho, were apparently careless in the preparation of an official calendar of eclipses, for one occurred without prior imperial warning. They are reported to have been executed for this in 2159 B.C.

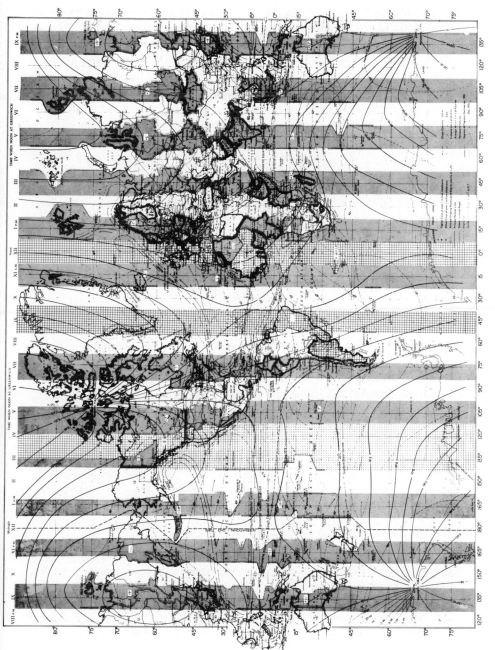

Figure 1.13   The time zones of the earth. Each zone is 15° of longitude wide, with local exceptions.

The modern calendar is based on one developed in ancient Rome; it may be traced back to the eighth century B.C. Originally there were 10 months in the calendar, each month beginning with a new moon. The 10 months were increased to 12 by the first century B.C. The average length of the 12 was $29\frac{1}{2}$ days, the period of time required for the moon to go through all its phases and return to the starting phase. Thus the Roman months were alternately assigned 29- and 30-day durations. Trouble rapidly developed: $29.5 \times 12 = 354$ days, not the approximate $365\frac{1}{4}$-day duration observed for the year. Consequently the local priests were authorized to insert a thirteenth month every third year or so to make things come out even. These were called "full" years; the 12-month years were "empty" years. The priests soon recognized the advantages of assigning a full year while their friends were in public office and empty years while others were in authority. Since the calendar varied from town to town, chaos rapidly developed; a traveler could go from year to year when moving from town to town. It may be assumed, however, that some of the politicians were more than content with the status quo.

1.27 Julian Calendar

In 46 B.C. Julius Caesar established a new calendar, containing among other things a month named July after him. In this Julian calendar there were 12 months, with an average length of $30\frac{1}{2}$ days (except for February, which was given 29 days). The average year world then be 365 days long. Since the tropical year was then believed to contain $365\frac{1}{4}$ days, an additional day was added to February every fourth year. Such years were called "leap years." The 31-day months were called "lucky" months; July became one of them. They alternated with the 30-day months. But supporters of the next emperor, Augustus, insisted on a lucky month named for him, so that August was now assigned 31 days at the expense of February, which was relegated to 28-day status (except in leap years).

Now $365\frac{1}{4}$ days is 11^m14^s longer than the tropical year. By A.D. 1582, therefore, a 14-day discrepancy between the two years had developed. The vernal equinox occurred on 11 March 1582 rather than the 25 March assigned to it by Julius Caesar.

1.28 Gregorian Calendar

To avoid, or at least reduce, this slippage, Pope Gregory XIII proclaimed the *Gregorian calendar*, the calendar presently employed in the western world.

One of the major features of this calendar is that the vernal equinox occurs on 21 March. This change was effected by having 15 October 1582 follow 4 October 1582. The period 5 October—14 October 1582 was simply eliminated. Another change was in the rule for leap years. In the Julian version every year divisible by 4 was a leap year; in the Gregorian calendar this is still true, except for the "century years." Only century years divisible by 400 are leap years in the Gregorian. Thus 1600 and 2000 are leap years but 1700 and 1900 are not, although they were in the Julian calendar.

The Gregorian average year is 365.2425 mean solar days long. It is therefore correct to 1 day in 3300 years. No day will have to be deleted for another 29 centuries.

Calendar reform can have serious economic consequences. England and her colonies in America did not adopt the Gregorian until 1752, when parliament decreed that 2 September 1752 was to be followed by 14 September 1752. Riots ensued, because landlords wanted a full month's rent and the tenants felt short-changed.

1.29 Julian Dates

Scientists are notoriously lazy. Many have avoided the complexities of the calendar through the use of Julian dates. In this system each day is assigned a number rather than a designation that includes a month and year.

The first day, JD1, is equivalent to 1 January 4713 B.C. It should not be concluded from this that this was the date of "creation." Rather, no historical records exist for earlier dates, so that negative dates are avoided. In this system 1 October 1965 is numbered JD2,438,736. A new Julian day starts at noon. Listings of Julian date numbers are to be found in the *ephemeris*.

Problems

1.1. A sounding rocket is launched eastward from Wallops Island, Virginia (latitude 37.5° N, longitude 75.7° W), at a zenith angle of 10°, to an apogee of 100 miles. Neglecting the precession and nutation of the rocket as well as aerodynamic forces, find the right ascension and declination of a point viewed by a directional Geiger counter mounted perpendicular to the spin axis of the rocket.

1.2. Compute the galactic latitude and longitude of a strong discrete radio source observed to be at $\alpha = 9^h52^m$, $\delta = +70°$ (1950).

1.3. Is the x-ray source in Scorpius ($\alpha = 16^h15^m$, $\delta = -15°$ [1950]) within 10° of the galactic equator?

1.4. A gravity-gradient-stabilized satellite passing vertically over New York City has an infrared detector aboard that is aimed at the Crab Nebula. Find the altitude angle and azimuth of the axis of the detector. The right ascension of the Crab is 5^h30^m; the declination is $+22°$.

1.5. A shipwreck survivor on a raft has a sextant and a perfect clock set to the standard time of the last time zone that the ship passed through. He notices that a certain star (whose right ascension he doesn't know) is on his meridian to the north at exactly 7:00 P.M. by the clock. Three days later, after drifting with the winds and currents, he finds that the same star is again on his meridian at 7:00 P.M. and that it is 1° higher above the horizon than when he observed it last.

(a) What is the net direction of drift of his raft during the three days? (North, northeast, east, southeast, south, etc.)

(b) How far (in degrees) did the raft drift in latitude during that time?

(c) How far (in degrees) did it drift in longitude?

(d) If the radius of the earth is 6370 km, what was the average speed of drift in kilometers per hour?

1.6. A certain star has an azimuth of 30° and an elevation of 45°; a second star has an azimuth of 45° and an elevation of 60°. What is the shortest arc through which a telescope must be moved to look at first one star and then the other?

1.7. The tracking station at Goldstone, California (latitude 35° N, longitude 115° W), is communicating with an interplanetary spacecraft, the coordinates of which are $\alpha = 19^h30^m$, $\delta = +30°$ (1950.0). Find the altitude angle and the azimuth of the antenna axis at Goldstone, at a sidereal time of 22^h30^m.

1.8. An observer aboard a ship at unknown position sees the star Castor ($\delta \simeq 30°$, $\alpha \simeq 7\frac{1}{2}$ hours) in the constellation Gemini cross his meridian to the north of his position at Greenwich mean time 2015 hours on 1 April. The elevation angle of Castor at the time of meridian transit is 75°. The equation of time is −4 minutes. Find:

(a) The ship's latitude.

(b) The ship's longitude.

(c) The standard time of the ship's time zone.

1.9. An astronomer at unknown north latitude sees a star on his meridian. Exactly 2 hours earlier the star was located at a north azimuth angle of 25° and had an elevation angle of 54°. Find the declination of the star.

1.10. An observer in Houston (latitude 30° N) observes a star on his meridian, north of his zenith, at 7:00 P.M. Exactly two hours later, at 9:00 P.M., he observes that the same star is 45° from the zenith. What is the declination of the star, and what is its north azimuth at 9:00 P.M.?

Chapter 2

CELESTIAL MECHANICS

2.1 Introduction

Celestial mechanics is that area of space science devoted to the motions of celestial bodies. It is not our purpose to attempt a thorough treatment of classical mechanics, or even a complete discussion of all the details of celestial mechanics. Instead we hope to outline some of the more pertinent features of the elegantly developed subject; the interested reader will find a bibliography that includes more detailed treatises at the end of the book.

Celestial mechanics is mostly the study of gravitation and its effects on bodies such as planets, satellites, and artificial space probes. It is true that forces other than gravitation enter into a complete description, such as atmospheric drag on orbiting bodies or the interactions of plasmas, such as the solar wind, with comets. These forces will be alluded to in the present chapter, but a more complete discussion of them is deferred to more appropriate chapters.

Johannes Kepler published his three laws of planetary motion in 1619. They constituted an analysis of Tycho Brahe's careful observations of the apparent motions of the planet Mars. Kepler's laws formed the basis for Newton's inquiries into planetary motions that today form the basis for classical mechanics. It is now our task to see how Kepler's laws follow from Newton's more general equations. Kepler's laws may be summarized as follows.

2.2 Kepler's Laws

We present them here in the order in which they are subsequently discussed.

Second Law

The orbit of a planet lies in a plane that passes through the sun, and the area swept over by a line joining the sun and the planet is proportional to the time elapsed.

27

First Law

The orbit of a planet is an ellipse; the sun is at one focus of each planet's elliptical orbit.

Third Law

The ratio between the square of the sidereal period of revolution P and the cube of the semimajor axis a is the same number for all planets in the solar system.

In treating the development of these laws from Newtonian mechanics we shall assume that each planetary orbit may be treated as a two-body problem; celestial objects other than the sun and the planet are assumed to exert negligible forces on these two bodies. The third law implies that the force of gravity is proportional to planetary mass. This circumstance supports our assumption that the other planets may be neglected, for their masses are all small compared to the solar mass. The largest planetary mass (that of Jupiter) is only about 0.1% of that of the sun. We shall return to a discussion of planetary masses later.

2.3   Derivation of Kepler's Laws

The coordinate system that we shall adopt is shown in Figure 2.1. Here $x^2 + y^2 + z^2 = r^2$, and the solar mass, denoted by m_1, is assumed to be so massive that the center of mass of the system is very close to the origin.

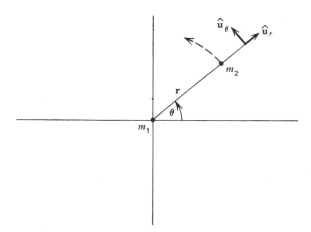

Figure 2.1 Coordinate system used for describing the motion of secondary mass m_2 about m_1, where m_1 is assumed to be very massive; $\hat{\mathbf{u}}_n$ and $\hat{\mathbf{u}}_\theta$ are unit vectors in the **n** and θ directions, respectively.

We also implicitly assume that we are dealing with a Newtonian frame of reference, one in which Newtonian mechanics is obeyed. We neglect for now the effects of relativity, and we also neglect the effects of the finite size and nonsphericity of planetary masses, treating them for the moment as point masses (mass at the planetary center). We shall treat cases later in which nonsphericity is important, as in the case of a satellite in a near-planet orbit.

Newton's law of gravitation may be expressed as

$$\ddot{\mathbf{r}} = -\frac{GM}{r^3}\mathbf{r}, \tag{2.1}$$

where $M = m_1 + m_2$ and is introduced to allow for the fact that the center of mass of the system is not quite at $(0, 0, 0)$. If $\mathbf{r}$ refers to the position vector of mass m_2, then (2.1) gives the acceleration of m_2 relative to the center of mass of the system; $\mathbf{r}$ is really the vector sum of the position vectors of the center of mass and of the particle m_2, both with respect to the origin. Also G is a constant, the value of which depends on the system of units employed, but it is assumed in Newtonian theory to be independent of position or of the configuration of any other masses in the universe.

2.4 Motion in a Plane

An interesting result emerges if we take the vector cross-product of $\mathbf{r}$ with the quantity $m_2\ddot{\mathbf{r}}$:

$$\mathbf{r} \times m_2\ddot{\mathbf{r}} = \frac{d}{dt}(\mathbf{r} \times m_2\mathbf{v}) = \frac{d}{dt}\mathbf{L}. \tag{2.2}$$

The product must be zero, since $\mathbf{r}$ and $\ddot{\mathbf{r}}$ are in the same direction. Thus

$$\frac{d\mathbf{L}}{dt} = 0; \tag{2.3}$$

$\mathbf{L}$ is a "constant of the motion." The quantity $\mathbf{L}$ is recognized as the vector angular momentum of the system. Since the angular momentum does not change in time, the orbit must be in an unchanging plane. The plane of the orbit will not change unless some external torque is exerted to disturb it.

2.5 Kepler's Second Law

Now in polar coordinates

$$\mathbf{r} = r\hat{\mathbf{u}}_r \tag{2.4}$$

and

$$\mathbf{v} = \dot{r}\hat{\mathbf{u}}_r + r\dot{\theta}\hat{\mathbf{u}}_\theta, \tag{2.5}$$

where $\hat{\mathbf{u}}_r$ and $\hat{\mathbf{u}}_\theta$ are unit vectors in the r and θ directions, respectively.

Therefore we may re-express our cross-product in (2.2) as

$$\mathbf{r} \times m_2\mathbf{v} = r\hat{\mathbf{u}}_r \times m_2\{\dot{r}\hat{\mathbf{u}}_r + r\dot{\theta}\hat{\mathbf{u}}_\theta\}, \tag{2.6}$$

or

$$\mathbf{r} \times m_2\mathbf{v} = m_2 r^2\dot{\theta}\hat{\mathbf{u}}_L, \tag{2.7}$$

where $\hat{\mathbf{u}}_L = \hat{\mathbf{u}}_r \times \hat{\mathbf{u}}_\theta$ is a unit vector orthogonal to $\hat{\mathbf{u}}_r$ and to $\hat{\mathbf{u}}_\theta$. Now from (2.7) and (2.2) we see that

$$r^2\dot{\theta} = h, \tag{2.8}$$

where h is a constant, such that $L = m_2 h$. We can see from Figure 2.1 that the quantity $\frac{1}{2}r^2\dot{\theta}$ is the rate at which $\mathbf{r}$ sweeps out an area; from (2.8), this rate is a constant. This constitutes a proof of Kepler's second law on the basis of Newtonian mechanics.

2.6   Conservation of Energy

We digress for a moment to see that energy is conserved in this central force problem. (We should note that we have implicitly assumed gravitation to be a *central* force, i.e., isotropic. This assumption is the simplest to make and appears to explain all observations that may be explained with Newton's mechanics.)

The equation of motion is

$$m_2\dot{\mathbf{v}} = -\frac{GMm_2}{r^3}\mathbf{r}. \tag{2.9}$$

If we now form the scalar product $m_2\mathbf{v} \cdot \dot{\mathbf{v}}$, we have

$$m_2\mathbf{v} \cdot \dot{\mathbf{v}} = -\frac{GMm_2}{r^3}(\mathbf{v} \cdot \dot{\mathbf{r}}), \tag{2.10}$$

or

$$\frac{d}{dt}(\tfrac{1}{2}m_2 v^2) = -\frac{GMm_2}{r^3}[\dot{r}\hat{\mathbf{u}}_r + r\dot{\theta}\hat{\mathbf{u}}_\theta] \cdot [r\hat{\mathbf{u}}_r]. \tag{2.11}$$

We therefore find, from (2.11), that

$$\frac{d}{dt}(\tfrac{1}{2}m_2 v^2) = +GMm_2\frac{d}{dt}\left(\frac{1}{r}\right), \tag{2.12}$$

so that

$$\tfrac{1}{2}m_2 v^2 - \frac{GMm_2}{r} = E, \tag{2.13}$$

where E, called the total energy of the system, is another constant. Thus we see that energy is conserved.

2.7   Kepler's First Law

The first law of Kepler follows from considering the equation of the orbit of m_2 about m_1. The equation of motion (2.9) may be resolved into components in $\hat{u}_r$ and $\hat{u}_\theta$, the terms of which may be separately equated. The radial component becomes

$$\ddot{r} - \frac{h^2}{r^3} + \frac{GM}{r^2} = 0. \tag{2.14}$$

We now let $\mu = GM$ and also let $r = 1/u$. In addition, we see that

$$\frac{dr}{dt} = -\frac{1}{u^2}\frac{du}{dt} = -\frac{1}{u^2}\frac{du}{d\theta}\frac{d\theta}{dt},$$

or

$$\dot{r} = -h\frac{du}{d\theta}. \tag{2.15}$$

Also,

$$\frac{d^2r}{dt^2} = -h\frac{d}{dt}\left(\frac{du}{d\theta}\right) = -h\frac{d^2u}{d\theta^2}\frac{d\theta}{dt}, \tag{2.16}$$

so that

$$\ddot{r} = -h^2u^2\frac{d^2u}{d\theta^2}. \tag{2.17}$$

Substitution of these quantities into (2.14) yields

$$\frac{d^2u}{d\theta^2} + u = \frac{\mu}{h^2}$$

for the equation of the orbit. Since μ and h are constants, this differential equation may immediately be solved for u; the result is that

$$u = A\cos(\theta - \omega) + \frac{\mu}{h^2}. \tag{2.18}$$

Alternatively (2.18) may be rewritten as

$$r = \frac{h^2/\mu}{1 + (Ah^2/\mu)\cos(\theta - \omega)}, \tag{2.19}$$

where A and ω are constants of integration. If we now let

$$e = \frac{Ah^2}{\mu}, \tag{2.20}$$

where e is a constant called *eccentricity*, (2.19) becomes

$$r = \frac{e/A}{1 + e \cos (\theta - \omega)}. \qquad (2.21)$$

Now the minimum value of r is reached when $\theta - \omega = 0$; it is called the distance of *perihelion*, r_p, where

$$r_p = \frac{h^2}{\mu(1 + e)}. \qquad (2.22)$$

The maximum value of r, called *aphelion* when a body is in orbit around the sun (*apogee* when it is in orbit about the earth), is achieved when $\theta - \omega = \pi$. It is

$$r_a = \frac{h^2}{\mu(1 - e)}. \qquad (2.23)$$

For closed orbits we may also define the quantity a, where

$$2a = r_p + r_a, \qquad (2.24)$$

and substitute these various factors into (2.21). The result is

$$r = \frac{a(1 - e^2)}{1 + e \cos (\theta - \omega)}, \qquad (2.25)$$

which we recognize as the polar equation of an ellipse (if e is less than unity). This constitutes a proof of the second law. It is truly remarkable, and a testimony to both Brahe and Kepler, that Kepler was able to see that the Martian orbit was elliptical rather than circular. This is because it turns out that the eccentricity of that orbit is only $e = 0.09$; r_p and r_a are not greatly different from each other, and the orbit is therefore nearly circular.

The shape of the earth's orbit may be found fairly readily from measurements taken of the apparent size of the sun throughout the sidereal year. As our planet gets closer to the sun, the size of the solar disc is increased slightly; the farther away we are, the smaller it is, with the average being approximately $\frac{1}{2}°$.

In general (2.25) is the equation of a conic. The type of conic depends on the value of e. Should $e = 1$, the orbit is parabolic; if it is greater than unity, it is a hyperbola. The special case $e = 0$ is for a circular orbit.

The constant ω, the *argument of perihelion*, is a phase angle that serves to denote the orientation of the major axis in the orbital plane. It is measured from the line of nodes (the line through the sun and through the points where the orbital plane intersects the ecliptic plane) to perihelion. In particular, the argument is measured from the *ascending node* (see Figure 2.2) to perihelion. The argument of perihelion is one of seven quantities, called the orbital

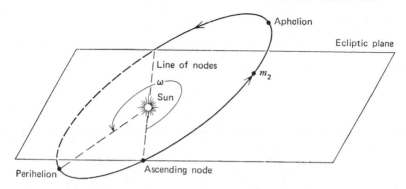

Figure 2.2   The orbital elements that describe the motion of a body within the solar system. The orbital and ecliptic planes intersect along the line of nodes.

elements, that completely specify the orbit. The orbital elements are described in Appendix A.

2.8 Kepler's Third Law

It still remains to deduce Kepler's third law. We may now write that

$$h^2 = \mu a(1 - e^2) \tag{2.26}$$

by adding our expressions for r_p and r_a and setting them equal to $2a$.
Also, we see that

$$\mu a(1 - e^2) = GMa(1 - e^2), \tag{2.27}$$

so that

$$h^2 = GMa(1 - e^2).$$

Alternatively,

$$(r^2\dot{\theta})^2 = GMa(1 - e^2). \tag{2.28}$$

Now if S is the area swept out by the vector $\mathbf{r}$, we therefore have that

$$\frac{dS}{dt} = \tfrac{1}{2}[GMa(1 - e^2)]^{1/2}. \tag{2.29}$$

We may integrate (2.29) and obtain

$$S = \tfrac{1}{2}[GMa(1 - e^2)]^{1/2}\, t + S_o, \tag{2.30}$$

where S_o is a constant of integration and t is time.
In one sidereal period P, m_2 moves completely around the orbit, returning to its starting place. If the orbit is elliptical, the area swept out during P is πab, where b, the semiminor axis, is

$$b = a(1 - e^2)^{1/2}. \tag{2.31}$$

We may therefore evaluate the expression (2.30) as

$$\pi a^2 (1 - e^2)^{1/2} = \tfrac{1}{2}[GMa(1 - e^2)]^{1/2}P. \tag{2.32}$$

Solving (2.32) for P, we find that

$$P = \frac{2\pi a^{3/2}}{[G(m_1 + m_2)]^{1/2}}, \tag{2.33}$$

Kepler's third law. This is known as the "harmonic law." Since m_1, the mass of the sun (in the solar system case), is so much larger than the planetary mass m_2, the latter may be neglected to a very good approximation. It is this approximation that was originally given by Kepler; all terms on the right-hand side of (2.33) except the semimajor axis are constants, independent of the particular planet under investigation.

Figure 2.3*a* Line-of-sight lunar gravity accelerations at 100 km altitude, compiled by P. M. Muller and W. L. Sjogren of Jet Propulsion Laboratory, California Institute of Technology, *Science*, August, 1968.

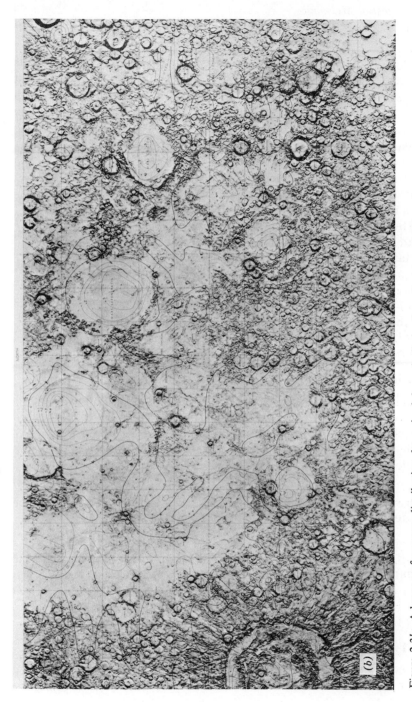

Figure 2.3b A lunar-surface mass distribution determined dynamically by the estimation of 580 surface mass points. Obtained by W. L. Sjogren, P. M. Muller, and P. Gottlieb, Jet Propulsion Laboratory, and L. Wong, G. Buechler, W. Downs, and R. Prislin, Aerospace Corporation 1969.

Kepler's third law may be used to measure the masses of those five planets that have negligible-mass satellites in orbit about them. The periods of revolution and mean distances are the measured quantities.

The earth's natural satellite, however, requires special treatment, since its mass is over 1 % of that of the planet. Both planet and satellite revolve about the common center of mass, called the *barycenter*. This motion of the earth causes the apparent position of nearby celestial objects (Mars, Venus, any planetoids) to undergo parallactic position shifts, on a *monthly* basis. Mars, for example, shifts about 17 arc-seconds each month (when closest to the earth) from the positions computed for its elliptical orbit about the sun. From this shift it is found that the lunar mass is about 1/81 of that of the earth, a result confirmed and refined by observations of lunar orbiting space-craft. (Indeed, these orbits have been so accurately measured that they have revealed the presence of "mass concentrations" beneath some lunar craters. These lumps of matter are called *mascons*.) (See Figures 2.3a, b.)

The masses of planets such as Venus and Mercury, which have no satellites, are more difficult to estimate precisely. Venus perturbs the orbits of both the earth and Mercury; the amount of the perturbation may be analyzed and it is found that a Venusian mass 0.82 as great as the earth satisfies the observa-tions. More accurate estimates may be made by analyzing the perturbations on the trajectories of spacecraft that fly close to the planet.

2.9   Artificial Earth Satellites

Kepler's laws, of course, may be used to describe situations other than the motions of planets about the sun. Thus the harmonic law applies to the motion of artificial satellites about the earth, for example. The mass of such a satellite is negligible compared to that of the earth. If P is expressed in minutes and a is in kilometers,

$$a = 332.3P^{2/3}. \tag{2.34}$$

This provides a quick way of estimating the mean altitude of a satellite, once the period of revolution P is determined. The mean altitude z is

$$z = a - R_e, \tag{2.35}$$

where R_e is the average radius of the earth. One can also approximate the maximum and minimum altitudes for such a satellite, if the eccentricity is known. For example, the minimum altitude would be $[a(1 - e) - R_e]$. It should be recalled, however, that this is an approximation that would be exactly true only for a spherical earth.

2.10   Type of Conic

We noted in Section 2.7 that the eccentricity of the orbit determined whether the orbit was to be elliptical or hyperbolic, say. In (2.20) we introduced the eccentricity as Ah^2/μ, a combination of other constants. Thus the eccentricity of an orbit has so far been discussed only in geometrical terms. We now show that e is physically determined by the total energy, E.

The energy equation (2.13) may be rewritten as

$$\tfrac{1}{2}m_2h^2u^2 - \mu m_2 u - E = 0 \tag{2.36}$$

at points in the orbit where the velocity is entirely transverse, since $v = hu$ at those points. We may solve (2.36) for u, finding that

$$u = \frac{\mu}{h^2}\left[1 \pm \left(1 + \frac{2Eh^2}{m_2\mu^2}\right)\right]^{\frac{1}{2}}. \tag{2.37}$$

The maximum value of u, u_{max}, is therefore

$$u_{max} = \frac{\mu}{h^2} + \frac{\mu}{h^2}\left[1 + \frac{2Eh^2}{m_2\mu^2}\right]^{\frac{1}{2}}, \tag{2.38}$$

which we may set equal to $A + \mu/h^2$, from (2.19). This leads to another expression for e when we employ (2.20); the new expression is

$$e = \left[1 + \frac{2Eh^2}{m_2\mu^2}\right]^{\frac{1}{2}}. \tag{2.39}$$

The eccentricity is indeed determined by the total energy. In fact, if

$E = 0$, the orbit is a parabola;

$E < 0$, the orbit is an ellipse (or circle);

$E > 0$, the orbit is a hyperbola.

2.11   The Vis-Viva Equation

Using (2.27), we may express the energy equation differently. That equation tells us that $h = [GMa(1 - e^2)]^{\frac{1}{2}}$; substitution of this into (2.39) leads to

$$E = -\frac{GMm_2}{2a} \tag{2.40}$$

upon rearrangement of terms. We may insert this value for E into the energy equation (2.13) and find that

$$v^2 = GM\left[\frac{2}{r} - \frac{1}{a}\right]. \tag{2.41}$$

Equation 2.41 is called the *vis-viva* equation. *Vis-viva* means "living force" in Latin and may be also translated as "kinetic energy." It is one of the most useful equations of celestial mechanics. Let us now see some illustrations of the usefulness of this equation.

For example, when $a = r$, we have a circular orbit. The vis-viva equation immediately tells us that the velocity of a body in a circular orbit, v_c, is $[GM/r]^{1/2}$. Another interesting case occurs when $a = \infty$; a body in such an orbit can "escape" from the central body denoted by M. The *escape velocity*, v_e, is also given by the vis-viva equation. It is $v_e = v_c\sqrt{2}$.

The velocity v_e represents the velocity required to travel "to infinity" from a planet or other massive body. It is the same velocity that a freely falling body would acquire under the influence of gravity, if it starts at infinity and strikes the planet. A spacecraft that approaches another planet from the earth will in general, therefore, have a speed in excess of the escape velocity for that planet. Consequently a reduction in speed will be required, should it be desired to place the spacecraft in orbit about the planet.

2.12   Some Applications of the Vis-Viva Equation

In the case of the earth v_c is about 5 miles per second. That is, a satellite that has this velocity will enter into a circular orbit. A satellite moving faster than this will find itself in an elliptical orbit.

Another example of the use of the vis-viva equation is in finding the apogee of a satellite injected into earth orbit with a speed of, say, 6 miles per second. It is most convenient to take the earth's radius as the unit of distance. In these units (2.41) now reads

$$v^2 = \left(2 - \frac{1}{a}\right), \tag{2.42}$$

if the velocity is expressed in terms of the circular orbit velocity. A velocity of 6 miles per second is 1.2 times the circular orbit velocity. We may now find a from $1/a = 2 - (1.2)^2$, so that we find $a = 1.79$ earth radii. This is the apogee distance from the center of the earth and it occurs halfway around the globe from injection, at an altitude above sea level of about 3200 miles. The orbit is an ellipse, with one focus at the center of the earth.

If we had been dealing with an interplanetary probe that was in orbit about the sun, the injection point would be perihelion, and aphelion would occur halfway around the orbit. The orbit would also be elliptical, but one focus of the ellipse would be the center of the sun rather than the center of the earth.

However, should the injection velocity be less than the circular orbit velocity, we would find that the elliptical orbit intersected the solid earth. This represents the trajectory of a ballistic missile, for example (Figure 2.4).

The magnitude of the velocity, as we have seen, determines the length of the semimajor axis.

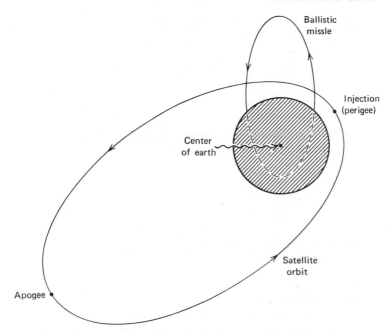

Figure 2.4   The type of orbit depends on the speed. Two orbits are shown. A satellite injected into orbit with a speed greater than v_c will have an elliptical orbit that has apogee halfway around the orbit from injection. A ballistic-missile trajectory is an ellipse that intersects the surface of the earth at two points, corresponding to the launch and impact sites.

2.13   Applications of Kepler's Laws

An example of the use of Kepler's laws is provided by the mission to Mars trajectory. Here we seek to determine the travel time for a spacecraft to make such a flight. We shall use the earth-sun distance as our unit of distance; it is called the astronomical unit (A.U.). Referring to Figure 2.5, we see that $2a = 2.52$ A.U. or $a = 1.26$ A.U.; the mean distance of the spacecraft from the sun is approximately the semimajor axis of the orbit.

We have chosen the particular orbit shown in the figure because it intersects the Martian orbit when Mars is at perihelion. Thus such an orbit requires the minimum energy expenditure. We have implicitly assumed here that the orbits of Mars and the earth are coplanar with the ecliptic, so that the "transfer orbit" is also in the ecliptic.

Now, applying Kepler's harmonic law,

$$P = \frac{2\pi a^{3/2}}{[GM]^{1/2}} = \frac{2\pi(1.26)^{3/2}}{0.0172} = 516 \text{ days}.$$

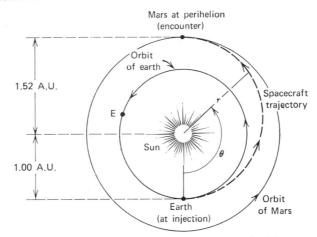

Figure 2.5   The "Mission to Mars" problem. All orbits are
assumed to be coplanar.

Thus the travel time, $P/2$, is 258 days (about $8\frac{1}{2}$ months). The earth will be
approximately at the point labeled E in the diagram, when the spacecraft
encounters Mars. We have used the so-called Gaussian value of G, which is
appropriate for these units. The numerical value is such that $G^{1/4} =$
0.017202099. It was determined by Gauss from observing the sidereal period
of the earth-moon system about the sun. This value of G defines the unit of
distance, the A.U., from Kepler's third law.

We may also find the equation of this minimum energy transfer orbit.
Since our spacecraft has a perihelion of 1.00 A.U. and an aphelion of 1.52
A.U. (Mars' perihelion distance), we have that

$$a(1 - e) = 1.00$$

and

$$a(1 + e) = 1.52.$$

These tell us that the eccentricity is $e = 0.21$. Thus the equation of the orbit is

$$r = \frac{1.20}{1 + 0.21 \cos (\theta - \omega)}.$$

A special and rather well-known case of these transfer orbits is the
"Hohmann transfer ellipse." It applies to transfer between two circular coplanar orbits.

2.14   Nonsphericity of an Attracting Body

We have thus far discussed the motion of celestial bodies as though they
were perfect spheres. In that case the center of mass of each may be considered as a point at the center of each object. There are many cases, however,

in which this approximation is not justified. None of the planets or their natural satellites is a perfect sphere; satellites in close planetary (or lunar) orbits will consequently experience forces that are noncentral in character; that is, they will experience forces that are position-dependent, not merely distance-dependent. The study of such forces is an extremely complex undertaking, and so we treat only a simplified case in subsequent sections.

2.15   Interplanetary Spacecraft Trajectories

In the preceding example our spacecraft was inserted into the transfer ellipse from an orbit for the earth that was assumed to be circular. The point of injection was chosen so as to take maximum advantage of the earth's speed of revolution about the sun.

A more realistic problem takes the eccentricity of the earth's orbit into account and also allows for injection at any point along the orbit. This situation is illustrated in Figure 2.6.

Since

$$\mathbf{v} = \dot{r}\mathbf{u}_r + r\dot{\theta}\mathbf{u}_\theta, \tag{2.5}$$

we may write

$$\tan \alpha = \frac{\dot{r}}{r\dot{\theta}}.$$

Also

$$v^2 = \dot{r}^2 + r^2\dot{\theta}^2 = GM\left(\frac{2}{r} - \frac{1}{a}\right);$$

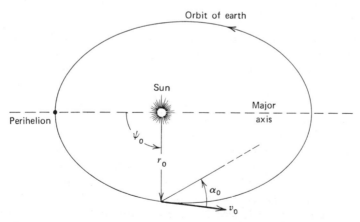

Figure 2.6   Insertion into interplanetary trajectory at an angle α_0 with respect to the earth's orbital velocity-vector $\mathbf{v}_0$.

hence

$$\dot{r} = \frac{a(1 - e^2)e \sin (\theta - \omega)\dot{\theta}}{[1 + e \cos (\theta - \omega)]^2} = \frac{e \sin (\theta - \omega)r\dot{\theta}}{1 + e \cos (\theta - \omega)}.$$

Using the relation

$$h^2 = GMa(1 - e^2),$$

we see that $h^2 = r^2 v_o^2 \cos^2 \alpha_o$, where v_o and α_o are the spacecraft's speed and direction (with respect to the tangent to the earth's orbit) at the point of injection, respectively.

Therefore we arrive at the results

$$\tan \alpha_o = \frac{e \sin \psi_o}{1 + e \cos \psi_o},$$

$$r_o = \frac{a(1 - e^2)}{1 + e \cos \psi_o},$$

and

$$v_o^2 = GM \left(\frac{2}{r_o} - \frac{1}{a} \right).$$

Given the distance r_o (from the sun) of the earth when the spacecraft is inserted into its trajectory, as well as v_o and α_o, we may use these relations to find ψ_o (the angle along the earth's orbit from perihelion to insertion), the semimajor axis a, and the eccentricity e of the spacecraft orbit. These results may be used, for example, to compute the period of the spacecraft's elliptical trajectory.

2.16 Attraction of a Point Mass by a Body of Finite Dimensions and Arbitrary Mass Distribution

The case of an artificial satellite in close orbit about a planet such as the earth provides an illustration of the problem. The earth is asymmetric; its daily rotation has produced an equatorial bulge and there is also some evidence (as we shall see) that more mass is located south of the equator than north of it. The presence of the bulge was established as early as 1745, when surveys and astronomical measurements showed that the degree of latitude increases in length from the equator to the pole; the difference between the equatorial and polar radii is 1/298 of the equatorial radius of the earth.

We again neglect the effects of atmospheric drag here, although this assumption is questionable at perigee distances of less than 200 miles. Drag forces are dependent on the cross-sectional area of the satellite, on the atmospheric density (a function in turn of altitude), and on the spacecraft velocity relative to the atmosphere, and they are also dependent on the mass of the satellite. We shall return to this topic in the next chapter.

For a planet such as the earth it is necessary to know the principal moments of inertia from other data in order to calculate the orbit of a satellite. Conversely, the observation of such an orbit provides information on the values of these moments, which may in turn be utilized to gain insight into the mass distribution of the planet.

We shall adopt the following coordinate system. Let X_o, Y_o, Z_o be the coordinates of the center of mass of the planet M, and let X, Y, and Z be the coordinates of the satellite m. Consider an element of mass dM of the planet; let ξ, η, and ζ be its coordinates with respect to X_o, Y_o, and Z_o. Also put $x = X - X_o$, $y = Y - Y_o$, and $z = Z - Z_o$, so that x, y, and z are the coordinates of m relative to the center of mass of the planet. Thus the distance between the satellite and the mass element, Δ, is given by

$$\Delta^2 = (x - \xi)^2 + (y - \eta)^2 + (z - \zeta)^2. \tag{2.43}$$

Now if ρ is the density of the planet (a function of the coordinates ξ, η, ζ), the gravitational potential external to the planet is given by

$$U = G \iiint_{(M)} \frac{\rho(\xi, \eta, \zeta)}{\Delta} \, d\xi \, d\eta \, d\xi. \tag{2.44}$$

This may be rewritten as

$$U = G \int \frac{dM}{\Delta}. \tag{2.45}$$

The force on the satellite m is given by the gradient of the potential; the components are

$$F_x = -m \frac{\partial U}{\partial x}, \qquad F_y = -m \frac{\partial U}{\partial y}, \qquad F_z = -m \frac{\partial U}{\partial z}, \tag{2.46}$$

where the potential U satisfies Laplace's equation everywhere outside the planet,

$$\nabla^2 U = 0. \tag{2.47}$$

We now have reduced our problem to that of obtaining U as a function of position. For example, for a spherically symmetric body,

$$U = \frac{GM}{r},$$

where $r^2 = x^2 + y^2 + z^2$.

2.17   Calculation of the Potential

The potential U may be found from (2.45). If we assume that the distance to the satellite from the center of mass of the planet is large compared with the dimensions of the planet, a power series solution for U is obtainable with the aid of the Legendre polynomials.

If
$$l^2 \equiv \xi^2 + \eta^2 + \zeta^2, \tag{2.48}$$

then, according to our assumption above, $l/r < 1$. We may rewrite our expression for Δ^2 (2.43) as

$$\Delta^2 = r^2 - 2(x\xi + y\eta + z\zeta)l^2$$

$$= r^2 \left[1 - 2\left(\frac{x\xi + y\eta + z\zeta}{rl}\right)\frac{l}{r} + \frac{l^2}{r^2} \right].$$

Now let

$$\frac{l}{r} = \alpha \tag{2.49}$$

and let

$$\frac{x\xi + y\eta + z\zeta}{rl} = q, \tag{2.50}$$

where the magnitude of q does not exceed unity. Now our expression for Δ^2 becomes

$$\Delta^2 = r^2(1 - 2q\alpha + \alpha^2). \tag{2.51}$$

The quantity $1/\Delta$ can now be expressed in terms of the Legendre polynomials, $P_n(q)$. We have

$$\frac{1}{\Delta} = P_0 + P_1\alpha + P_2\alpha^2 + P_3\alpha^3 \cdots + P_n\alpha^n + \cdots. \tag{2.52}$$

The values of the Legendre polynomials may be computed from

$$P_n(q) = \frac{1 \cdot 3 \cdot 5 \cdots (2n-1)}{1 \cdot 2 \cdot 3 \cdots n}\left[q^n - \frac{n(n-1)}{2(2n-1)}q^{n-2} \right.$$

$$\left. + \frac{n(n-1)(n-2)(n-3)}{2 \cdot 4 \cdot (2n-1)(2n-3)}q^{n-4} - \cdots \right]; \tag{2.53}$$

for example, the Legendre polynomials of order 0 through 6 are as follows:

$$P_0(q) = 1,$$
$$P_1(q) = q,$$
$$P_2(q) = \tfrac{1}{2}(3q^2 - 1),$$
$$P_3(q) = \tfrac{1}{2}(5q^3 - 3q),$$
$$P_4(q) = \tfrac{1}{8}(35q^4 - 30q^2 + 3),$$
$$P_5(q) = \tfrac{1}{8}(63q^5 - 70q^3 + 15q),$$
$$P_6(q) = \tfrac{1}{16}(231q^6 - 315q^4 + 105q^2 - 5).$$

Thus the potential U may be written as

$$U = G \int \left[1 + P_1(q) \frac{l}{r} + P_2(q) \frac{l^2}{r^2} + P_3(q) \frac{l^3}{r^3} + \cdots \right] dM. \quad (2.54)$$

This integral (2.54) may be written as

$$U = U_0 + U_1 + U_2 + U_3 + \cdots,$$

a sum of terms that we shall not attempt to calculate here; the reader is referred to texts in celestial mechanics for the details. The result of the calculation is that

$$U_0 = \frac{GM}{r},$$

the result that would be obtained if all the mass of the planet were concentrated at its center of mass. Another result is that $U_1 = 0$, but

$$U_2 = -\frac{G}{r^3} \left[\frac{(C - A + B)}{2} (\tfrac{1}{2} - \tfrac{3}{2} \sin^2 \lambda) - \tfrac{3}{4}(A - B) \cos^2 \lambda \cos^2 \Psi \right], \quad (2.55)$$

where λ is the latitude and Ψ is the longitude. Also, A, B, and C are the moments of inertia about the x-, y-, and z-axes. For a planet that is rotationally symmetric about the z-axis, $A = B$. Hence U_2 reduces, for spheroidal and "pear-shaped" planets, to

$$U_2 = -\frac{G}{r^3} [(C - A)(\tfrac{1}{2} - \tfrac{3}{2} \sin^2 \lambda)]. \quad (2.56)$$

The expression for U_3 contains 10 integrals, while that for U_4 contains 15. We therefore terminate the series with U_2 and write the potential U of a spheroid as

$$U = \frac{GM}{r} \left[1 + \frac{B_2}{r^2} (-\tfrac{1}{2} + \tfrac{3}{2} \sin^2 \lambda) \right], \quad (2.57)$$

where $B_2 \equiv -(C - A)/M$. B_2 may be evaluated for a planet such as the earth by observation of the precession of the equinoxes or of the orbits of artificial satellites. The analysis of these observations has given rise to the conclusion that slightly more mass is present south of the equator than north of it.

We should note, however, that it now appears that higher-order terms are necessary in order to specify the potential with sufficient accuracy. Expansions up through U_4 are almost adequate for satellites in orbit about the earth, but those that take U_5 into account may yield better agreement with observation.

2.18   Perturbations of the Orbit

If we take the gradient of the potential, we find an expression containing a radial component plus terms that are inversely proportional to r^5, but are nonradial. As we indicated previously, we are therefore dealing with a *noncentral* force field.

We may define a torque $\mathbf{N}$ on the satellite. This is $\mathbf{N} \equiv -\mathbf{r} \times \nabla U$. When the indicated vector algebra is performed, we find that

$$\mathbf{N} = \frac{3Gz}{r^5}(C - A)(y\mathbf{i} - x\mathbf{j}); \qquad (2.58)$$

$\mathbf{i}$ and $\mathbf{j}$ are unit vectors along the x- and y-axes, respectively. This external torque causes the plane of the satellite orbit to change. We see that the torque will vanish only if $C = A$; if the planet is spherically symmetric the plane of the orbit is unchanging.

This last statement is not quite correct. The torque given by (2.58) will also vanish if the orbit is in the equatorial plane of the planet ($z = 0$). However, in that case the force of gravity will still be larger than for a spherically symmetric planet.

There is also another special case, motion along the y-axis, that does not lead to a twisting of the orbital plane. In this case $y = 0$, and the torque is directed along the y-axis. The result of this is that the orbit turns in its own plane, so that the perigee point rotates. This turning of the orbit in its own plane will also occur for all polar orbits.

2.19   Numerical Values

We expect that in general the application of the torque $\mathbf{N}$ on the satellite will lead to precessional effects on the orbit. This is because the torque tends to pull the plane of the orbit into the equatorial plane of the earth.

The equatorial bulge of the earth produces gravitational forces on a satellite in a nonequatorial orbit that are unequal and also noncentral. That is, neither the attractive force due to the bulge matter on the side of the planet near the satellite nor the force exerted by the far-side material is directed toward the planetary center; both are directed toward the bulges.

The precession that results is such that the orbit moves westward; if the orbit crosses the equatorial plane at a point P at a given moment, it will cross at some point P' to the west of this at the end of one period. Thus the plane of the satellite's orbit continually shifts westward. The amount of the shift is given by the rate of change of the orbital element Ω (see Appendix A). The shift is westward because we have assumed the satellite to have been launched *eastward* into orbit, to take advantage of the earth's rotation. The precessional effects due to the nonsphericity of the earth are not too difficult to evaluate explicitly.

It turns out that if we define a *disturbing function* R, such that $R = U -$

GM/r, the change of orbital elements depends on R and on the elements themselves. For example, the rate of change of Ω, the motion of the ascending node of the orbit, is given by

$$\dot{\Omega} = \frac{1}{na^2(1 - e^2)^{\frac{1}{2}} \sin i} \frac{\partial R}{\partial i}, \tag{2.59}$$

where n is the angular frequency. The other orbital elements also change; an illustration is provided by the rate of change of the argument of perigee, which is

$$\dot{\omega} = \frac{\cos i}{na^2(1 - e^2)^{\frac{1}{2}} \sin i} \frac{\partial R}{\partial i} + \frac{(1 - e^2)^{\frac{1}{2}}}{na^2e} \frac{\partial R}{\partial e}. \tag{2.60}$$

Since R is in the form of a series, the accuracy of the expressions for $\dot{\Omega}$ and $\dot{\omega}$ will be determined by the number of terms carried through.

The rates given in (2.59) and (2.60) are readily estimated for a spheroidal planet such as the earth. The rates are given (to first order in J) by

$$\dot{\Omega} = -\frac{Jn \cos i}{a^2(1 - e^2)^2} \tag{2.61}$$

and

$$\dot{\omega} = \frac{Jn(2 - \frac{5}{2} \sin^2 i)}{a^2(1 - e^2)^2}. \tag{2.62}$$

Here, as in Appendix A, i is the inclination of the satellite orbit to the celestial equator, a is the semimajor axis of the orbit in earth radii, and n is the mean angular velocity (degrees per day) of the satellite in its orbit. For the earth the quantity $J(= -\frac{3}{2}B_2/R^2)$ is about 1.624×10^{-3}.

Now for a polar orbit ($i = 90°$), $\dot{\Omega} = 0$, so that there is no precession of the line of nodes. The line of apsides, however, does regress or advance, since $\dot{\omega} \neq 0$.

For an orbit that is inclined 45° to the equator and that has a sidereal period of 2 hours with an eccentricity of 0.2, we find that

$$\dot{\Omega} = -3.37° \text{ per day},$$
$$\dot{\omega} = +3.58° \text{ per day}.$$

Thus the perigee point moves forward in the orbit, along the direction of the satellite's motion. It returns to its starting place in 101 days. Also, the ascending node completes a revolution in 107 days. We see that a satellite in orbit about the earth passes through a considerable volume of space.

Figure 2.7 shows the perigee motion for the OGO-2 satellite (also called POGO-1). This spacecraft was launched into a near-polar orbit (inclination, 87.3°) on 14 October 1965; it acquired useful data on the geomagnetic field until 2 October 1967. The perigee altitude was 410 km; apogee occurred at 1510 km. The figure shows a plot (for the first few months from launch) of

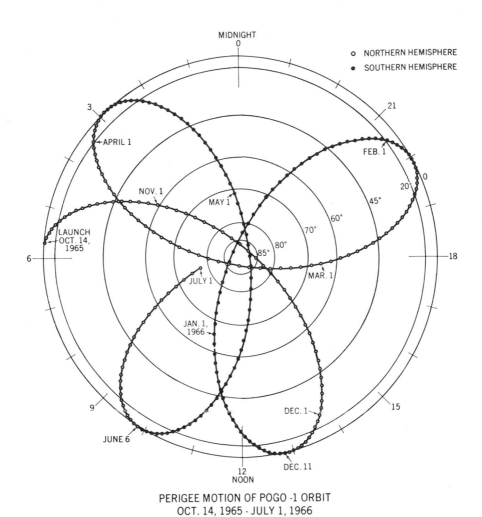

MIDNIGHT
0

o NORTHERN HEMISPHERE
● SOUTHERN HEMISPHERE

PERIGEE MOTION OF POGO -1 ORBIT
OCT. 14, 1965 - JULY 1, 1966

Figure 2.7 Motion of perigee point of first polar-orbiting geophysical observatory, POGO-1. This satellite was in a near-polar orbit. The view is from above the North Pole and the locus of perigee is shown against the parallels of latitude (concentric circles) and local time (azimuth scale).

the locus of perigee for OGO-2 as seen from above the north pole. The concentric circles are parallels of geographic latitude; the azimuth scale is hours of local time. The OGO-2 orbit at any epoch would project onto this diagram as a thin ellipse passing through the point given for perigee and the parallel of latitude equal to its inclination (87.3°). Thus for the first 10 days from launch, each observation of the geomagnetic field equatorward of 60° latitude occurred between 4 and 6 o'clock (A.M. and P.M.) local time.

2.20   The Lagrangian Points

It is not possible, in general, to solve the three-body problem in closed form. There are, however, some special cases that are amenable to solution in analytic form. Among these are the five points called the "Lagrangian points." These are the points where there are no net forces on a particle.

Let us for illustration look at the earth-moon system. The third body, the mass of which is negligible compared to the other masses, might be an artificial satellite or a particle of interplanetary dust. We wish to compute the positions of the five Lagrangian points for this system. Three of them are on the earth-moon line; they are known as the synodic solutions, for they revolve about the earth just as the moon does. The other two, $L/4$ and $L/5$, are at the apices of equilateral triangles whose base is the earth-moon line. It is relatively simple to compute the positions of $L/1$ and $L/2$, and it is instructive to do so.

Since the system is in rotation (see Figure 2.8), all that we have to do is equate the gravitational and centrifugal forces:

$$\frac{GM_e}{(r-a)^2} - \frac{GM_m}{a^2} = (r-a)\dot{\theta}_M{}^2. \tag{2.63}$$

In (2.63) a is the distance of $L/1$ from the moon; it is also equal to the distance

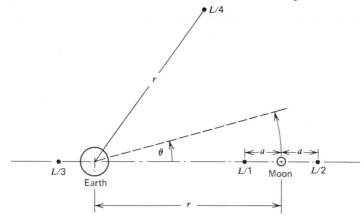

Figure 2.8   The Lagrangian points of the earth-moon system. It is not clear at this writing whether interplanetary dust has in fact concentrated at points $L/4$ and $L/5$ of this system.

of $L/2$ from the moon, which is taken to be revolving around the earth with an angular velocity $\dot{\theta}_m$. Since the angular momentum h_m of the moon is just $r^2\dot{\theta}_m$, we may rewrite (2.63) as

$$GM_e\left[\frac{1}{(r-a)^2} - \frac{M_m}{a^2 M_e}\right] = \frac{(r-a)h_m{}^2}{r^4}. \qquad (2.64)$$

To simplify (2.64), let α be the ratio of the masses M_m/M_e and let $y \equiv a/r$ be the distance of $L/1$ in dimensionless units. Equation 2.64 becomes

$$\frac{1}{(1-y)^2} - \frac{\alpha}{y^2} = \frac{(1-y)h_m{}^2}{GM_e r}. \qquad (2.65)$$

Since y is expected to be small, on physical grounds, we may use the expansion

$$\frac{1}{(1-y)^2} = 1 + 2y + 3y^2 \cdots \qquad (2.66)$$

and retain only the first two terms on the right-hand side. Also, for a circular orbit, the sidereal period P is just

$$P = \frac{2\pi r^2}{h_m}, \qquad (2.67)$$

whereas the harmonic law (2.33) tells us that

$$P^2 = \frac{4\pi^2 r^3}{GM_e}, \qquad (2.68)$$

in the present case. If we square (2.67) and set it equal to (2.68), we can show that the quantity $(h_m{}^2/GM_e r)$ that appears on the right-hand side of (2.65) is equal to unity. Putting our expansion (2.66) into (2.65), we may now transform the latter expression into

$$y^3 - \frac{\alpha}{3} = 0. \qquad (2.69)$$

We may now evaluate (2.69) numerically to find α. The masses are such that $\alpha = 1/81$; this yields $a = 61{,}000$ km on either side of the moon. Of course, if we had used more terms of the expansion (2.66), we would have a more accurate result. It should be noted that α does *not* correspond to the neutral gravity point in the system.

It is interesting to inquire whether the Lagrangian points have real physical significance. It appears that they do; the 14 so-called Trojan satellites have been observed in the sun-Jupiter system at $L/4$ and $L/5$ (see Figure 2.9). These points are particularly important because calculations indicate that they are *stable*; a particle placed at $L/4$ or $L/5$ will experience a restoring force if its orbit is slightly perturbed. The synodic solutions $L/1$, $L/2$, and $L/3$ are not stable. The Trojan satellites are thought to have originated in the asteroid belt. The 14 satellites were captured into $L/4$ and $L/5$ by the perturbing mass of Jupiter, and now precede and follow Jupiter in its orbit about the sun.

Many ground- and space-based attempts have been made to detect dust particles concentrated at $L/4$ and $L/5$ in the earth-moon system. The results are presently conflicting; some observers report having seen faint light patches (presumably sunlight scattered from concentrations of interplanetary dust) at $L/4$ and $L/5$, with integrated brightnesses just detectable by the naked eye.

It is possible that the gegenschein (Section 10.20) is due to dust concentrated at $L/3$ in the sun-earth system. This interpretation, however, is open to question.

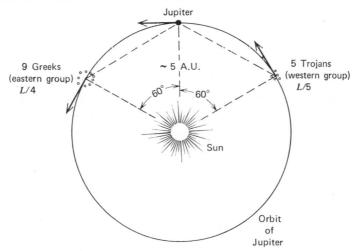

Figure 2.9   Approximate locations of the 14 Trojan Satellites of Jupiter; they slowly oscillate about the Lagrangian points $L/4$ and $L/5$ of the Jupiter-sun system. By modern convention those to the east of Jupiter are named for Greek warriors; those to the west, for the Trojans. (There is, however, one "spy" for the other group in both the Greek and Trojan "camps.")

Problems

2.1. If a particle describes a circle passing through the origin, find the law of force (depending on the distance alone) under which it moves.

2.2. If a particle describes an ellipse with the origin at the center, find the law of force under which it moves.

2.3. Find the velocity (relative to the earth) with which a spacecraft will get to Mars from the earth on a minimum-energy-transfer orbit.

2.4. A satellite is injected into an orbit at an altitude of 1000 km. At the point of injection the satellite is moving horizontally with a velocity of 8 km per second. Take $G = 6.67 \times 10^{-8}$ dynes-cm^2 per gm^2 and the mass of the earth as 6×10^{27} gm.
   (a) What is the semimajor axis of the orbit?
   (b) What are the apogee and perigee distances?
   (c) What is the eccentricity of the orbit?
   (d) What is the orbital period?
   (e) What are the velocities at perigee and apogee?

2.5. What kind of harmonic law of planetary motions would you have if the gravitational force varied inversely as the cube of the distance? Would Kepler's second law still hold?

2.6. There is a class of comets called "sungrazers"; their perihelia may be considered to be negligible. If the aphelion of one such comet is 200 A.U. (take 1 A.U. $= 1.5 \times 10^{13}$ cm), how often will it graze the sun?

2.7. (a) Compute the minimum speed for a spacecraft to escape from the solar system if it is launched at 1 A.U. from the sun.
   (b) Ignoring the effects of the earth's atmosphere, the moon, and the other planets, do the same as in (a) except launch the probe from the earth.

2.8. An interplanetary probe of negligible mass is observed to have an aphelion distance of 1.145 A.U. and is at 0.988 A.U. from the sun at the point of closest approach. Find
   (a) The semimajor axis of the orbit.
   (b) The eccentricity of the orbit.
   (c) The sidereal period of the probe.
   (d) The speed of the probe at perihelion.

2.9. A solar probe is injected into an elliptical orbit with an eccentricity of 0.5. Find the ratio of the speed of the spacecraft at perihelion to its speed at aphelion.

2.10. A small spacecraft finds itself trapped at the Lagrangian point $L/2$ of the sun-earth system. If the orbit of the earth is taken to be a circle, will the vehicle ever be in the shadow of the earth?

2.11. A satellite launched into earth orbit has an orbital inclination of 33°. The eccentricity of the orbit is 0.166, and the period is 126 minutes.
(a) At what rate and in what direction does the ascending node move?
(b) At what rate and in what direction does its perigee move?

2.12. Show that the speed v of a particle in a hyperbolic trajectory at a distance r from a planet of mass M is

$$v^2 = GM\left[\frac{2}{r} + \frac{1}{a}\right],$$

where a is the semitransverse axis.

2.13. A satellite is to be injected into a circular orbit of radius $3R$ about a planet of mass M and radius R. At the point of injection the booster has a malfunction, and the satellite velocity vector at burnout (at $3R$) has magnitude $\frac{2}{3}\sqrt{GM/R}$ and is inclined outward 30° from horizontal. What are the orbital parameters (semimajor axis, eccentricity, apogee, and perigee)? (*Hint*: Angular momentum is mh, where $h^2 = GMa(1 - e^2) = |\mathbf{r} \times \mathbf{v}|^2$.)

Chapter 3

THE EARTH'S ATMOSPHERE

3.1 Introduction

We live at the bottom of an "ocean" of air that is in many respects just like the seas that we are more familiar with, although the analogy is not exact. For example, the temperature and pressure of the atmosphere depend critically on depth (altitude), and so does the pressure in the ocean (the sea temperature also varies, but not as sharply).

The environment of our planet provides a basis for the understanding of the other planets in the solar system. We discuss the atmosphere of the earth in this chapter, including in our discussion the variation of parameters such as pressure, temperature, composition, and degree of ionization. Other phenomena, such as airglow and Van Allen radiation, are discussed in following chapters.

3.2 Nomenclature and Temperature Profile

The atmosphere of the earth has several reasonably distinct regions with different atmospheric properties at various altitudes. These are shown in Figure 3.1a. The nomenclature used here has been introduced as result of the study of the temperature profile of the atmosphere and is now generally adopted.

Temperature in the atmosphere is measured with a variety of techniques. At low altitudes (up to about 25 km) it is measured with balloon-borne thermistors. Higher-altitude determinations are conducted with sounding rockets as the lifting vehicles. Here the speed of sound may be determined through the use of grenades that are ejected from the rocket at various altitudes along its trajectory. The speed of sound, $V_{s'}$, is a temperature-dependent quantity; the relationship is

$$V_s = \left(\frac{\gamma R}{M} T\right)^{1/2}. \tag{3.1}$$

Here γ is the ratio of specific heats, T the atmospheric temperature, and M the mean molecular weight. The quantity R is the universal gas constant, 8.31×10^3 joules per °K-kg.

54

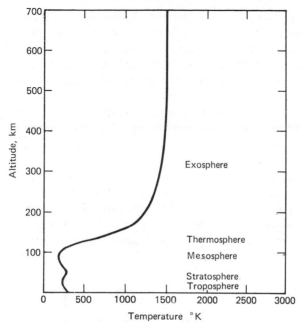

Figure 3.1a Nomenclature and the temperature profile adopted for the "standard atmosphere" adapted from a report of the U.S. Committee on Extension to the Standard Atmosphere (COESA), 1962.

At the very highest altitudes temperature may be inferred from studies of atmospheric drag on artificial satellites. The drag force depends on the density, which in turn is related to the temperature through a quantity called the *scale height*, H. We shall return to a discussion of H shortly.

Atmospheric drag forces are dissipative in character; they are velocity-dependent. The drag force on an orbiting spacecraft is imparted by elastic collisions with the molecules of the residual atmosphere. If the air is at rest and the satellite is orbiting at a speed v, the drag force is proportional to $(\frac{1}{2}\rho v^2)$, where ρ is the atmospheric density at the satellite. The drag is also proportional to the "presentation area" of the satellite; if the surface is a sphere, the presentation area is just one quarter of the total surface area. The coefficient of proportionality is called the *drag coefficient*, C_D. C_D depends on the geometry of the spacecraft and usually must be determined empirically. It turns out, however, that C_D is usually nearly equal to unity for a spherical spacecraft, assuming that the dimensions of the craft are large compared with a mean free path for an atmospheric constituent. If the dimensions are small compared to the mean free path of the gas molecules, and if the molecules either adhere to the surface or are totally reflected, then $C_D \sim 2$.

The lowest altitude region of the earth's atmosphere is called the troposphere ("turning sphere"); it is the region where clouds and weather are formed and is, as the name suggests, quite turbulent. It terminates at the *tropopause*, where the temperature halts its steady decrease with increasing altitude. The altitude and temperature of the tropopause fluctuate a little in time, but it nominally occurs at about 15 km with a temperature of around $-60°C$ at midlatitudes. It is interesting that the tropopause temperature is lowest around the equator ($\sim -100°C$) and warmest at the poles (perhaps $-40°C$). The tropopause altitude, on the other hand, is lowest over the poles (~ 6 km) and highest over the equator (~ 18 km).

The stratosphere is found above the tropopause. The name implies that it is nonturbulent. It is far less turbulent than the troposphere, but modern investigations have shown it to be not strictly calm. The temperature increases slowly with altitude through the stratosphere. The stratosphere terminates at the *stratopause*, where the temperature dependence on altitude once again reverses sign. The stratopause is usually somewhere around 50 km of altitude. Figure 3.1*b* shows the mean temperatures that have been adopted for planning purposes as part of the "standard atmosphere."

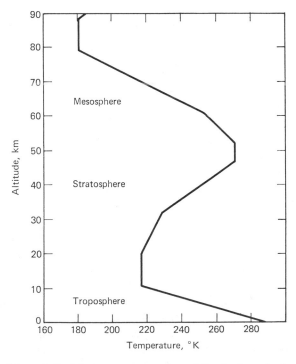

Figure 3.1*b*   The temperatures adopted for the "standard atmosphere." Adapted from 1962 COESA Report.

The stratosphere is thickest over the poles and may (at times) be essentially nonexistent over the equator. This is because of the altitude variation of the tropopause. If it is indeed true that the stratosphere may be of zero thickness at the equator, the terrestrial atmosphere actually has two "stratospheres," on either side of the equator. Research is now under way, using radioisotopes as tracers, on the mechanisms by which stratospheric air is circulated across the equator, a process that does appear to operate.

The temperature decreases with increasing altitude throughout the *mesosphere*, the next higher region. This temperature decline terminates at about 80 km, where the *mesopause* occurs. Clouds are occasionally observed even as high as the mesopause. They may be observed after dark on the earth's surface, because the sun's rays still illuminate them. Such clouds are called "noctilucent clouds" (Figures 3.2a, b) and their composition, altitude, and rate of occurrence are presently the object of study. They are so thin that stars may be easily seen through them; thicknesses of ~ 1 km have been deduced.

The *thermosphere* receives its name from the fact that very high (kinetic) temperatures may be found there. The temperature increases with altitude through the thermosphere, apparently reaching values as high as nearly 2000°K during those periods when the sun is most disturbed.

The highest altitude region, the *exosphere*, is really the "fringe of space." Some of the molecules comprising our atmosphere may escape into interplanetary space from the exosphere. The mean free paths are no longer than the scale height in the exosphere; the base of the exosphere is taken to be that altitude where the mean free path is equal to H and is usually found at about 400 km of altitude. Temperature is relatively independent of altitude in the exosphere.

3.3   Temperature Distribution in the Troposphere

We may inquire into the reasons for this odd-shaped temperature profile. In the troposphere we see at once that the atmosphere must be unstable (turbulent), since it is hottest at the bottom, rather than the top. The heat source here is solar radiation that is absorbed by the planetary surface; the warm surface heats the troposphere from below. We therefore expect that the major heat transfer process responsible for cooling the troposphere is convection; radiation is less important in the lower atmosphere than convection, at least up to the troposphere.

A thermally insulated 15-km column of air heated at the bottom (say with a heating element at a temperature T) would after some time become isothermal at T. We note, however, that the troposphere exhibits a temperature "lapse rate" of about −6.5°C per km up to the tropopause. There must therefore be a heat sink somewhere in the troposphere in order for the atmosphere

Figure 3.2 Two photographs, taken at high latitudes, near local midnight. The wavelike structures in the noctilucent clouds are sometimes observed to move at 400 miles per hour, giving an indication of wind speeds at 80-km altitudes. Reproduced from *Particles in the Atmosphere and Space* by R. D. Cadle, New York: Reinhold, 1966.

to cool in this manner. We shall return to the identification of the heat sink after first calculating the cooling by convection.

The lapse rate may be ascribed to convective expansion of a rising parcel of air. If we make the adiabatic assumption (that there is no heat input except from the ground and that there is no heat "sink"), we may estimate what the lapse rate should be. Although the very fact that cloud formation takes place tells us that this assumption is hardly valid; considerable heat is released wherever water-vapor condensation occurs. We also assume that the troposphere is composed of an ideal gas.

In the adiabatic case

$$PV^\gamma = A = \text{constant},\tag{3.2}$$

where P is the pressure and V the volume of our parcel of air. To digress for a moment, $\gamma = \frac{5}{3}$ for an ideal (monatomic) gas; it is $\frac{7}{5}$ for diatomic gases, and is more generally $(g + 2)/g$, where g is the number of degrees of freedom for the molecule. It turns out that $\gamma = 1.4$ is a reasonable value for "air" for most purposes.

The ideal gas law tells us that

$$PV = nRT,\tag{3.3}$$

where n is the number of moles, or

$$V^\gamma = \left(\frac{nRT}{P}\right)^\gamma,\tag{3.4}$$

so that

$$\frac{T^\gamma}{P^{\gamma-1}} = B, \text{ a new constant.}\tag{3.5}$$

Therefore

$$T = BP^{\gamma-1/\gamma},\tag{3.6}$$

the adiabatic lapse rate (where we should recall that the pressure P depends on the altitude). Convection results in a power law dependence of the temperature on pressure.

We remarked earlier that there must be a heat sink somewhere in the lower atmosphere. The requisite sink is provided by the molecules of water vapor in the troposphere. Water vapor is quite opaque in the infrared; it strongly absorbs heat energy from the ground and reradiates some of this in the "wings" of the pressure-broadened spectral lines. Now very little water vapor lies at higher altitudes than the tropospheric region (and that present is not subject to pressure broadening), so that the reradiated energy escapes directly to space and therefore is not available for tropospheric heating.

In fact, H_2O is only one of the triatomic molecules that are good infrared radiators. Others found in the terrestrial atmosphere are CO_2 and O_3. Relatively complicated molecules like these make the radiative heat transfer process important. We may say, therefore, that the presence of triatomic molecules (even in relatively small concentrations) determines the temperature profile of a planetary atmosphere.

We should note that a calculation based on dry-air convection above, as described in (3.6), leads to a lapse rate of perhaps $-9°C$ per km. The presence of water vapor and its condensation, however, reduces the lapse rate to the observed value.

3.4   Temperature of the Stratosphere

The stratosphere is expected to be stable, since it is heated from the top. Measurements indicate that it is indeed much more stable than the turbulent atmosphere below it. Its relatively warm temperature is caused by the presence of a layer of ozone, O_3. This gas absorbs solar ultraviolet rays and thus heats up its surroundings. (O_3 also protects us at sea level from this harmful radiation.)

The stratosphere, however, is not completely stable. Sudden "explosive warmings" are occasionally observed. As seen in Figure 3.3, the temperature

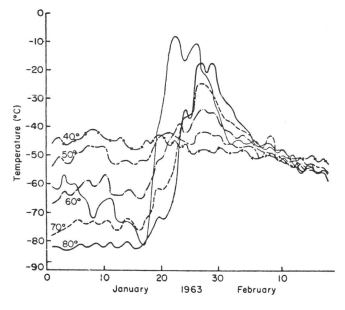

Figure 3.3   An explosive warming of the stratosphere at the 10-millibar pressure level ($\sim$100,000 feet) in January 1963. The event is shown at several latitudes that range from 40° N poleward; the data are averaged over longitude. In addition, the light curve illustrates the data from one station, at 90° W longitude. From W. L. Webb, *Structure of the Stratosphere and Mesosphere,* Academic Press, Inc., 1966.

of large regions of the stratosphere may increase by $\sim30°C$ in 1 week. The warmings have so far been observed only in the northern hemisphere and most often in January. They are presumably connected with disturbances originating below (in the troposphere) and propagating through the tropopause. Mechanical energy (turbulence) may thus be converted to heat in such events.

3.5 Temperatures of the Mesosphere and Thermosphere

Above the stratopause the air cools again through convection and through radiation from the ozone layer. This atmospheric cooling terminates at the mesopause (which is also the altitude where the atmosphere becomes ionized, rather than a mixture of neutral gases). Temperatures at the mesopause are the coldest found in our planet's atmosphere.

A substantial *positive* temperature gradient exists in the earth's atmosphere at greater altitudes, at heights between ~80 and ~400 km. The explanation for this warming with altitude is based on the facts that (a) the air becomes quite rarefied with height, (b) there are few if any triatomic molecules present in the upper atmosphere, and (c) solar radiation of small penetrating power (the extreme ultraviolet, $\lambda < 1750$ Å) heats the atmosphere from above. Additional heating arises from the exothermic chemical reactions entered into by the ions and electrons that are present.

As we have seen, the temperature profile in the lower atmosphere is determined by the heating (from the planetary surface below) and cooling that occur through the processes of convection and infrared radiation by triatomic molecules. In the upper atmosphere, however, convection is inhibited because the heat input is from above, rather than from below. Even if convection were to occur, it would be inefficient because of the reduced pressures. In addition, single atoms are present, rather than polyatomic molecules; substantial cooling via infrared radiation does not take place. Hence the only mechanism that remains for cooling the high atmosphere is downward conduction. It is this conduction that results in the increase of temperature with increasing altitude.

The high temperatures found in the upper reaches of our atmosphere do not imply that there is a large heat reservoir there, especially in view of the small densities that exist at those altitudes. Instead, the high temperatures (nearly 2000°K at some times) merely testify to the inefficiency of the process for heat removal from the region. Only about 10^{-6} of the solar energy supplied to the earth is absorbed in the thermosphere; most of this energy is absorbed on the ground.

The temperature rise eventually ceases at extreme altitudes, and the temperature at the top of the atmosphere is essentially independent of altitude. The exosphere is isothermal. The reason for this isothermal character is that the mean free paths are very long at great heights. The thermal conductivity is quite high in the exosphere.

Any gradient of temperature above ~400 km requires the introduction of heat at higher levels and therefore a transport by conduction throughout the upper thermosphere. No known solar ultraviolet absorption processes occur at these heights, and only particles entering the atmosphere from beyond could supply such heating. However, as remarked above, the mean free paths are too large to supply high-altitude heat, even if such particles were found to exist.

3.6   Temperature Variability

The temperature of the upper thermosphere varies between 900° and 1700°K. The temperature is strongly dependent on solar activity, reaching its highest values when the sun is most disturbed. An active sun produces much more ultraviolet than does the quiet sun. Since solar activity follows an 11-year cycle, the temperature at high altitudes follows a similar cycle.

The upper atmospheric temperature, quite unlike its low-altitude counterpart, varies considerably throughout the day. At high altitudes afternoon temperatures may exceed predawn temperatures by a factor of 1.3. An equivalent excursion at the ground would be 100°K! The sensitive response of temperature to diurnal change in the solar heat input is a consequence of the low density and correspondingly low heat capacity of the upper atmosphere.

There are other variations in the high-atmosphere temperature. One such variation appears related to the 27-day solar rotation period; the exospheric temperature varies by ~200°K with this period. Also, for some as yet unexplained reason, the high-altitude temperature varies by as much as 1000°K during a geomagnetic "storm." A storm is a relatively major fluctuation in the earth's magnetic field; at sea level the field may decrease by ~1% in an hour or less and may not recover to its prestorm value for a week or so. Geomagnetic storms are believed to be due to the effects of interplanetary magnetic fields, as transported by plasma issuing from the sun, when they interact with the earth's magnetic field (see Chapter 9).

3.7   The Pressure Profile

We have remarked before that the pressure is a function of the altitude h. If we assume that the pressure balance is such that the pressure just equals the weight of overlying air, so that an element of pressure is given by the weight of an air parcel, we may write an equation of hydrostatic equilibrium

$$dP = -\rho g \, dh, \qquad (3.7)$$

for a parcel of air of density ρ, unit cross-sectional area, and thickness dh. We have assumed that g, the acceleration due to gravity, is independent of h in the altitude region of interest.

The ideal gas law (3.3) may be rewritten as

$$P = nkT, \tag{3.8}$$

where k is Boltzmann's constant, 1.38×10^{-23} joule per °K. Now the density ρ is

$$\rho = nm, \tag{3.9}$$

where m is the mass of the average air "atom." Substitution of these last two relations into (3.7) results in

$$\frac{dp}{p} = -\frac{mq}{kT} \, dh. \tag{3.10}$$

We may integrate (3.10) and find that

$$P = P_0 e^{-h/[kT/mg]}, \tag{3.11}$$

or

$$P = P_0 e^{-h/H}. \tag{3.12}$$

Equation 3.12 is the familiar "barometric law." The quantity P_0 is a constant of integration; it is also the pressure at sea level, where $h = 0$. Figure 3.4

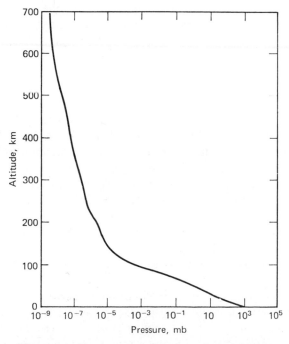

Figure 3.4   Pressure as a function of altitude in the standard atmosphere. Courtesy U.S. Committee on the Extension to the Standard Atmosphere (COESA), 1962.

shows how closely the real atmosphere approximates the barometric law. Below ~125 km the exponential atmosphere is evidently a good fit.

We introduced the parameter H in (3.12); the definition is that

$$H = \frac{kT}{mg}.\tag{3.13}$$

H is called the *scale height* of the atmosphere; it is that distance over which the pressure falls by a factor of e.

The barometric law shows us that the vast majority of the atmosphere is found at low altitudes. This fact reveals something about the *twinkling* of celestial objects. Mars sometimes subtends the tiny angle of only 4 seconds of arc, and yet it doesn't twinkle. Suppose that we make a generous allowance and say that the atmospheric irregularity responsible for the twinkling is as much as 1 mile away from the observer. Since 4 seconds = 2×10^{-5} radian, the size of the irregularity can at most be about 1 inch! In other words, very tiny atmospheric irregularities are responsible for twinkling.

3.8   Scale Height

Equation 3.13 defines the scale height, which is one of the most useful parameters in the study of planetary atmospheres. For example, if the scale height is truly constant throughout the altitude range of interest, we know that the atmosphere is isothermal and also that the composition is constant (or else m would change). The scale height therefore represents a convenient way of characterizing the atmosphere.

There are many facets of the scale height. If we set $mgH = kT$, we have something resembling the conservation of energy in a gravitational field. This tells us that an atom of mass m at a temperature T will have enough kinetic energy to reach an altitude H above the starting point. It happens that H becomes very large in the exosphere; hydrogen (a major constituent of the exosphere) can thus escape. At low altitudes (near sea level) the average mass for air is about 29; this leads to a scale height H of 8 km. H is quite constant up to an altitude of about 120 km, since $H = 7 \pm 1$ km up to this height; air motions must provide good mixing of the atmospheric constituents. Figure 3.5 illustrates the dependence of H on altitude.

At greater altitudes there is no longer good mixing of the atmospheric constituents, and *diffusive separation* sets in. That is, at the reduced pressures found at $h > 120$ km, diffusion proceeds quite rapidly; the mixing is ineffective. Thus gravitational diffusive separation becomes efficient. The lighter gases float on top of the heavier ones. At these elevated altitudes, therefore, each species has a different value of H. (This is because the mean free paths become sufficiently great at 120 km that the various species move about as independent particles.)

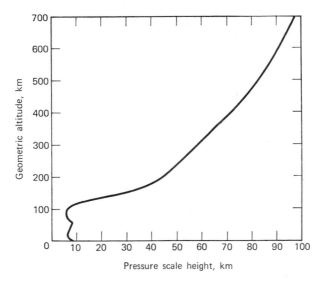

Figure 3.5   Pressure scale height H as a function of geometric altitude. Courtesy U.S. Committee on the Extension to the Standard Atmosphere (COESA), 1962.

3.9   Density Variation

A good rule of thumb is that up to 100 miles of altitude the density changes by a factor of 10 every 10 miles. Thus the density at $h = 20$ miles ($\sim$100,000 feet) is only 1% of the density at sea level; 99% of the atmosphere is found below 20 miles of altitude. From (3.12) we see that the pressure falls by a factor of 10 in 2.13 scale heights, which leads to this rule of thumb. Figure 3.6 displays the variation of number density of the neutral atmosphere with altitude up to $\sim$700 km.

3.10   Present Atmospheric Composition and its Evolution

Near sea level the principal permanent gases now in the earth's atmosphere have the characteristics given in the Table 3.1 (upper three lines). There are a few others present in trace amounts. There are also some variable-concentration gases. The latter are shown in the lower three lines of the table. The upper part of Table 3.1 tells us the average mass of an "air" molecule. It is about 29, so that $H = 8$ km, as we noted previously. In fact, the air at the lower altitudes is evidently so well mixed that $H = 7 \pm 1$ km all the way up to an altitude of 120 km.

At heights in excess of 120 km the picture becomes more complicated, for each species assumes its own scale height. H_i, under the action of diffusive

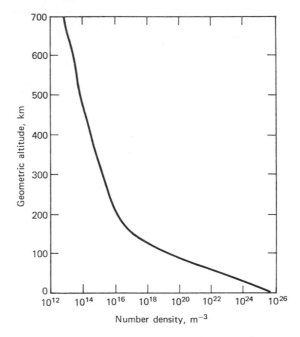

Figure 3.6 Number density n as a function of geometric altitude. Only neutral particles are considered. Courtesy COESA, 1962.

separation. The value of H_i depends on solar activity, because the thermopause temperature depends on this activity through the intensity of the received solar ultraviolet radiation. In addition, the thermopause temperature is affected by the solar rotation period of 27 days, for this temperature changes by perhaps 200°K in that period. Also the temperature is markedly changed during geomagnetic "storms" (severe magnetic disturbances); temperature changes of as much as 1000°K have been detected during such

TABLE 3.1. PERMANENT GASES IN THE EARTH'S ATMOSPHERE

Gas	Molecular Mass	% by Volume
N_2	28.016	78.110
O_2	31.999	20.953
Ar	39.942	0.943
H_2O	18.005	0–7
CO_2	44.009	0.01–0.1
O_3	47.998	0–0.01

storms. One speculation was that this heating is caused by atmospheric absorption of hydromagnetic waves, but detailed calculations show that this is not the case, and the heating remains an unsolved problem.

Figure 3.7 shows the various species found at altitudes between 120 km and 700 km. The logarithmic slope of each line is proportional to the individual scale heights H_i.

Hydrogen is normally found floating above the helium layer, as one might expect from diffusive separation, at altitudes in excess of 1300 km (the reverse situation exists at lesser heights). Thus a *geocorona* of hydrogen atoms envelops the earth; it apparently extends to great distances from the planet (see Section 5.9). The atoms of the geocorona move on escape trajectories or in ballistic orbits, or, because of (infrequent) collisions with other exospheric constituents, move in satellite orbits and thus remain in the planetary vicinity. If in ballistic motion, they eventually return to the earth.

The question of the origin of this atmospheric composition is interesting. Presumably all of the planets were formed at about the same time, from the same chemically homogeneous medium, the primordial solar nebula. But Mars and Venus, as we shall see, seem to have atmospheres that are primarily composed of CO_2, while the envelopes of Jupiter and Saturn contain H_2 and CH_4. The earth is unlike any of these.

The earth is deficient, with respect to the sun, in elements that include H_2 and He, but that also include C, N, and the noble gases (Ne, Ar, Kr, and Xe). All these are gases at a few hundred degrees (centigrade). The earth's relative concentration of nonvolatile elements, such as Na, Mg, and Al, appears to resemble that of the sun. We therefore conclude that the earth was sufficiently warm at some previous epoch in its history to drive off the volatile elements, and that the remaining traces of these substances are due to outgassing of the planetary interior.

Indeed, planetary outgassing is believed to be the source of our present-day atmosphere. Volcanoes provide the most striking illustration of this outgassing. The original atmosphere most likely consisted of water vapor and hydrogen, with heavier gases such as NH_3 and CH_4 (or CO_2 and nitrogen) also present. Either mixture is consistent with laboratory experiments on the effects of electric discharges ("lightning") on such atmospheres; large hydrocarbon molecules and amino acids result from the discharges.

In any event, it now seems clear that the oxygen now present is due to photosynthesis by plants, and has only accumulated over the last billion years or so. On the earth, the study of beds of iron ore has revealed that the red beds, those that are rich in oxidized iron (the ferric form), mark the advent of oxygen in our atmosphere. The earliest continental red beds are less than 2×10^9 years old; the oldest such beds are apparently less than half the age of the earth.

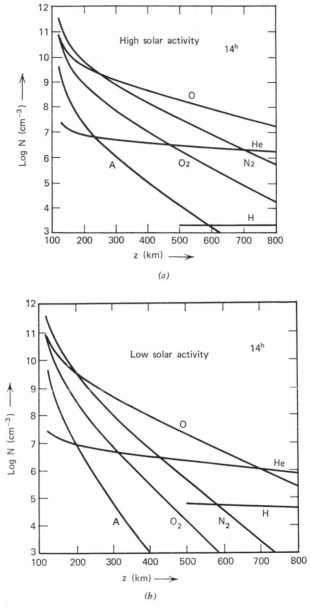

Figure 3.7 The number densities of various constituents of the earth's atmosphere for 1400 local time and two intensities of solar activity. Courtesy COESA, 1962.

3.11   Thickness of the Atmosphere

We next ask: What is the total number of atoms, n_T, present in a cylindrical column (of terrestrial air) that extends to infinity from the surface of the planet? The cross-sectional area of the cylinder is taken as unity. To answer this question we may integrate the barometric law (3.12) over altitude. Our expression becomes

$$n_T = \int_0^\infty n_0 e^{-h/H} \, dh, \qquad (3.14)$$

where the scale height H is assumed to be a constant equal to 8 km. The quantity n_0 is related to the surface pressure P_0, for the number of atoms at sea level is just

$$n_0 = \frac{P_0}{kT}. \qquad (3.15)$$

The result of the integration is that

$$n_T = n_0 H. \qquad (3.16)$$

Thus, if the atmosphere were *isobaric* (so that the density would be independent of altitude), all of the earth's atmosphere would be found in a layer just 8 km thick. (We have, of course, assumed here that the earth is flat, not spherical.) This thin layer confirms our earlier remark that the vast majority of the atmosphere is found at low altitudes. It is interesting to evaluate (3.16) numerically, for n_T gives us the total number of atmospheric atoms between the observer and the universe. The evaluation is left as an exercise for the student.

For investigations such as the study of cosmic rays, it is useful to express the thickness of the atmosphere in different units. The surface pressure on earth is 1013 millibars; this is equivalent to a mass of 1033 gm per cm², the mass of a 76-cm unit column of mercury. This latter figure is the amount of atmospheric mass above 1 cm² of the earth's surface. It is also interesting that sea-level pressure is equal to a 10-meter-high column of water; the atmosphere is equivalent in thickness to 31 feet of water. Humans are therefore rather like fish that live at the bottom of a 31-foot-deep lake, the waters of which seriously distort their view of the universe. But a balloon at 130,000 feet is only about 1 inch from the surface of this "lake"; a sounding rocket may be considered actually to "pop through" the surface, while a satellite "lives" above the surface.

3.12   The Ionosphere

We have so far been discussing the neutral atmosphere. We turn now to the ionosphere, that region of ionized gases electrically charged by the action of solar radiation and found at great altitudes. The ionosphere was first alluded to as a possibility by Gauss in 1841. It was postulated by Balfour Stewart in 1883, in order to explain some of the observed variations in the geomagnetic field. Finally, the presence of the ionosphere was established by Kennelly and by Heaviside in 1902. These workers were able to explain the long-range propagation of radio waves found by Marconi in the previous year, by using the ionosphere as a high-altitude "reflector." Kennelly and Heaviside named it the "electrified layer"; later usage has resulted in a contraction to the "E-layer." Actually, as we shall see, the ionosphere doesn't "reflect" anything; incident radio waves are *refracted* back down to the ground by the ionized medium.

Measurements ("soundings") may be made of the density of ionization in the ionosphere. This may be accomplished by transmitting frequency-modulated pulses of electromagnetic energy toward the ionosphere. The frequency at which the energy is "reflected" back toward the transmitter depends on the density of ionization, as we shall see. The transmitters may be located on the ground (such instruments are called "ionosondes") or on orbiting earth satellites that are called "topside sounders."

These soundings have now revealed that there are at least three layers of ionization in the ionosphere. The one below the E-layer is called the D-layer, while the one above is known as the F-layer; these names were given by Appleton in the mid-1920s, as he developed magnetoionic theory.

3.13   Effect on Scale Height

It is interesting to see the effect of the presence of ionization on scale height, as compared to that found in a similar but neutral planetary atmosphere.

If the atmosphere is ionized, we may write that the scale height of a species, H_i, is given by

$$H_i = \frac{kT_e + kT_i}{(m_i + m_e)g}. \tag{3.17}$$

In (3.17) the subscripts e and i refer to the electron and ion, respectively.

Suppose now that the kinetic temperatures of the electrons and ions are equal, so that

$$T_e = T_i = T. \tag{3.18}$$

Equation 3.17 now becomes

$$H_i = \frac{2kT}{m_i g}, \tag{3.19}$$

since the mass of the electron is negligible compared to that of the ion.

An examination of (3.19) reveals that the scale height has seemingly been *doubled*; H_i is twice that found in a neutral atmosphere. This follows if the electron and ion temperatures are really equal. It appears, however, that the electron temperature is frequently several times that of the ions. Photo-ionization constitutes a source of photoelectrons with energies of some tens of electron volts, well in excess of the thermal energies of the other ionospheric constituents. Thus the calculation of electron temperatures in the ionosphere involves a consideration of equilibrium between the rate at which ambient electrons in the ionosphere are heated by collisions with photoelectrons and the rate at which they are cooled by collisions with ions and neutrals.

3.14 Ionospheric Electric Fields

The apparent doubling of the scale height results from the electric fields present in the ionosphere. The scale height for electrons is of course very great; charge separation therefore occurs and an electric field results. In effect, we may think of a "layer" of electrons "floating" at great heights above the region of positive charge concentration.

To evaluate the electric field strength E, which is directed *upward* toward the electrons, we may write

$$eE = \tfrac{1}{2}m_i g, \tag{3.20}$$

since the doubled scale height is equivalent to saying that the "effective" gravity is half that of that pulling on the neutral gas. Therefore, if $T_e = T_i$, we have

$$E = \frac{mg}{2e} \tag{3.21}$$

as our explicit expression for the charge separation field strength.

Another way of writing the effect on the scale height is to say

$$H_i = \frac{kT}{mg - eE}, \tag{3.22}$$

where $E = mg/2e$ cancels half the gravity on, say, the oxygen ion.

This field is so strong that it is "responsible" for the exosphere being composed of hydrogen. The value of E is so great that the protons are pulled up and "float" on top of the other atmospheric constituents. (We shall return to the origin of hydrogen and helium in Section 3.18.)

This redistribution of the positive charges, however, tends to reduce these electrostatic forces. For example, the insertion of the high-altitude proton layer tends to reduce the scale height for singly-ionized oxygen atoms down toward the value for H found for neutral monatomic oxygen.

This idea may be fixed with a detailed example. Assume that there is a trace of a heavy gas (O^+, let us say) in an ionized hydrogen ionosphere, and that we wish to find the scale height H_0 of the heavy species in this rather artificial atmosphere. We may write

$$H_0 = \frac{kT}{m_0 g - eE},$$

where m_o is the mass of the oxygen atom. Now if m_p is the mass of the proton,

$$E_{H^+} = \frac{m_p g}{2e},$$

so that

$$H_{O^+} = \frac{kT}{m_o g - em_p g/2e}.$$

Since the mass of oxygen is 16 times the mass of the proton, we find that

$$H_{O^+} = \frac{kT}{m_0 (\frac{32}{31})g}.$$

This tells us that

$$H_{O^+} = (\tfrac{32}{31})H_O.$$

Thus the scale height of the ionized oxygen atoms is very nearly equal to that of the neutral gas, rather than the doubling we previously estimated.

A calculation (which we shall leave as an exercise) of the inverse problem, in which we have a trace of ionized hydrogen in a gas of ionized oxygen atoms, would show that the proton scale height H_p is

$$H_p = -\frac{16}{7}\frac{kT}{m_0 g} = \frac{-kT}{7m_p g}.$$

Thus $H_p = -\frac{1}{7}$ the scale height for neutral hydrogen. The minus sign, when interpreted in the framework of the barometric law, reveals, as expected, that the protons rise above the oxygen.

These ideas on ionospheric electric fields may be checked experimentally by releasing an ionized glowing cloud at rocket altitudes (several hundred kilometers). The presence of an electric field will alter the scale height of the ions in the cloud. This may be observed from the ground; the optical emission as a function of altitude will be altered. It appears at this writing that barium will serve nicely. If it is released when it is dark on the ground. but sunlight is present at the release altitude, the gas will be photoionized and also glow.

3.15 Ionization Profile

A midlatitude vertical profile of the ionization density (number of ions per cubic centimeter) is shown in Fig. 3.8 for the earth's atmosphere. The most pronounced feature is the "layering" of charge density above 80 km. The

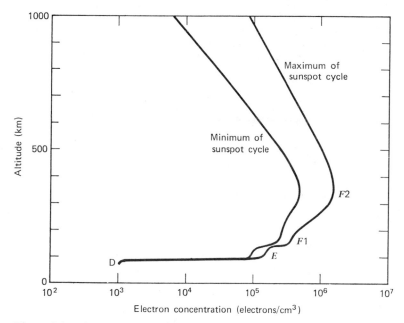

Figure 3.8 The concentration of electrons in the earth's ionosphere. The D-layer disappears at night, and the F1- and F2-layers coalesce in the absence of sunlight. These data apply at midlatitudes.

ionization trails off at extreme altitudes, forming a "magnetospheric plasma" of relatively low-energy ions around the earth.

The magnetospheric plasma distribution depends on local time and the state of geomagnetic activity. When there is only moderate activity, however, the distribution is as shown in Fig. 3.9 for times near noon. The shaded, inner region, called the plasmasphere, contains plasma with number densities that range from 10^3 to 10^2 protons cm^{-3}. The outer, unshaded region is known as the plasma trough; particle densities there range from about 1 to 10 protons cm^{-3}. The boundary between the two regions has been named the "plasmapause." The plasmapause is only about 0.15 earth radius thick and is aligned along the geomagnetic field lines. It is as though the ionosphere terminates abruptly at $\sim 4R_E$ (over the equator); the ionization decreases sharply there. It is interesting that this upper region of the F-layer changes its boundary altitude with local time; a pronounced evening bulge exists. Plasma is found between 4 and $5R_E$ shortly after dusk (see Fig. 3.10).

The ionosphere has a lower boundary also. At night virtually no ionization is present below about 80 km; the D-layer disappears at night. Also, the F1- and F2-layers more or less coalesce during the night. It is also interesting that there are occasionally "clouds" of ionization, called "sporadic E," at

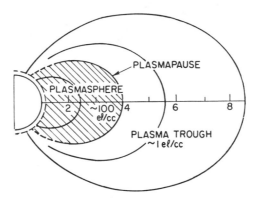

Figure 3.9 The configuration of magneto-spheric plasma as deduced from whistler measurements made on the earth. The data refer to 1400 *LT* and a planetary magnetic index that varies between 2 and 3 (moderate activity). From Carpenter, in *Journal of Geophysical Research*, 1966.

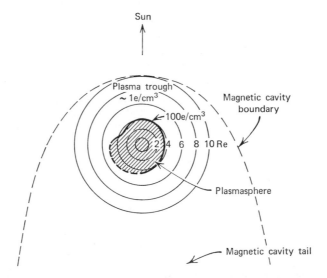

Figure 3.10 Schematic view of the distribution of thermal ions in near-earth space deduced by Carpenter in 1966 from whistler data. The equatorial plane is shown. The distribution applies for moderately disturbed geomagnetic conditions ($Kp = 2$ to 4).

low E-layer altitudes. These occur more frequently than previously thought, and their origin is presumably different from the ionization found at other altitudes.

It is not at all clear why the E- and F1-layers do not disappear at night, even during the long polar night. In fact, they continue to exhibit their usual diurnal behavior during this long absence of sunlight.

Ionization due to meteors has been suggested for the continued existence of the E-layer. Convective upward (outward) movement of the ions along the geomagnetic field lines into regions where the density is too low for recombination might be responsible for the durable F1-layer. Other possibilities that have been mentioned include solar wind particles penetrating into the magnetospheric "tail" of the earth and then coming back upstream, as it were, into the polar regions. It is also possible that aurora-like low-energy particles continuously bombard the polar caps. We shall allude to this again in the chapter on aurorae.

Sporadic E may be due to the heavy ions deposited at high altitudes by meteors. If a wind shear exists at sporadic E altitudes, the heavy ions may form a layer that is highly conductive.

3.16 Layer of Charge

It is of interest to investigate the origin of the layering of charge. S. Chapman was the first to discuss this problem.

The ions are produced by energetic (ultraviolet and x-ray) photons from the sun that photoionize the atoms in the earth's atmosphere. Because of the rapid increase with depth of atom density, the photons produce more and more ions at greater depths. Finally, however, a depth is reached at which so many solar photons have been absorbed that production begins to fall off with further increases in depth. A layer of production thus results.

We may make these ideas more precise. If the differential solar energetic photon flux at the top of the earth's atmosphere is F photons per cm²-second-cps, and if we for simplicity's sake restrict ourselves to monochromatic radiation incident at a zenith angle θ_0, we may compute the photon flux that survives down to an altitude h.

From the barometric law the number density of atmospheric atoms as a function of h is given by

$$N(h) = N_0 e^{-h/H} \text{ atoms cm}^{-3}, \tag{3.23}$$

where, as before, H is the scale height. Now if α_v is the absorption cross-section of a given atmospheric species for an incident frequency v, the photon flux of frequency v at h will be

$$F(h) = F \exp \left[\frac{-\alpha_v}{\mu_0} \int_h^\infty N(h')\, dh' \right], \tag{3.24}$$

where $\mu_0 = \cos\theta_0$. The integral represents the total number of atoms traversed. The expression (3.24) may therefore be rewritten as

$$F(h) = F \exp\left[\frac{-\alpha_v N(h)H}{\mu_0}\right]. \tag{3.25}$$

Let β_v be the absorption cross-section for photoionization. If photoionization were the only absorption process, we would then have $\beta_v = \alpha_v$. Thus $q(h/\mu_0)$, the number of ionizations per cm³-second, would be the product of the flux at h times β_v, multiplied by the number density of atoms at h. Thus q is

$$q(h/\mu_0) = \beta_v N(h) F \exp\left[-\frac{\alpha_v}{\mu_0} N(h)H\right]$$

$$= \beta_v N_0 F \exp\left[-\frac{h}{H} - \frac{\alpha_v N_0 H}{\mu_0}\exp\left(-\frac{h}{H}\right)\right]. \tag{3.26}$$

To find the altitude, $h_{\max}$, at which the rate of production is maximal, we set

$$\frac{dq}{dh} = 0. \tag{3.27}$$

This results in

$$\exp\left[\frac{+h_{\max}}{H}\right] = \frac{\alpha_v N_0 H}{\mu_0}, \tag{3.28}$$

so that $h_{\max}$ varies throughout the day as μ_0 changes. We can also see that at this altitude of maximum production, $h_{\max}$, the maximum rate of ion production is

$$\frac{F\beta_v\mu_0}{e\alpha_v H}.$$

In fact, when $\beta_v = \alpha_v$ and the sun is at the zenith (so that $\mu_0 = 1$), the maximum rate of ion production is just $(1/e)$, the incident solar flux, divided by the scale height. This is a result that we might guess on purely physical grounds.

It still remains, however, to describe the formation of a layer of charge. Let q_m be the maximum rate of ion production, which occurs when $\mu_0 = 1$; h_M is the altitude of maximum production under those circumstances. We wish now to consider q relative to q_m. We find that

$$q(h/\mu_0) = q_m \exp\left[1 - \left(\frac{h - h_M}{H}\right) - \frac{1}{\mu_0}\exp - \left(\frac{h - h_M}{H}\right)\right]. \tag{3.29}$$

This is the number of ions *produced* per cm³ per second. Ions are also *lost*, however, through recombination of the electrons and ions. If we assume the ionosphere to be in equilibrium, the loss rate equals the production rate. We also assume that the recombination rate is proportional to the number density of ions multiplied by the number density of electrons, so that the loss rate, $\bar{q}$, becomes

$$\bar{q} = \alpha_{rec} N_i N_e, \tag{3.30}$$

where α_{rec} is the *recombination coefficient*, which may be measured in the laboratory for a given atomic species. Furthermore, if we have an atmosphere such that appreciable negative ion formation does not occur, $N_i = N_e$, and we may rewrite (3.30) as

$$\bar{q} = \alpha_{rec} N_e^2. \tag{3.31}$$

In line with our assumption of equilibrium, we have

$$q(h/\mu_0) = \alpha_{rec} N_e^2. \tag{3.32}$$

This tells us that the density of electrons, N_e, is given by

$$N_e(h/\mu_0) = \left(\frac{q_M}{\alpha_{rec}}\right)^{1/2} \exp\frac{1}{2}\left[1 - \frac{(h - h_M)}{H} - \frac{1}{\mu_0}\exp\frac{(h - h_M)}{H}\right]. \tag{3.33}$$

It is instructive to examine (3.33) at altitudes close to h_M. When $(h - h_M)$ is small, (3.33) may be expanded in a power series, which results in

$$N_e(h/1) = \left(\frac{q_M}{\alpha_{rec}}\right)^{1/2}\left[1 - \frac{(h - h_M)^2}{4H^2}\right], \tag{3.34}$$

where we have again taken the sun to be directly overhead.

We note that (3.34) reveals a parabolic variation of electron density with altitude, at least when $(h - h_M)$ is small compared to H. We thus see that a layer of charge does indeed form. In fact, several layers may form, since α_v is dependent on the wavelength of the incident radiation. There are several wavelengths in the solar spectrum that are important for the atoms found in our atmosphere. We may compile a table of the (believed) causative agents for the various layers (Table 3.2).

TABLE 3.2. CAUSES OF VARIATION IN ELECTRON DENSITY LAYERS WITH ALTITUDE

Layer	Altitude (km)	Atmospheric Composition	Causative Radiation
D	60–85	NO^+	Lyman alpha (1216 Å)
E	85–140	O_2^+, NO^+	Soft x-rays
F1	140–200	O^+ (at top)	HeII (304 Å)
F2	200–400	O^+, N^+	HeII

3.17   Ionospheric Hydrogen and Helium

The hydrogen in the exosphere is believed to result from the water vapor found at lower altitudes. From studies of noctilucent clouds, which are apparently dust particles coated with ice, H_2O molecules are present at altitudes of the order of 80 km. Photodissociation of these molecules produces neutral hybrogen. The neutral hydrogen may be transformed into protons through a charge exchange reaction with hydrogen,

$$H + O^+ \rightarrow H^+ + O.$$

The electric fields in the ionosphere then may pull the ionized hydrogen upward, as we have shown.

The atmosphere (and the oceans too, for that matter) of the earth is believed to have come from outgassing of terrestrial rocks. The necessary heat would be supplied by the sun, internal radioactivity, and possibly other sources.

The helium in the atmosphere, however, originates in the radioactive decay of elements in the planetary crust. This helium represents a real problem. The helium influx to the atmosphere from the crust of the earth averages 2×10^6 atoms cm^{-2} sec^{-1}. However, the thermal loss rate (averaged over the solar cycle) is only 6×10^4 atoms cm^{-2} second^{-1}. The discrepancy is not understood.

3.18   Ionosondes

It is interesting to see how ionic densities are measured. This amounts to understanding the principles underlying the operation of an ionosonde.

We start with two of Maxwell's equations,

$$\nabla \times \mathbf{H} = \mathbf{j} + \frac{\partial \mathbf{D}}{\partial t} \tag{3.35}$$

and

$$\nabla \times \mathbf{E} = - \frac{\partial \mathbf{B}}{\partial t}. \tag{3.36}$$

In these equations $\mathbf{j}$ is the current density, $\mathbf{D}$ is the displacement vector ($= \epsilon \mathbf{E}$), $\mathbf{B}$ is the magnetic induction, and $\mathbf{H}$ is the magnetic field strength. Notice that for an electromagnetic wave propagating through a vacuum, (3.35) simplifies to

$$\nabla \times \mathbf{H} = \epsilon \frac{\partial \mathbf{E}}{\partial t}. \tag{3.37}$$

Let us assume that a radio frequency wave of angular frequency ω is incident on the ionospheric plasma, and let us further assume that the ions are too massive to move appreciably under the action of the wave, but that the electrons may move. Now the current density becomes

$$\mathbf{j} = ne\mathbf{v}, \tag{3.38}$$

where $\mathbf{v}$ is the velocity of the electrons. The acceleration $\mathbf{a}$ of the electrons by the sinusoidal radio wave is just

$$\mathbf{a} = \frac{e}{m} \mathbf{E}_0 \sin \omega t, \tag{3.39}$$

where $\mathbf{E}_0$ is the amplitude of the wave and e and m are the charge and mass of the electron, respectively. We may integrate (3.39) to find the instantaneous velocity of the electrons in terms of the wave parameters. The result is

$$\mathbf{v} = \int \mathbf{a} \, dt = - \frac{e}{m} \frac{\mathbf{E}_0}{\omega} \cos \omega t. \tag{3.40}$$

The current density expression (3.38) now becomes

$$\mathbf{j} = - \frac{ne^2}{m\omega} \mathbf{E}_0 \cos \omega t. \tag{3.41}$$

Since $\mathbf{D} = \epsilon\mathbf{E}$,

$$\mathbf{D} = \epsilon\mathbf{E}_0 \sin \omega t, \tag{3.42}$$

so that

$$\frac{\partial \mathbf{D}}{\partial t} = \epsilon\omega\mathbf{E}_0 \cos \omega t. \tag{3.43}$$

Hence (3.35) becomes

$$\nabla \times \mathbf{H} = - \frac{ne^2}{m\omega} \mathbf{E}_0 \cos \omega t + \epsilon\omega\mathbf{E}_0 \cos \omega t, \tag{3.44}$$

or

$$\nabla \times \mathbf{H} = \left[\epsilon\omega - \frac{ne^2}{m\omega} \right] \mathbf{E}_0 \cos \omega t. \tag{3.45}$$

At high frequencies the second term in (3.45) approaches zero, so that $\mathbf{E}$ and $\mathbf{H}$ are related as they are usually in a vacuum at high frequencies [this relation is summarized by (3.37)].

However, when

$$\epsilon\omega = \frac{ne^2}{m\omega}, \tag{3.46}$$

$\nabla \times \mathbf{H} = 0$; propagation through the ionized medium does not occur and our externally incident wave is "reflected." This happens at a critical frequency (called the *plasma frequency*, ω_p) such that

$$\omega_p = \left[\frac{ne^2}{\epsilon m} \right]^{\frac{1}{2}}. \tag{3.47}$$

We see that the plasma frequency depends on the electron density. There are several ways of describing this "reflection" at a critical frequency dependent on the electron density. For example, the displacement current is equal and opposite to the conduction current in the medium at ω_p; there is no net flow of current at this frequency.

Another way of looking at this is to calculate the index of refraction, k, through calculating the speed of the wave in the medium. When a non-conductor is placed in an electric field $\mathbf{E}$, the distribution of the electric charges that constitute the atoms and molecules of the medium is altered to produce an internal field that opposes the original field. This is called *polarization*; dipole moments are induced in the atoms, or, if already present, are aligned by the field.

In a dielectric, the field that would exist in the absence of polarization is called the electric induction $\mathbf{D}$; the polarization field is denoted by $4\pi\mathbf{P}$, and the resultant field is denoted by $\mathbf{E}$. We have

$$\mathbf{E} = \mathbf{D} - 4\pi\mathbf{P}.$$

When the medium is isotropic, the three vectors are in the same direction, and for small fields $|\mathbf{E}|$ is proportional to $|\mathbf{D}|$, so that $D = \epsilon E$ and $E = 1 + 4\pi(P/E)$. The index of refraction k is given by $\epsilon^{1/2}$ at frequencies far removed from those where significant absorption of energy by the medium occurs.

Suppose now that there are N oscillators per unit volume and that each one may be represented by an electron controlled by an elastic restoring force and a small damping force proportional to the electron's velocity $\dot{\mathbf{r}}$. The equation of motion of the electrons is

$$m\ddot{\mathbf{r}} + g\dot{\mathbf{r}} + \beta\mathbf{r} = e\mathbf{X} + e\mathbf{E},$$

where $\mathbf{r}$ is the displacement from a mean position and $\mathbf{X}$ is a field due to the dipoles. Indeed, the right-hand side of the equation of motion is just $(\mathbf{E} + \mathbf{P})e$, and since everything is in the $\mathbf{r}$ direction, we may drop the vector signs.

The polarization P is equal to the dipole moment per unit volume, so that $P = Ner$. We may substitute this into the equation of motion, along with $E = E_0 e^{i\omega t}$, and we may assume that P has the same frequency ω as E.

It turns out that a good approximation to k is given by

$$k \equiv \frac{c}{v} = \left(1 - \frac{ne^2}{m\omega^2}\right)^{1/2} \tag{3.48}$$

in a plasma of this kind (where the ions are very massive). Thus the index of refraction is zero when $\omega = \omega_p$. When we recall Snell's law of optics, we see that a vertically incident wave will undergo *total reflection* when $k = 0$ (the process, however, is really refraction by $180°$ at $k = 0$).

We should also note in passing that $k \leqslant 1$ is perfectly permissible. This merely says that the phase velocity in the medium is greater than the speed of light in a vacuum, but the group velocity (the speed at which energy is transported) is always $\leqslant c$. A physical "reason" for the increased phase velocity in the medium is that the wavelength increases in the medium, but the frequency is unchanged. We see from (3.48) that *imaginary* values of k are possible; this occurs at frequencies below the plasma frequency ω_p. This situation corresponds to energy absorption by the electrons; the cloud of electrons heats up.

There is still another way of saying this in terms of the bulk properties of the ionized medium. The dielectric constant ϵ is related to the index of refraction k through $\epsilon = k^2$, at least at frequencies far removed from the critical frequencies of the medium, so that absorption is negligible. In "normal" (i.e., nonionized) media the dielectric constant is positive. In ionized media, however, ϵ may be zero (when $\omega = \omega_p$), or it may even be negative. Thus the dielectric constant depends very much on the frequency of the incident electromagnetic radiation. Since the velocity of propagation is determined by ϵ, we expect *dispersion* to occur; high frequencies will travel faster in the medium.

This idea underlies the explanation for "whistlers." On a long-distance telephone line one occasionally hears a sharp click if lightning strikes. This is followed by a whistler (Figure 3.11), which is a sound whose pitch (frequency) starts off high and then rapidly falls, in periods of the order of

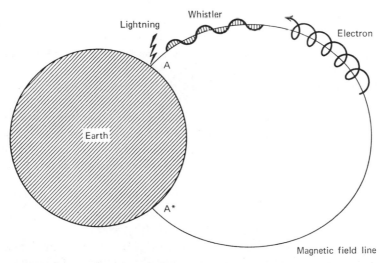

Figure 3.11   A whistler, a circularly polarized wave generated by lightning, propagates along a field line (and can interact resonantly with a charged particle moving along the line). Points A and A* are said to be *conjugate*.

seconds. It may then be followed some seconds later, by another, fainter whistler, where the frequencies are even more spread out in time. They may also be heard on radio receivers tuned to very low frequencies. As many as 20 whistlers may be detected after each disturbance.

It is thought that the original lightning stroke disturbance propagates along the magnetic field lines through the ionosphere to the "conjugate point," from which the signal bounces back. Conjugate points are those where a given field line enters the ionosphere; there are two such points (one in each hemisphere) for each field line. Dispersion occurs along this (ionized) path; the original "click" is decomposed, as it were, into its component frequencies by this dispersion. An analysis of the time spread or dispersion will yield information on the average electron density n along the path. Therefore whistlers tell us about electron densities at great heights, for some of the field lines may extend out to several earth radii from the planetary surface.

3.19   Conductivity of the Ionosphere

The study of the propagation of electromagnetic radiation in the ionosphere of the earth is greatly complicated by the presence of the geomagnetic field. For one thing, the propagation depends on the polarization of the wave because of the Faraday effect (see Section 14.4). For another, the conductivity of a plasma that has magnetic field lines embedded in it is not isotropic. In general the conductivity is a tensor rather than a scalar; the calculation of ionospheric currents, through the relation $J = \sigma E$, will therefore lead to anisotropic currents.

In the absence of a magnetic field the conductivity of an ionized medium may be calculated in a relatively straightforward manner. This may be accomplished by solving the differential equations of motion for the electrons and ions, taking into account the damping of the oscillations of the charged particles by neighboring particles, when they are accelerated by an electromagnetic wave of angular frequency ω. It is clear that the conductivity will be proportional to the electron density, and there will be an inverse dependence on the cross-section for collisions between the electrons (and ions) and the atmospheric constituents. It is beyond the scope of this book, however, to perform a detailed calculation of σ_0, the so-called "zero-field conductivity."

The result of such a calculation is that

$$\sigma_0 = ne^2 \left[\frac{1}{m_e(\nu_e - i\omega)} + \frac{1}{m_i(\nu_i - i\omega)} \right], \qquad (3.49)$$

where ν_e is the collision frequency for electrons. It is a temperature-dependent quantity that is equal to the sum of the collision frequencies of electrons with neutral atoms and with ions. It thus also depends on the densities of the neutrals and ions and, hence, on the altitude. The quantity ν_i is the collision

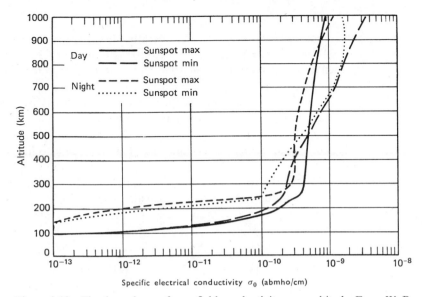

Figure 3.12 The dependence of zero field conductivity σ_0 on altitude. From W. B. Hanson in *Satellite Environment Handbook*. Reprinted with the permission of the Stanford University Press. Copyright 1965 by the Board of Trustees of the Leland Stanford Junior University.

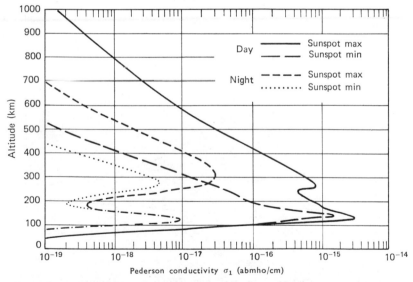

Figure 3.13 The dependence of the Pedersen conductivity σ_1 on altitude, as calculated by Hanson. Comparison with Figure 3.12 shows that $\sigma_1 < \sigma_0$. From W. B. Hanson, in *Satellite Environment Handbook*. Reprinted with the permission of the Stanford University Press. Copyright 1965 by the Board of Trustees of the Leland Stanford Junior University.

frequency of the ions; v_i is independent of temperature if the collision cross-section depends on $1/v$, where v is the ion velocity. It is possible to evaluate (3.49) readily when ω, the driving frequency, is zero, and thus to determine the dependence of σ_0 on the altitude h.

The conductivity in the direction of the field lines is unaffected by the presence of the magnetic field. This conductivity is equal to σ_0.

The conductivity *across* the field lines, called the Pederson conductivity, σ_1 (Figures 3.12, 3.13, and 3.14), is less than σ_0. This is because charges "prefer" to follow the field rather than cross it, a result that follows from the Lorentz expression for the force exerted by a magnetic field on a charged particle. The Pederson conductivity may also be calculated; if $\omega_e(= Be/m_e)$ and $\omega_i(= Be/m_i)$ are the cyclotron frequencies of the electron and ion, respectively,

$$\sigma_1 = ne^2 \left[\frac{v_e - i\omega}{m_e[(v_e - i\omega)^2 + \omega_e^2]} + \frac{v_i - i\omega}{m_i[(v_i - i\omega)^2 + \omega_i^2]} \right]. \qquad (3.50)$$

Another conductivity, which takes the Hall effect into account, must be evaluated in ionospheric problems. (The Hall effect occurs in conductors

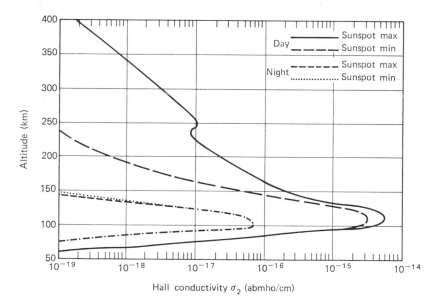

Figure 3.14 Dependence of the Hall conductivity σ_2 on altitude. The curves presented in this and Figures 3.12 and 3.13 are based on the assumptions that the frequency ω of the applied electric field is negligible, compared with the various collision frequencies, and that the magnetic field strength at sea level is 0.5 gauss. After Hanson in *Satellite Environment Handbook*. Reprinted with the permission of the Stanford University Press. Copyright 1965 by the Board of Trustees of the Leland Stanford Junior University.

when a magnetic field is present that is transverse to the current flow parallel to the electric field-strength vector). The Hall effect produces an electric field orthogonal to both the original electric field and the magnetic field. The Hall conductivity of a plasma, σ_2, is given by

$$\sigma_2 = ne^2 \left[\frac{\omega_i}{m_i[(v_i - i\omega)^2 + \omega_i^2]} - \frac{\omega_e}{m_e[(v_e - i\omega)^2 + \omega_e^2]} \right]. \qquad (3.51)$$

The Hall effect may produce *horizontal* electric fields through polarization of the atmosphere; these fields are in addition to the vertical polarization fields discussed in Section 3.14. We see that σ_1 and σ_2 may be called "transverse conductivities"; they refer to particle motion across the field lines and are in general less than σ_0.

Finally, there is another conductivity that is frequently useful in ionospheric studies. It is the Cowling conductivity, σ_3, where

$$\sigma_3 = \sigma_1 + \frac{\sigma_2^2}{\sigma_1}. \qquad (3.52)$$

We see that $\sigma_3 \gg \sigma_1$. This is most easily understood at the geomagnetic equator (where the field lines are all horizontal) in a horizontally stratified ionosphere. The Pederson conductivity is less than σ_3 because the (vertical) Hall current horizontally polarizes the medium; the conductivity is therefore enhanced.

The conductivity therefore depends on direction with respect to the magnetic field, as well as on altitude (particle density). Since the conductivity depends on field direction, it also depends on latitude (see Section 9.4).

Problems

3.1. (a) Derive the relationship between the acceleration of an orbiting spacecraft, a_D, and the local atmospheric temperature. Assume that the spacecraft is spherical, with a radius R.
   (b) If the spacecraft is orbiting in a stationary exosphere of hydrogen atoms (assumed isothermal at a temperature $T = 1000°K$), how much greater is a_D when the atmospheric temperature is suddenly raised to $1500°K$?
   (c) Discuss the ways in which atmospheric drag measurements may be used to develop a theory of exospheric winds. Include in your discussion the effects, if any, caused by ionization of the atmosphere and the electrical potential of the spacecraft.

3.2. An observer in a valley at 5000 feet of altitude notices that snow just covers the top of a nearby mountain. The valley temperature is $+68°F$. How high above sea level is the mountain top?

3.3. Calculate the adiabatic lapse rate for a planetary atmosphere where the equation of state is given by $P(V - b) = nRT$; b is a constant.

3.4. Suppose that our atmosphere is completely at rest and in diffusive equilibrium. Find the altitude at which $10\% \ O_2$ and $90\% \ N_2$ would be found.

3.5. If the temperature were $1000°K$ at a height of 400 km, what fraction of the molecules with molecular mass 2 will have at least the escape velocity?

3.6. The scale heights of planetary atmospheres are rarely constants. A better approximation is to express the scale height as a linear function of altitude, so that $H = H_0 + \alpha h$, where α and H_0 are constants. Find the pressure as a function of height in a planetary atmosphere where the scale height changes in this manner.

3.7. In a ionosphere of oxygen atoms there exists a trace of ionized hydrogen. Find the proton scale height.

3.8. (a) If the vertical potential gradient near the ground is measured as 150 volts meter^{-1} and the conductivity is 1.5×10^{-16} (ohm cm)$^{-1}$, find the vertical conductor current.

(b) Find the potential difference between the earth and the atmosphere at a height of 20 km for the following representative observations of conductivity.

Height (km)	σ (10^{16} ohms^{-1} cm^{-1})
0	2.5
2	3
4	6
6	10
8	20
10	26
12	38
14	60
16	70
18	85
20	110

(c) If the conductivity is assumed to be uniform between 20 and 80 km, find the potential difference between the surface and the lower edge of the D-layer.

3.9. During the eclipse of 31 August 1932 observations of ion density in the E-layer at the time of maximum eclipse indicated a decrease in concentration of electrons from 6.3×10^4 cm^{-3} to 5.0×10^4 cm^{-3} in 10 minutes. If we assume that there was no ion production during this 10-minute period, what is the effective electron recombination coefficient? Find the rate of ion production following the eclipse when observations indicated a maximum concentration of 9×10^4 electrons cm^{-3}.

3.10. Show that (3.48) is valid. Take $g = 0$; no damping (absorption of energy) exists.

3.11. Show that the approximate dependence of the speed of light c on altitude z in the earth's atmosphere is given by

$$\frac{dc}{dz} = -\frac{c(c_0 - c)}{c_0 \rho} \frac{d\rho}{dz},$$

where c_0 is the speed of light in a vacuum and ρ the atmospheric density (*Hint:* Express the index of refraction in terms of ρ and also express the radius of curvature of a horizontally moving light ray in terms of the gradient of ρ. At heights that are small compared to the earth's radius dc/dz may be expressed in terms of the height and the result rearranged to yield the given dependence.)

3.12. The base of the exosphere is called the *critical* level; it is that height at which a fast neutral particle moving upward has a probability of e^{-1} of escaping from the atmosphere without having any collisions. Show that for an isothermal exosphere the density at the critical level is such that the mean free path for a fast particle in the horizontal direction equals the vertical scale height.

3.13. A certain planetary atmosphere is composed entirely of atoms of mass m, where m is very large compared to the electron mass. The ions formed by photoionization have a single positive charge, and the electrons formed by photoionization immediately become attached to neutral atoms to form a negative ion. If the planetary atmosphere is isothermal at temperature T, and the planetary gravity is g, what is the scale height of the ions (negative and positive) in terms of the neutral scale height? (Neglect electron mass compared to atom mass.)

3.14. An ionizing photon flux F_0 is vertically incident on a planetary atmosphere composed of a gas for which the absorption cross section is σ, the neutral scale height is H, the surface neutral number density is n_0, and the recombination coefficient is α. Find the photoionization production rate per unit volume and the equilibrium ion number density at the altitude where the photon flux is $1/e$ of the incident flux F_0. How many ions are produced per second in a column of unit cross-sectional area above this level?

3.15. Consider a very massive planet whose atmosphere is composed entirely of a neutral perfect gas of mass M_0. Assume that the atmosphere is isothermal with temperature T_0, and that the variation of gravity with altitude is important. Also assume that the surface gravity is g_0 and that the atmosphere is in hydrostatic equilibrium. Obtain an expression for the pressure as a function of height h above the surface if the radius of the planet is h_0. How does the scale height vary with height (does it increase, decrease, approach a constant, approach 0, etc.)?

3.16. Consider an ionosphere in which doubly-ionized atomic oxygen (O^{++}) is the principal constituent, and in which there is a trace of CO_2^+. Assume that the ion and electron temperatures are the same, and that the variations of temperature and gravity with altitude can be ignored. What is the scale height of the O^{++}, what is the scale height of the CO_2^+, and how do these scale heights compare with the neutral scale heights for both species?

3.17. Assume that an ionosphere has a perfectly vertical magnetic field of magnitude B, and that the conductivity is anisotropic but uniform (i.e., σ_0, σ_1, and σ_2 do not vary with altitude or horizontal position.) Choose a coordinate system with z-axis vertical, x-axis to the south, and y-axis to the east. Let $\sigma_1 = 3\sigma_2$ and $\sigma_0 = 8\sigma_1$.

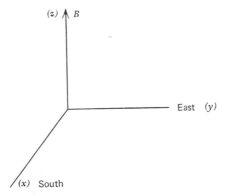

Figure 3.15

(a) If the southward electric current component, the eastward electric current component, and the vertical (up) electric current component are all equal, what are the southward and eastward electric field components in terms of the vertical electric field component E_z? (Your answers should not contain σ_0, σ_1, or σ_2.)

(b) If the eastward electric current component is equal to the vertical current component, but the southward current component is zero, what are the southward and eastward electric field components in terms of the vertical electric field component E_z?

3.18. A planetary atmosphere is composed of two un-ionized gases, which we designate as gas A and gas B. The molecular weight of gas B is twice that of gas A. At a height $h_0 = 20$ km, the number densities of the two gases are equal. Assume that the temperature is constant in altitude, and that the scale height of gas A is 10 km.

(a) What is the ratio of number densities of gas A to gas B at the surface of the planet ($h = 0$)?

(b) What is the ratio of number densities of gas A to gas B at $2h_0$ (40 km)?

(c) What are the answers to (a) and (b) if gas A is un-ionized and gas B is singly ionized? [Assume that $T_e = T_i$ and that $n_{A\,(ion)}(h_0) = n_B(h_0)$.]

(d) For the original assumption that both gases are un-ionized and the number densities are equal at $h_0 = 20$ km, what is the ratio of the total number of atoms of gas A above 20 km to the total number of atoms of gas B above 20 km?

3.19. A planetary atmosphere has a constant scale height of H_0, with surface number density n_0. A photon flux F_0 of a certain wavelength λ strikes the top of the planetary atmosphere vertically. Let the absorption cross-section for photons of wavelength λ be σ_λ cm^2.

(a) Find the flux of photons of wavelengths λ at two scale heights above the surface.

(b) Find the ratio of the flux at any height h to the flux at a point one scale height higher $(h + H_0)$.

Chapter 4

THE OTHER PLANETS
AND THEIR ATMOSPHERES

There is, at this writing, only one measurement *in situ* of a planetary atmosphere other than that of the earth. That is, the only information (scale heights, say) we have on the atmospheres of the other planets (aside from Venus) is presently restricted to an analysis of the electromagnetic radiation reaching our instruments either on earth or on spacecraft that have flown close by the other planets.

4.1 Planetary Distances and the Astronomical Unit

Since planetary temperatures are deduced from measurements used with radiation theory, we must first learn how far away the planets are. We may measure planetary distances from the sun in terms of the earth's distance from the sun, and then evaluate the latter quantity astronomically in absolute units. The mean distance between the sun and the earth is called the astronomical unit (A.U.).

Planetary distances may be derived from measurements of the sidereal periods P and of the synodic periods S. The sidereal period is the time, as seen from the sun, required for the planet to complete its orbit (about the sun) with respect to the "fixed" stars. P is the "actual" period of revolution. The synodic period, on the other hand, is the time required for the planet to resume the same configuration with respect to the sun. S is thus a function of the observer and in our case is measured from the earth; the lunar synodic period, for example, is the time between "full moons." The quantity S is that which we observe; the more useful P is deduced from it.

Astronomically there are two "kinds" of planets, superior and inferior. Superior planets are more distant than the earth from the sun, while inferior planets are closer than we to the sun. It is possible to distinguish observationally between the two kinds: inferior planets, for example, are never observed to be in "quadrature" (where the sun-earth-planet angle is 90°). The maximum angles from the sun, or "elongations," for Mercury and Venus are 28 and 48°, respectively. Superior planets never pass between the earth and sun; they are never in "inferior conjunction."

For an inferior planet the following relation holds:

$$\frac{1}{S} = \frac{1}{P} - \frac{1}{E},$$

(4.1)

where E is the length of earth's year, P is the sidereal period, and S is the quantity to be measured (the synodic period). (For convenience we may set $E = 1$ and measure S and P in years.) Equation 4.1 follows from Kepler's third law; we note that the inferior planet moves faster than the earth. In fact, the inferior planet moves $(1/P)$ revolutions each day (on the average), while the earth moves $(1/E)$ revolutions in pursuit. Since the planet gains one whole revolution on the earth in one synodic period, it gains $(1/S)$ per day, so that $(1/S)$ is equal to the difference between $(1/P)$ and $(1/E)$, as expressed in (4.1). The problem may be turned around for a superior planet, where the earth plays the inferior planet role. For a superior planet the corresponding relation is (see Figure 4.1)

$$\frac{1}{S} = \frac{1}{E} - \frac{1}{P}.$$

(4.2)

Jupiter, for example, which is a superior planet, is observed to have a synodic period of 1.094 years. Therefore its sidereal period is 11.86 years.

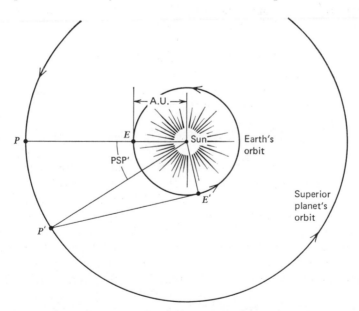

Figure 4.1 Determining the distance of a superior planet from the sun in terms of the astronomical unit, A.U., using observations made from the earth.

Mercury, an inferior planet, is found to have $S = 116$ days. Thus $P = 88$ days for Mercury.

Kepler's third law relates the mean distance of the planet from the sun, a, to the sidereal period P. Thus, (2.33) may be used to find the planetary distances once the synodic periods have been measured. As we noted earlier, it is possible to express planetary distances in terms of the astronomical unit. For inferior planets

$$a = \text{A.U.} \times \sin \theta, \tag{4.3}$$

where θ is the angle of greatest elongation. The problem is only slightly more complex for superior planets. Referring to Fig. 4.1, we see that we can determine the angles PSP' and ESE', since we know the sidereal period of the planet. Position TES is called *opposition;* $SE'P$ is *quadrature*, and angle PSP' is proportional to the elapsed time between these two configurations. In the figure the distance SE is of course the A.U. Now

$$ESE - PSP = \theta. \tag{4.4}$$

We also see that

$$a = \frac{\text{A.U.}}{\sin (\pi/2 - \theta)}. \tag{4.5}$$

It is interesting that Copernicus was able to establish the relative planetary distances to within 2% of today's accepted values.

We must still determine the A.U. itself, in absolute units. This may be crudely done by measuring the solar parallax p. A point on a limb of the solar disc is displaced by an angle of about 18 arc-seconds ($= p$) when the point is viewed from two observatories diametrically across the earth from each other. If D is the diameter of the earth,

$$\text{A.U.} = 206265 \left(\frac{D}{p}\right) \text{ miles} \tag{4.6}$$

when p is measured in arc-seconds and D in miles. Our problem therefore reduces to finding the earth's diameter D. Eratosthenes was able to do this in about 230 B.C. (Figure 4.2). He observed that on the first day of summer there was no wall shadow in a deep well at Syrene (modern Aswan). However, at a city 5000 "stadia" almost due north (Alexandria) there was a wall shadow in a well there, at angle of about 1/50 of a circle, or 7° to the vertical. Since he "knew" the sun to be so far away that the sun's rays were sensibly parallel at both places, we must have that the earth's circumference C is

$$C = \left(\frac{360°}{7°}\right) 5000 \text{ [stadia]}. \tag{4.7}$$

All we have to do is find out how long a Greek "stade" was; C/π will then

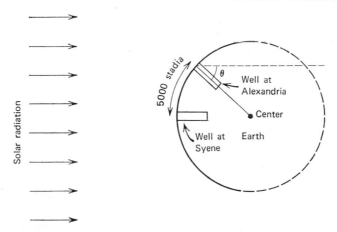

Figure 4.2 Schematic diagram, showing Eratosthenes' method for measuring the diameter of the earth more than two thousand years ago. The shadow angle θ is approximately $7°$ at the summer solstice.

yield D. The length of the stade has remained in doubt to the present day, because three different stadia were in use. One, the traveler's stade, was ~0.1 mile long. However, modern surveyors can accurately establish the distance between the two observatories in kilometers. The result of this work is that the A.U. = 1.49598×10^8 km.

There are some errors inherent in the procedure outlined above for finding the value of the A.U. from measurements of the diameter of the earth and of the solar parallax. It is difficult to measure p precisely, for it is never clear that the two observatories are really looking at exactly the same point on the solar disc. Therefore a more accurate technique is to measure the earth–Venus distance by radar (allowing, somehow, for the effects of the two planetary ionospheres). This distance is expressible in terms of the A.U., which may then be found from the measurement.

4.2 The Solar Constant

In order to estimate the heat received by a planet through radiation from the sun, we must first determine the effective temperature of the sun, $T_\odot$. This may be accomplished through the use of Stefan's law and by measuring the "*solar constant*" at the earth.

Stefan's law is

$$\frac{dQ}{dt} = S\sigma(T_i^4 - T_o^4) \tag{4.8}$$

for an idealized heated solid known as a "black body." In this expression the left-hand side is the number of ergs per second radiated by a body of surface area S to its surroundings. The absolute temperature of the black body is T_i; T_o is the temperature of the surroundings. The quantity σ is a universal constant, the value of which is 5.67×10^{-5} erg cm^{-2} second^{-1} °K^{-4}.

Stefan's law may be obtained by integrating Planck's law of radiation over all wavelengths. This latter expression is

$$R(\lambda)\, d\lambda = \frac{2hc^2}{\lambda^5}\left[\frac{1}{\exp\left(hc/\lambda kT\right) - 1}\right] d\lambda \qquad (4.9)$$

(ergs per $d\lambda$ wavelength interval per second per cm^2 and into unit solid angle). Planck developed this relationship in 1901 by assuming that physical systems have only discrete energy levels and that the emission or absorption of radiation is associated with transitions between two of these levels. The energy lost or gained by the system is just $h\nu$; the energy of a quantum depends on its frequency.

Now we recognize that

$$\frac{dQ}{dt} = \int_0^\infty R(\lambda)\, d\lambda. \qquad (4.10)$$

As indicated above, this integration yields Stefan's law.

We define the solar constant Q_0 to be the amount of radiant solar energy incident on a unit cross-sectional area per unit time, above the earth's atmosphere. It may be measured with sounding rockets. Once Q_0 is found, it may be multiplied by the surface area of the spherical shell (radius, 1 A.U.) surrounding the sun to determine the total power radiated by the star. This may be set equal to Stefan's expression for the energy radiated by the sun (4.8). Thus

$$Q_0 \cdot 4\pi \,(\text{A.U.})^2 = 4\pi R_\odot{}^2 \sigma T_\odot{}^4,$$

where $R_\odot$ is the radius of the sun. This yields that $T_\odot = 5778°$K; the "effective solar temperature" (the temperature of the visible disc or photosphere) is nearly 6000°K. There is perhaps a $\pm 10°$ uncertainty in this, principally determined by the error in measuring Q_0 in the rocket experiments. We should note that $T_\odot$ refers to an average over the whole solar disc, including the darker limbs. If we were to restrict ourselves to studies of the center of the disc, the effective temperature would be higher: 6000°K.

It is possible that Q_0 is time-dependent, for the sun may produce more radiant energy during the active phase of its 11-year cycle, particularly in the x-ray portion of its spectrum. We should also note that we have assumed the sun to be a black body so that the emission is a smooth function of the wavelength λ. This, of course, is not true. In particular, there are many lines present at short wavelengths (the ultraviolet and x-ray regions), so that the solar spectrum does not resemble a black body at all at those wavelengths. However, this spectral region represents only a very small fraction of the total energy output of the sun.

4.3 Radiation

By planetary temperature T_p, we shall usually mean the temperature of the outer regions of the atmosphere, the regions that are in equilibrium with the sun. Some people call T_p the "effective temperature" of the planet. There are some exceptions to this, however, that we shall presently discuss.

In equilibrium a planet reradiates as much energy as it absorbs from the sun. We continue to assume here that each radiating body acts as a perfect black body, an idealized object capable of absorbing all the electromagnetic radiation falling on it. The equilibrium assumption that we have made is but a special case of *Kirchhoff's radiation law*, which says that the radiancy of any body at any temperature is equal to a fraction of the radiancy of a black body at that temperature, the fraction being the absorptivity (or emissivity) at that temperature. Thus it tells us that a good absorber is also a good emitter.

The radiancy of a body is the amount of radiant energy emitted per second per unit surface area S. The emissivity of a black body is $e = 1.00$ (by definition), and it is a result of Kirchhoff's law that the absorptivity of that black body is also unity. Thus (4.8) should be rewritten (to include non-blackbodies) as

$$\frac{dQ}{dt} = Se\sigma(T_i^4 - T_o^4). \tag{4.8a}$$

In the case of real bodies e is a function of the wavelength. It is this latter fact that makes the greenhouse effect (Section 4.6) possible; this fact is also utilized in the design of passive thermal control systems for spacecraft. We shall ignore the "nonblack" properties of the sun and planets in the following analysis of planetary temperatures, but it is interesting to look at Table 4.1, which summarizes how e can be different from the absorptivity a.

In the case of gold, for example, the absorption of sunlight is only moderate but the emittance is very low. This is because in the case of solar radiation 94% is received at wavelengths less than 2 μ but the radiated energy is primarily at wavelengths in excess of 10 μ. Thus the temperature of a gold-plated earth satellite would have to be very high to dissipate the absorbed heat.

4.4 Planetary Temperatures

Planetary temperatures may be estimated with Stefan's law and the assumption of equilibrium. We may set (the heat radiated by the planet) = (heat from the sun intercepted by the planet) − (heat reflected from the planet).

TABLE 4.1. HEAT SHIELD MATERIALS OR COATINGS

Material	Approximate a	e^*	Remarks
White organic paint	0.2–0.3	0.9	300°C maximum temperature
Plastics	0.2–0.9	0.8–0.9	a depends on filler and color of plastic
Glass	Transparent	0.5–0.8	Good surface for most metal coatings
Black organic paint	0.9	0.9	400°C maximum temperature
Aluminum paint	0.3–0.4	0.3–0.4	Easily applied; durable
Gold	0.3–0.4	0.02–0.05	Soft unless protected

* The emittance of metals increases with alloying and at elevated temperature. The values here are for 0–300°C.

Hence, if R_p is the planetary radius, and $R_\odot$ the solar radius, we may write

$$4\pi R_p{}^2 \sigma T_p{}^4 = \sigma T_\odot{}^4 \left[\frac{4\pi R_\odot{}^2}{4\pi a^2} \right] \pi R_p{}^2 (1 - A). \qquad (4.11)$$

In this expression the planet is assumed to be at a mean distance a from the sun. The quantity A is called the *albedo;* it is the fraction of incident energy scattered and reradiated with no change in wavelength, or, more simply, reflected.

When we speak of the isotropic albedo we are referring to a quantity that has come to be known as the *Bond albedo*, A. This is different from the *geometrical albedo*, p; p is the amount of energy reflected back in the direction of the observer. Some information may also be gleaned from the wavelength dependence of the albedo, A or p; Rayleigh scattering goes as λ^{-4} while scattering from large particles will give rise to a different spectral shape for the albedo. Thus the size of the particles in a planetary atmosphere may be inferred from photometric (i.e., spectral) data on the albedo.

It is interesting in (4.11) that we have assumed that the planet radiates thermally equally in all directions. This assumption is certainly unrealistic; the daylit side will be hotter and therefore will radiate more. However, this condition may be met if the planet is an excellent thermal conductor. It may also be approximately met if the planet is a rapid rotator, and if it has a dense atmosphere so that heat may readily be transported to the night side.

4.5   The Subsolar Temperature

The subsolar point is on the planetary surface and on a line between the centers of the sun and the planet. The temperature of a square centimeter located at this point, T_{sp}, is again derivable from an application of Stefan's law and the assumption that the subsolar point is in radiative equilibrium with the sun. T_{sp} is the most useful temperature for the planets that are slow rotators. The expression for T_{sp} is

$$T_{sp} = T_\odot \left(\frac{R_\odot}{a}\right)^{1/2} (1 - A)^{1/4}, \tag{4.12}$$

where A, as before, is to be measured. The values of A are summarized in Table 4.2.

TABLE 4.2

Planet	Albedo, A	Remarks
Mercury	0.06	Low A indicates little atmosphere
Venus	0.76	Brightly reflecting clouds
Earth	0.39	Obtained from moonshine
Mars	0.15	Thin atmosphere
Jupiter	0.51	Organic atmosphere
Saturn	0.50	Organic atmosphere
Uranus	0.66	Unknown atmosphere
Neptune	0.62	Unknown atmosphere
Pluto	0.16	Small planet, highly eccentric orbit

4.6   The Greenhouse Effect

A planetary atmosphere may be relatively transparent to incident solar radiation at short wavelengths. The earth has such an atmosphere. Thus the sun directly heats the planetary surface. However, the warm surface radiates longer wavelengths (such as the infrared), and the atmosphere is quite opaque to much of this long-wave radiation. Thus the heat energy is "trapped" by the atmosphere, tending to increase the temperature over that of a black body as predicted by (4.11). This atmospheric effect is called the "greenhouse effect." It is the same effect that is responsible for the warmth inside greenhouses and also for the heating of the interiors of automobiles far above the atmospheric temperature when the car is left out in the sun for a long time. The windows of the auto (rolled up so that cool air may not enter) are transparent to solar radiation, but not to the longer waves reradiated by the interior.

4.7   Molecular Spectra and Planetary Temperatures

The optical methods that we have so far discussed for determining planetary temperatures lead to some basic uncertainties in the interpretation of the results. When we use the black body radiation laws in this fashion, we are not really sure just what it is that T_e refers to; assumptions have to be made regarding things like planetary conductivity or atmospheric density before the *effective temperature* may be understood as the average temperature of the visible surface.

If, however, we measure the *spectrum* of the planet, rather than merely measuring the total energy radiating from it, an analysis of the planetary spectrum may resolve the uncertainties. The analysis will result in a determination of the temperature T of the molecule that causes the emission or absorption spectrum. If we localize the place on the planet from which the light emerges, we can more safely say that we know the temperature at that place.

In general, molecular spectra can be divided into three spectral ranges that correspond to the three different types of transitions possible between molecular quantum states. The spectra consist of both individual lines and "bands". Bands appear to most spectroscopes to be composed of continuums, but instruments of high resolving power show that each molecular band consists of closely packed spectral lines.

The three different spectral ranges are due to *rotation spectra* (equally spaced single lines), *vibration-rotation spectra* (complicated line systems), and *electronic spectra* (bands). The first named generally falls at the longest wavelengths, while electronic spectra occur at much higher frequencies such as the ultraviolet.

Atmospheric temperatures usually affect only the rotational spectra of molecules, because the spacing between rotational energy levels is usually small compared with the product of Boltzmann's constant and the temperatures commonly experienced in planetary atmospheres ($\sim10^{2}°$K). Thus the *Boltzmann factor*, $\exp{(E_1 - E_2)}/kT$, is apt to be significant; even high-quantum-number states have reasonable probabilities of being occupied. This probability applies equally to all the spectral components of the state. Hence the number of rotational lines detectable will be large; this number provides a measure of the temperature.

Rotational spectra, unfortunately, tend to fall in the far infrared region. For example, the frequencies v absorbed or emitted by a *rigid rotator* (i.e., a diatomic molecule whose interatomic spacing is kept rigidly fixed, so that no vibration may occur) are given by $v = hJ/4\pi^2 I$, where I is the moment of inertia about an axis perpendicular to the line joining the two atoms, and J is the angular momentum quantum number ($\Delta J = \pm 1$ are the only allowed transitions). In emission spectra J refers to the initial state; J is the final state in absorption lines. Now $I = \sum mr^2$ is of the order of 10^{-40} gm-cm² for most molecules, so that for J somewhere between 1 and 10 we find λ somewhere between 3 and 30 μ. Rotation-vibration spectra generally occur at shorter wavelengths, but the temperature dependence is less important.

The far infrared is difficult to measure with ground-based instruments, even if the intensities are sufficient to permit a spectral analysis. This difficulty has two basic causes. First, the earth's atmosphere is an effective absorber in that region (see Chapter 15). Second, even if our atmosphere were otherwise transparent in this spectral region but contained the same kinds of molecules as those in the planetary atmosphere under investigation, it would not be easy to separate the planetary line from its terrestrial counterpart. Only the relative motion of the two planets, which in turn displaces the planetary line through the Doppler effect, permits the separation to be made, and the displacement is apt to be small.

Nevertheless it is to be expected that this area of planetary research will become most active and fruitful as large telescopes are lifted to the top (and beyond) of the earth's atmosphere. This is because space-borne telescopes will be above the absorbing atmosphere of the earth.

4.8 Mercury

This, the planet closest to the sun, has a very eccentric orbit, $e = 0.206$. In angular size mercury is only about $\frac{1}{60}$ the diameter of the sun, or about 30 arc-seconds, when it transits across the solar disc. It is so small that many sunspot groups in the solar surface appear larger during such a transit.

The sidereal period of revolution is 88 days, and until recently it was thought that the rotational period was "locked into" the revolution period so that the two were equal. This was deduced from optical observations of the faint surface markings.

However, radar has now been brought to bear on measuring the period of rotation. The 1000-foot antenna at Arecibo, Puerto Rico, has been used in Doppler measurements of surface features. The result is that the rotation period is 59 days (or just about two-thirds the period of revolution); the sense is prograde. A spherical planet as close to the sun as is Mercury should, given sufficient time, always present the same face to the sun at perihelion, much as our moon does to the earth. If the matter within Mercury is not spherically distributed, torques will be exerted, as we saw in Chapter 2. (These torques, however, have a strong inverse dependence on distance; an eccentric planetary orbit may cause a "loss of lock" as the planet gets farther away [nears aphelion].) Mercury may be spinning with an angular velocity precisely 1.5 times its orbital mean motion because the long axis of the presumably permanently deformed planet is pointed toward the sun at each perihelion passage. The particular ratio of spin rate to revolution rate implies that the three principal moments of intertia—A, B, and C—are related as

$$\frac{B-A}{C} \sim 10^{-8}.$$

This relatively rapid rotation also affects atmospheric calculations. The subsolar temperature computed from the measured value of A in (4.12) is $+770°F$. This is a very high temperature, and it is therefore doubtful that much of a planetary atmosphere could survive; the molecules would escape. This idea is substantiated by optical observations, although some observers have reported seeing a faint "haze" that could be caused by a residual atmosphere of heavy gases, such as CO_2. The surface pressure, however, could not be in excess of 3 millibars, only 0.3% of the value found at the surface of the earth. And if the surface pressure is so low, the night-side temperature "must" be quite cold—perhaps only 4° or 5°K.

But radio astronomy indicates otherwise. The received radio power from Mercury when only one quarter of the disc was sunlit is consistent with a surprisingly hot $+350°F$. Many suggestions were immediately made to attempt to explain the surprisingly high temperatures. Some of these were as follows.

A temperature this high on the night side could be evidence for a dense atmosphere that readily conducts heat from the daytime side of the planet. It could also, however, be evidence for a hot planetary interior. The heating might be caused by radioactivity. Cautious observers noted that we must always bear in mind one other possibility: there may be some feature of planets and their atmospheres that we are thus far unaware of. In the case of Mercury this feature is the nonsynchronous rotation of the planet.

Mercury is small; a diameter somewhat smaller than the radius of the earth is obtained from the observation that it subtends only 30 arc-seconds. The mass is also small (5% of that of the earth), so that the average density turns out to be a little higher than that of the earth. In fact, Mercury, Venus, and Earth, the three planets closest to the sun, have the highest densities in the solar system (except possibly Pluto, but measurements on Pluto are presently conflicting).

The eccentric orbit leads to an advance of the perihelion point. Some 530 seconds of the 573 arc-seconds per century observed for the precessional rate of the perihelion point are accounted for by the (Newtonian) attractions of the other planets, mainly Venus because it is closest, but the remainder must be explained by non-Newtonian mechanics. It might be thought that special relativity would suffice to explain the 43 seconds, for at perihelion the planet has the highest velocity and hence the mass departs detectably from m_0, the rest mass. It turns out, however, that special relativity predicts only 7 arc-seconds per century. The additional advance of 36 seconds is predicted by the general theory of relativity

As we saw in Chapter 2, the precession of perihelion might also be due to a slight nonsphericity of the sun; this possibility is difficult to evaluate observationally However, Dicke has recently found that the equatorial radius differs from the polar radius by some 35 km, or about 5 parts in 10^5. This degree of nonsphericity could give rise to the remaining 36 seconds of arc per century observed for the advance of Mercury's perihelion point, without invoking relativity (see Section 16.10).

4.9 Venus

This planet has been called "earth's twin," but this is a misnomer indeed, for only the mass and diameter are at all earthlike (Figure 4.3).

Figure 4.3 The third planet from the central star, taken at synchronous altitude (5.5 planetary radii). This planet is known to some as the blue planet; its albedo is variable because of the white "cloud" patterns observed in its atmosphere. It is not clear, from this picture, whether life exists on that planet. Courtesy National Aeronautics and Space Administration.

To begin with, the Venusian year is considerably shorter than ours, for Venus completes its orbit around the sun in only 225 days. Next, the planet is *hot*. Also the Venusian atmosphere of clouds prevents us from seeing the surface of the planet, so that the size of the planet itself below the clouds has been in doubt. Recent radar data indicate 6050 ± 5 km for the radius.

Infrared and radio measurements of Venus have been made, both from the earth and from the spacecraft Mariner 2, which flew by Venus at a distance of six planetary radii in 1962. Venus became the first planet visited by a spacecraft from earth, when the Soviet capsule Venera 4 soft-landed in 1967, two days before Mariner 5 flew by at an altitude of 2480 miles. Several debates were settled by Venera 4, a 1-meter-diameter, 383-kg spherical capsule intended to have landed with the aid of a parachute. A lively debate continues, at this writing, as to whether the spacecraft did in fact function properly all the way down to the planetary surface.

According to Venera 4, the surface is at a temperature of 270°C, nearly the melting point of lead. This temperature is ~100°C less than deduced from the microwave observations of the (average) planetary temperature, and it is therefore possible that Venera 4 landed on a feature ~20 km above the mean level of the Venus surface. Venera 5 and Venera 6 were crushed after they entered the Venusian atmosphere in 1969. (They had been designed to withstand 25-atm pressures.) It is possible, therefore, that Venera 4 did not survive to the surface.

In addition, Venera 4 reported that the atmospheric composition is over 93% CO_2, with perhaps a trace of water vapor. That CO_2 is present in the Venusian atmosphere was generally accepted previously. Venera 4 also detected $(1 \pm 0.6\%)$ oxygen, before telemetry reception ceased during its descent to the planet.

The detailed composition of the clouds remains in doubt. It is now clear that CO_2 and H_2O are present. The amount of CO_2 in the clouds was inferred from the earthbound measurements of the absorption of solar infrared; the result was some 5000 times the amount present in our atmosphere. There is about 0.2 meter-atmosphere of CO_2 present in our atmosphere (a meter-atmosphere is the number of molecules present in a 1-meter path at standard temperature and pressure). If the number of meter-atmospheres is interpreted to be the product of the pressure times the scale height, a typical terrestrial value of a major constituent (since $H \sim 7$ km) is several thousand meter-atmospheres.

It was also inferred from the width of the infrared absorption lines (say at $\lambda = 10.4\ \mu$) that the surface pressure on Venus must be very great. It was assumed here that the broadening is due to pressure; one estimate was that the surface pressure of CO_2 alone is somewhere between 10 and 20 atm. Venera 4 measured 18.5 ± 3.5 atm.; if the spacecraft did actually land

~20 km "up," the surface pressure is about 100 atm, since a scale height of ~12 km is consistent with the data.

Water vapor is another possible constituent of the atmosphere. Water vapor absorbs electromagnetic radiation at $\lambda = 1.34$ cm, but it does not absorb at $\lambda = 1.90$ cm. Therefore the ratio of the microwave fluxes at these two wavelengths should provide data on the presence of H_2O vapor. This measurement was conducted from Mariner 2 and no intensity difference was found, indicating at most ~20 gm cm^{-2} of water was present. However, balloon observations of infrared absorption by Strong had revealed the opposite result. He measured the amount of $\lambda = 1.13\,\mu$ absorbed from the solar spectrum by the Venusian atmosphere and did find absorption; hence water. The indicated amount is about 20 μ precipitable, or about 2×10^{-3} that present (on the average) over the earth. This result presumably applied to the atmosphere above the cloud tops, leading to the suspicion that the total amount of H_2O in the atmosphere could be quite large. The conflict between the two results is presently unresolved. Venera 4 did detect the presence of water; somewhere between 0.1 % and 0.7 % of the atmosphere below 26 km is composed of water. This amounts to 20–140 gm cm^{-2}, or about 10^{-4} of the amount found on the surface of the earth.

The presence of CO_2 (and possibly of H_2O as well) has led to the speculation that volcanism caused the thick atmosphere to form. A continuing "barrage" of large volcanoes would be consistent with the high temperature and the atmosphere as we now know it; it will be interesting to learn whether gases such as H_2S are present in the atmosphere.

It is also possible to infer the temperature of the cloud tops from infrared absorption and reflection. This has been done from Mariner 2 and from the earth. Since CO_2 absorbs at $\lambda = 10.4\,\mu$, but not at $8\,\mu$, the temperature inferred from the energy reflected at those wavelengths must come from a region above the altitudes where CO_2 is found (the clouds). The result is $-43°C$ above the cloud tops on Venus. The Venera 4 capsule measured a lapse rate of ~10°C per km at altitudes between 26 km and the surface; the entire descent took place on the night side of the planet (Figures 4.4 and 4.5).

Refraction in the Venusian atmosphere, according to an occultation experiment conducted with Mariner 5, leads to some odd effects. The refraction is so strong that light travels around the globe! This statement ignores absorption, of course.

It is interesting to speculate on the nature of the Venusian surface, if we suppose it subjected to a pressure of $\geqslant 15$ atm and temperatures near 300°C. We might have pools of molten aluminum, bismuth, lead, tin, and magnesium. Conceivably, under these conditions, things like liquid acetic acid and liquid benzene may be found on the surface. We might also suspect the

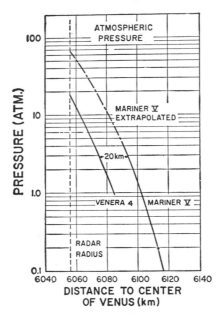

Figure 4.4 Atmospheric pressure data from Venera 4 and Mariner 5. The profiles are based on the assumption that 90% of the Venus atmosphere is CO_2. The two sets of data agree if the altitude scales are displaced by a total of 20 km. From R. Jastrow, *Science* **160**, 1408 (1968). Copyright 1968 by the American Association for the Advancement of Science.

existence of hurricane-like winds, in order that the day and night temperatures be about equal. There are reports that Venera 4 was severely buffeted during its descent to the surface. These strong winds presumably cause a great deal of surface erosion. (There is radar evidence, however, that great mountain ranges exist on the planet.) The flowing dust may cause some heating through friction. Venus would not at this point seem destined to be a resort area.

The length of the Venusian day defied measurement until recently. This was again because of the cloud blanket; the surface and any markings on it are hidden from view. Radar, however, penetrates the clouds. Doppler measurements indicate that the "day" is even a little longer than the year. The indicated duration is 243.09 ± 0.18 days (compared to the 225-day year), and in a most surprising *retrograde* direction. All the other planets rotate and revolve in the same sense. Indeed, it even appears that Venus may always present the same hemisphere to the *earth;* this will be true if the period is 243.16 days, retrograde.

It appears that the motion of the earth may control the spin angular velocity of Venus. One may wonder how the torque, due to the earth's interaction with the deformation presumed to be permanently present in Venus, could stabilize a resonant rotation rate in the presence of the enormously larger solar torque on that same permanent deformation. The answer is that the large solar torque has zero average over a synodic period; the earth's torque is weak by comparison, but the phenomenon of resonance greatly amplifies its effect.

We should note that Mariners 2 and 5 found no evidence for a planetary magnetic field around Venus, the same result reported by Venera 4. The magnetometer on Mariner 5, however, apparently did detect a fluctuation associated with the interaction of the solar wind with the planet. We shall return to this topic when we discuss the geomagnetic field.

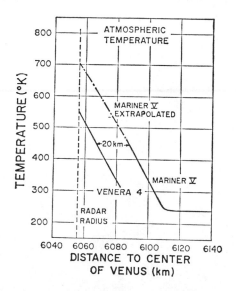

Figure 4.5 Atmospheric temperature data for Venus from two spacecraft. The data again agree if the altitude scales are displaced by 20 km. Note the differences from earth's profile. Convection apparently ceases 50 km above the surface, where the atmosphere becomes isothermal. From R. Jastrow, *Science,* **160,** 1408 (1968). Copyright 1968 by the American Association for the Advancement of Science.

4.10   Mars

Since someone is always writing about the red planet, Mars is known to some as the "newspaper planet." Unlike Venus, it is considerably smaller than the earth; the Martian diameter is equal to the earth's radius, and we can see the surface. Moreover, the Martian day is only a half-hour longer than that of the earth's (Figure 4.6).

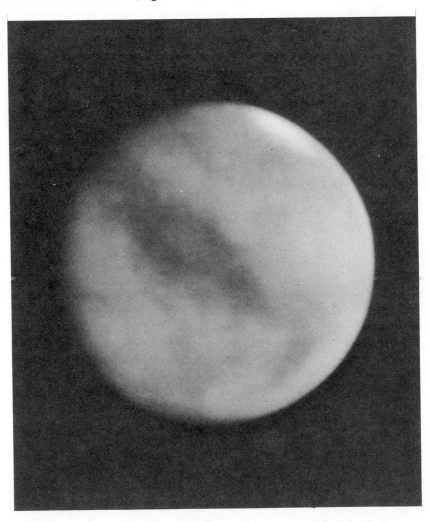

Figure 4.6   One of the best photographs of Mars to be taken from the earth. It was obtained with a 60-inch telescope. Courtesy Hale Observatories, California Institute of Technology.

Flights to Mars require considerable advance planning, for the synodic period is 780 days; the planet is closest to us every 2.1 years (the distance of closest approach is by no means the same, however).

We know the atmosphere must be thin, for we can see the surface most of the time and we can also measure the manner in which the planet occults various stars. If there were no atmosphere, the star would be "blotted out" sharply as the limb of the planet intervened between the observer and the star. An atmosphere, however, will make the occultation more gradual.

Mariner 4 came within four planetary radii of the surface of the planet in 1965, and then passed behind it, so that the planet in effect occulted the spacecraft. The attenuation of the telemetry signal by the planetary atmosphere afforded an opportunity to measure the density of the Martian atmosphere. This is because the atmosphere refracted the signals. At high altitudes the refraction is caused by the Martian ionosphere; the derived index may be interpreted in terms of the electron density. At lower altitudes (as the spacecraft neared the limb of the solid planet) the refraction is due to the neutral atmosphere. In practice the refraction itself is not measured. The phase shift of the radio wave as it passes through the medium is the measured parameter. The phase shift has different signs for ionized and neutral media. Such measurements can be performed with great precision. It is possible to relate the phase shift to refractive index.

By fitting the observations to a barometric law, it is possible to deduce both a surface pressure and a scale height (Figure 4.7). It is necessary first, however, to assume an atmospheric composition. An atmosphere of 100% CO_2 with a surface pressure of only 5 millibars appears to fit the observations.

The concentration of CO_2 had been previously measured spectroscopically from the earth, and it turned out to approximate to the *total* particle concentration found from the Mariner 4 occultation experiment. Mariners 6 and 7, which flew to about 3400 km from the center of Mars during July and August 1969, determined atmospheric parameters near the planetary equator and over the polar regions, respectively.

The surface temperature deduced in 1965 from the measurement of scale height was 180°K, in agreement with some of the astronomical determinations using (4.10) (Figure 4.8). Mariner 7 measurements, conducted four years later, indicated a surface temperature nearly 70°K warmer than this near the southern pole of Mars. This result assumes the composition to be 90% CO_2. It is interesting that the atmosphere is thin and the planet is cold; outgassing of the rocks should be slow when little heat is available.

The Martian surface gravity is only about one-third that of the earth. Therefore a surface pressure of 5 millibars corresponds to an atmospheric thickness of ∼15 gm per cm², a small percentage of that found at the earth's surface.

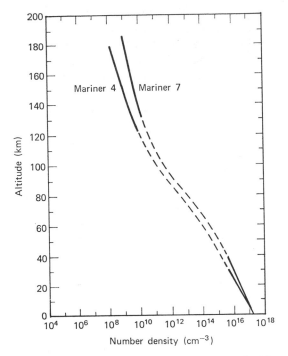

Figure 4.7 (*a*) Number density versus altitude. The Mariner 4 (1965) data were obtained near 50° S at 1:00 P.M. local time in late winter. The solar zenith angle was 67°. The Mariner 7 (1969) data were obtained near 58° S at 2:30 P.M. local time in early spring. The solar zenith angle was approximately 56°. From Kliore et al., *Science*, **166,** 1396 (1969).

The nature of the surface of Mars is interesting. The television pictures taken by Mariner 4 revealed the surface to be rather moonlike, with great numbers of craters. Since Mars is much closer to the asteroid belt than is our moon, comparative crater counts on the two bodies may shed light on the question whether the craters are of impact origin; the sporadic meteor flux may be higher near Mars than it is at 1 A.U.

Mars is unlike the moon in at least one respect: the coloring of the surface changes with the Martian seasons. There are two "caps" of white material over the poles; these shrink in that planet's summer and advance with the advent of winter. This has been interpreted as a thin (perhaps only 0.1 mm or so) layer of frost. It must be thin, because it would not shrink so quickly in the summer if it were anything like the thickness of ice over the polar regions of the earth.

That the caps consist of frozen H_2O rather than CO_2, say, is inferred from spectroscopic observations of a small amount of water vapor on Mars, conducted from the earth. The average amount in the atmosphere of Mars is perhaps 5–10 μ of precipitable H_2O, and it appears to depend on the Martian seasons; as much as 50 μ has been detected in the winter hemisphere.

As we have remarked in connection with Venus, water absorbs in the infrared. To separate any Martian absorption from telluric lines, the Doppler effect may be used, since the two planets move with respect to each other. This motion shifts the wavelength of the Martian line. More direct measurements may be made with infrared radiometers aboard flybys.

The atmosphere of Mars is much drier than the earth's. At midlatitudes on the earth there is an average of 23 μ of precipitable water at 240 millibars (35,000 feet of altitude) and an average of 13 μ at 60,000 feet.

In addition to the polar effects, there are other color changes as well. Most of Mars is orange-red. It might therefore be rather desertlike. Some people have suggested that the color may be due to iron oxide, but there are

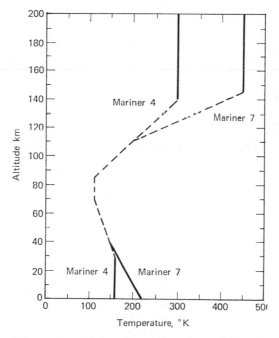

Figure 4.8 Temperature versus altitude above the surface of Mars, as deduced from occultation experiments conducted on radio emissions from Mariner 4 and Mariner 7. From A. Kliore et al., *Science*, **166**, 1396 (1969).

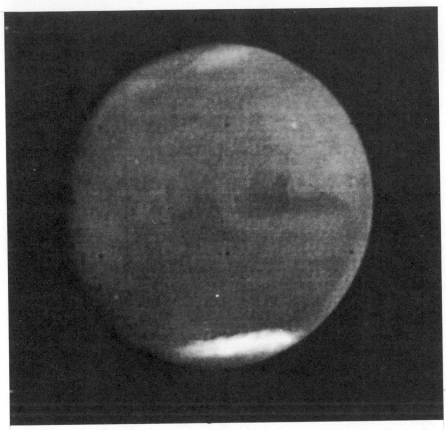

Figure 4.9   This picture shows the equatorial region covered in the near encounter phase of the Mariner 6 pass. *Meridiani Sinus* and *Sabaeus Sinus* compose the prominent dark feature parallel to the equator in the right-hand part of the photograph. The bright area at the top is *Cydonia*. The irregularity of the edge of the south polar cap could be due to topographical relief. The six small, very dark spots on the face of the planet are reference points in the focal plane of the camera. Courtesy Jet Propulsion Laboratory, California Institute of Technology.

also darker gray-green regions called "maria." The maria, in spite of their name, are *not* bodies of water, since bright reflected sunlight is not detected from them. The color of the maria does change with the seasons, becoming greenest in the spring. The polarization of light reflected from the maria is suggestive of low-order plant life such as algae and mosses. Furthermore, Sinton has detected absorption bands at infrared wavelengths in the maria; organic substances could explain the observation, leading to the suggestion that low-grade life may exist on Mars. It is possible, however, that the color

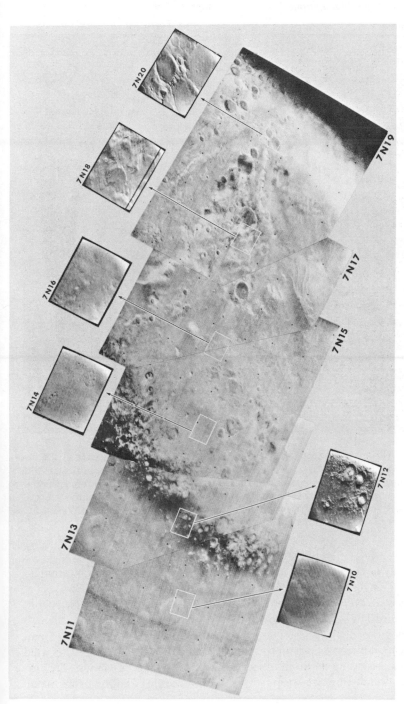

Figure 4.10 Spectacular view of the south polar cap of Mars, including the pole itself, was obtained by the Mariner 7 television cameras on 4 August, 1969. The edge of the cap is visible over a 90° span of longitude (from 290° E to 20° E), and the cap itself is seen over a latitude range from its edge at 60°, southward to and beyond the pole. The evening terminator is at the right edge of the five-picture mosaic. The five wide-angle pictures were taken 84 seconds apart, from left to right. Six narrow-angle frames cover, at high resolution, small areas within the overlapping portions of the mosaic. In wide-angle frames 7N11 and 7N13, where the local sun angle was about 53°, craters are visible both on and off the polar cap. In narrow-angle frame 7N12 the cap edge is seen in fine detail. The bright deposit on the cap is believed to be frozen carbon dioxide a few feet deep. Craters on the cap edge show "snow"-covered, southward-facing slopes and exposed northward-facing slopes.

Figure 4.11 Slightly above center is one of the largest clearly defined craters in the far encounter series taken by Mariner 6. Made visible by a white rim and central spot, this crater measures some 300 miles in diameter and is larger than any crater found on the moon. The crater has been identified surprisingly as *Nix Olympics*, long familiar to astronomers as a feature that becomes progressively brighter throughout the Martian day. Courtesy Jet Propulsion Laboratory, California Institute of Technology.

changes may be caused merely by blowing dust on the surface that alternately exposes and covers darker material.

The mass of Mars (only about one-ninth that of earth) is obtained from Kepler's third law, as applied to the two Martian moons, Phobos ("fear") and Diemos ("panic"). These two are so little that their sizes are too small to be measured by telescopes on the earth; a diameter of $\sim$10 km is estimated for Phobos, assuming its Bond albedo to be like that of our moon. Diemos may be only half this size, using the same procedure. Phobos orbits at an altitude of about 3715 miles above the surface (about equal to the diameter of the planet); it is the only natural satellite in the solar system that revolves about its primary in less time than the primary requires for a complete rotation.

This circumstance had even led some to suggest that Phobos is artificial. One of the Mariner 7 photographs revealed that Phobos is ellipsoidal, with a

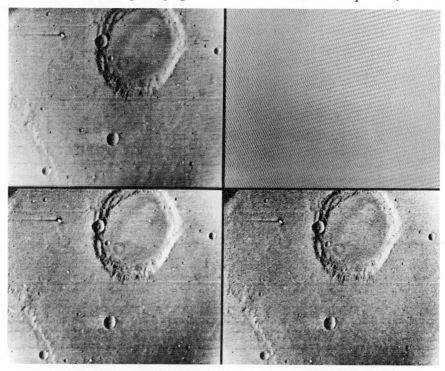

Figure 4.12 Example of preliminary computer enhancement of Mariner 6 near-encounter picture 18 of Mars. A portion of this frame, as originally recorded, is shown at the upper left. Apparent in it is a faint "basket weave" pattern due to electronic "pickup" in the sensitive preamplifier of the camera system, as well as a general "softness" due to the limited resolution of the Vidicon image tube. Computer analysis reveals the "pickup" pattern shown at the upper right. When the appropriate numerical value, determined from this pattern, is subtracted from each of the 658,240 elements of the picture, the result is as shown at the lower left. Two further computer programs may then be used to compensate for the smearing due to the Vidicon tube, with the result shown at the lower right. The final processing procedure will involve more refined versions of these steps, as well as programs designed to remove the numerous small blemishes, correct for electronic and optical distortions of the image, and correct for the photometric sensitivity of each picture element of the Vidicon tube. The computer will also be used to combine the digital and analog video data into a single photometrically accurate picture. Courtesy Jet Propulsion Laboratory, California Institute of Technology.

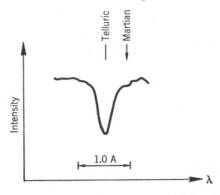

Figure 4.13 Tracing of the infrared spectrum of Mars, showing H_2O vapor feature measured from earth. From L. Kaplan, G. Munch, and H. Spinrad, *The Astrophysical Journal*, University of Chicago Press, 1963.

major diameter of 14 miles and a minor dimension of 11 miles, and has a very low albedo, perhaps only ~5%. The large size argues against an artificial origin for that satellite.

It is interesting that Jonathan Swift "predicted" the existence of these two satellites in 1726 in his book, *Gulliver's Travels* This was some 150 years before their discovery by Hall in 1877.

No planetary magnetic field or zone of trapped radiation was detected near Mars during the flight of Mariner 4. These observations are at least consistent with our present ideas of planetary fields, as we shall see in Chapter 9 (on geomagnetism).

4.11   The Asteroids

The asteroids are small bodies that are found primarily near the ecliptic plane and have mean distances from the sun such that they spend most of the time between the orbits of Mars and Jupiter. Some 2700 asteroids are now known, and it has been estimated that perhaps 50,000 are optically observable with the 200-inch Mount Palomar telescope. A few asteroids have diameters of the order of 100 km, but most of them have diameters as small as a few kilometers; probably there are also large numbers of microasteroids covering the entire range below the observed sizes.

There are two views on the origin of the asteroids. One is that they represent the fragments of a broken-up planet. This view, the more popular of the two, is largely based on the fact that Bode's law (see Section 4.15) "predicts" a planet at ~2.8 A.U., where the asteroids are mostly found. However, it appears that the combined mass of all the asteroids may well be less than 0.1 % of the mass of the earth, itself a relatively small planet.

The other view has been put forth by H. Alfvén and his collaborators; it is that the asteroidal belt represents an intermediate stage in the formation of planets. This latter view thus links the present conditions in the asteroidal region with the epoch in which the earth and the other planets were accreting from interplanetary grains.

An analysis of asteroidal material taken from those asteroids that come close to the earth (perhaps secured through a spacecraft-asteroid docking maneuver) could make possible a reconstruction of the essential features of the earth, the moon, and the earth-moon system as they were in the past, according to Alfvén. Among the most important data to be acquired are those concerning the chemical composition and its variation over the asteroid, so that information on the variation of isotopic abundance in the solar system at different distances from the sun will become available. These data would supplement the lunar samples returned by the astronauts. Like the lunar samples, asteroidal samples may provide undisturbed records of the history of the solar system; geological processes on planets such as the earth, Venus, and Mars have obliterated the records there.

4.12   Jupiter

Jupiter is easily the most massive planet in the solar system; the mass of this red, yellow, and tan object is about 300 times that of the earth and about 0.1 % the mass of the sun. It is also the largest planet physically. It has an equatorial diameter some 11 times the earth's diameter (Figure 4.14).

Jupiter, unlike the inner planets we have so far discussed, spins rapidly. The day (at the equator) is only about 10 hours long, and this rapid rotation is presumably responsible for the very large equatorial bulge. The oblateness amounts to 1 part in 15.

It is interesting that the rotational period depends on the Jovian latitude, also quite unlike the planets previously discussed. Near the poles the day is about 6 minutes longer than at the equator; equatorial material gains one whole lap approximately every 100 rotations on polar material. This means that we are viewing a gaseous atmosphere that contains strong winds and (presumably) shear layers. The shear layers are inferred from the "bands" that are observed and that are believed to be at different altitudes. They are thought to be the result of the wind profile on the planet.

There are other markings on the planet. The most notable is the Great Red Spot, which is dull red in color and some 30,000 miles long by 7500 miles wide (much larger than the earth). It reached its present prominence about the year 1831, but was visible even before that. The spot seems to move about with respect to the rest of the planet. The nature of the spot remains a mystery.

Figure 4.14   Photograph of Jupiter, taken with 60-inch telescope. Courtesy Hale Observatories, California Institute of Technology.

Along with hydrogen, ammonia and methane have been identified spectroscopically in the planet's atmosphere. The ammonia must be mostly in the solid form, since the subsolar temperature is some 70°C below ammonia's freezing point of −68°C.

Since the average density of Jupiter, $M/\frac{4}{3}\pi R^3$, is only 1.34 gm per cm³, the planet is presumed to consist mainly of light elements. It seems improbable that there is a core of metals, as there is inside the earth. Hydrogen is the most likely candidate for the major constituent. Indeed, it is thought by some that the composition of Jupiter is much like that of the primordial material that went into forming the solar system.

But the hydrogen, somewhere deep within the planet, is exposed to pressures of the order of 1 million atm; the density of the hydrogen near the center may be ~5 gm per cm³. Quantum mechanics tells us that at high pressures the wave functions representing the electrons from the various atoms begin to overlap significantly.

One consequence of this high-pressure situation is that some electrons are "squeezed out" of the valence band "by" the exclusion principle. The exclusion principle of Pauli operates for particles such as electrons. These electrons, in the classical analogy, are forced up into the conduction band. Hence at these very high pressures hydrogen becomes *metallic;* that is, a good conductor. All of this happens, we should note, even though the atoms do not touch (classically).

The fact that high-pressure hydrogen is a good conductor may help to explain what is perhaps the most astonishing feature of Jupiter: its radio emission. *Jupiter emits 100 times as much radio per unit area as does the quiet sun.*

Our present-day ideas of planetary magnetic fields hold that a rapidly rotating planet that contains a large, conducting core may have a strong magnetic field associated with it (see Chapter 9). As we have seen, Jupiter does spin rapidly and may have a core of metallic hydrogen. If Jupiter does in fact have a strong planetary magnetic field, charged particles may be trapped within the ordered field region around the planet and may radiate at radio wavelengths.

Jupiter is some 5 times as far distant from the sun as the earth. Therefore it should only receive about 4% of the solar heat received by the earth and should be quite cold. And as we have noted, cloud-top subsolar temperatures of only about 135°K are observed at optical wavelengths.

At radio wavelengths, however, the picture is quite different. Any object at an absolute temperature greater than zero will emit radiation; the flux and spectral distribution will be governed by the Planck law. At long (radio) wavelengths we may employ the Rayleigh-Jeans approximation to the Planck law, and define a radio flux S as

$$S(\lambda) = \frac{2kT_b\Omega}{\lambda^2},\qquad(4.13)$$

where λ is the wavelength, k is Boltzmann's constant, Ω is the solid angle subtended by the source at the observer, and T_b is a quantity called the "brightness temperature." T_b is the temperature of an equivalent black body that radiates S watts per square meter-hertz at a wavelength λ meters; it characterizes the thermal radio emission of an object. If the angular extent Ω of the source is small compared with the beam width of the antenna, S is sufficient to specify the strength of the radiation as the power per unit area per unit interval of frequency that is incident on the earth. If, however, the

TABLE 4.3. TEMPERATURE OF JUPITER AT VARIOUS RADIO WAVELENGTHS

Wavelength	T_b	Remarks
3 cm	135°K	Agrees with optical
10 cm	600°K	Temperature of Venusian "surface"
74 cm	50,000°K	Ten times photosphere temperature
15 meters	10^{12}°K	!

source is diffuse, we speak of the *brightness b* [watts per square meter per hertz per steradian]; $S = b\Omega$.

Radio observations of Jupiter indicate that S is approximately independent of wavelength. The experimental value is $S = 6 \times 10^{-26}$ watts per square meter per hertz. Table 4.3 summarizes the observations.

We are forced to conclude either that the longer-wave, decameter radiation from Jupiter is nonthermal or that Jupiter is hotter than the sun. We may eliminate the latter possibility, since the planet is not self-luminous in the optical.

Radio emission from Jupiter (Figure 4.15) was first discovered by Burke and Franklin in 1955. The decametric radiation occurs in bursts; the average duration of a burst is about half a second. The bursts themselves occur in groups that may last for a few hours. There seems to be an inverse relationship between the frequency of occurrence of these bursts, and solar activity. Thus there was a minimum in Jupiter's emissions during the 1958 maximum of solar activity. It is now believed that at least some of these fluctuations

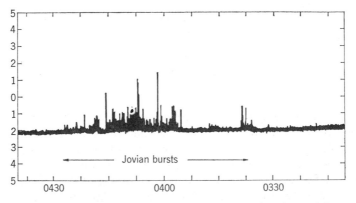

Figure 4.15   Example of the sporadic radio emission detected from Jupiter. Adapted from A. G. Smith and T. D. Carr, *Radio Exploration of The Planetary System*, D. Van Nostrand Company, Inc., 1964.

are of interplanetary origin; they may be due to fluctuations in the solar wind, so that the interplanetary medium causes radio "scintillations" (just as turbulent cells in our atmosphere cause poor "seeing" in the optical).

It is interesting to consider the energy in each burst, assuming that it is generated at Jupiter. Allowing for the Jupiter–earth distance, the total *radiated* energy is of the order of 10^{18} ergs. Now if we rather arbitrarily assume that the causal process for this emission has an efficiency of perhaps 10^{-4} or 10^{-5} (this is the ratio of radiated energy to causal-event energy), the total energy in each "event" is 10^{22} to 10^{23} ergs. Since 1 ton of TNT releases 4.2×10^{16} ergs, we are dealing with "events" that yield something like 1–10 megatons. And there may be several "events" per hour.

On the earth, the only events of this magnitude are the detonation of thermonuclear devices and the eruption of volcanoes. For example, the "loudest sound heard by man" resulted from the Krakatoa eruption of 1883. That volcano created a pressure wave in the atmosphere that circled the earth some three or four times. The pressure wave contained 10^{24} ergs of energy.

The decimeter radiation flux from Jupiter is much more intense than the decametric radiation found in the bursts, and is stable in time. (The radiation, except for the infrared, below a wavelength of about 4 cm is approximately that expected for a $\sim 135°K$ black body.) This decimetric radiation is strongly polarized, such that the plane of the electric vector is $10°$ or less from the rotational equator. The amount of polarization is about 30%. The fact that the plane of polarization varies by $\pm 10°$ with the equatorial rotation period of Jupiter is evidence that the magnetic axis is inclined to the rotational axis by about $10°$.

It is now believed that the decimeter source is to be found in a zone of magnetically trapped charged particles in the space around Jupiter. This trapped radiation is presumably rather like the Van Allen radiation around the earth, but richer in relativistic electrons and larger in physical extent. The reason for this belief is that the emission appears actually to come from the space surrounding the planet, rather than from the disc itself. In this picture the relativistic particles emit synchrotron radiation by virtue of the (presumed) planetary magnetic field. We shall return to synchrotron radiation when we discuss radio astronomy.

The emitting region is perhaps 60 Jovian radii in total width. It is also asymmetric, a situation that reminds us of the distribution of Van Allen particles around the earth. It is possible to estimate the planetary magnetic field from measurements of the radio spectrum, assuming that the spectrum is due to the synchrotron process (see Chapter 14). It is also assumed for this purpose that the electron density is equal to the density of Van Allen electrons. One estimate is that the required field strength is 1 gauss at three planetary radii; this indicates a planetary magnetic moment some 3×10^4 times that of the earth.

Jupiter has 11 moons (in addition to the so-called Trojan satellites trapped at the Lagrangian points $L/4$ and $L/5$; see Section 2.20). One of the moons, Io, has a remarkable effect on the decametric radio emissions from Jupiter. The "signals" are strongly correlated with Io's position in its orbit about Jupiter. Since Io is the satellite closest to the planet, it is thought that the moon is actually passing through the trapped radiation; the perturbing effect of Io's passage on the trapped orbits somehow causes enhanced radio emission to occur. In any event, the expectation is that the Jovian decametric radiation is also a result of trapped radiation around the planet.

Radio waves have also occasionally been observed to emanate from the other planets, but with nothing like the Jovian intensity. For example, emissions have been observed from Venus. It is believed, however, that these bursts are from Venusian *lightning strokes*, since the bursts have a frequency spectrum like the radio spectrum of terrestrial lightning. The latter frequency spectrum falls much less sharply with increasing frequency than the Jovian radio spectrum does. Finally, each Venusian burst involves only 10^{-9} of the energy released in a Jovian burst.

Before leaving Jupiter, we should discuss one more topic. We noted that Jupiter is cold in the optical. It is, however, warmer in the infrared—warmer than it should be from solar heating. We shall return to a discussion of this when we consider exotic astronomy, but let us note for now that Jupiter seems to have a source of internal energy. This may be a result of gravitational contraction.

4.13   Saturn

The average density of Saturn is so low that, like a famous soap, it would float in water. The density is only 0.71 gm cm^{-3}; we conclude that it must be composed largely of hydrogen, as in Jupiter. Also like Jupiter, Saturn is a rapid rotator; the equatorial day is only 10^h14^m long. Since the solar constant at Saturn is only 1% of the value at 1 A.U., Saturn should be (and is) a very cold place. The subsolar temperature is only $120°K$. Thus most of the ammonia is frozen out; the remaining methane is presumably responsible for the greenish color of Saturn, since methane absorbs the solar red and yellow wavelengths.

Saturn is most famous for its rings (Figure 4.16). The outer diameter of the rings is 171,000 miles, while the inner edge is only 6000 miles above the planet's surface. Saturn's axis is tilted about $27°$ from the plane of the ecliptic, so that we get to see the rings from practically all aspects. The thickness of the rings must be less than 20 miles; they can't be seen edge-on.

We know that the rings are not continuous rigid bodies; the inner regions

Figure 4.16 Ground-based photo of Saturn, taken with the 200-inch telescope on Mount Palomar. Courtesy Hale Observatories, California Institute of Technology.

rotate faster than outer zones, in apparent agreement with Kepler's third law. It is presumed that the rings are composed of particles about the size of sand or gravel particles, with a total mass no greater than one-fourth the mass of our moon. The particles may be coated with ices, a possibility suggested by their albedo.

All three rings are within the *Roche limit* for Saturn (see Section 8.6). Roche showed that if a satellite approaches too close to a planet, the tidal action of the planet tears the satellite into pieces, provided that the satellite is not too dense. For a satellite whose density is the same as Saturn's the limiting distance (Roche limit) from the center of the planet, below which the tidal action becomes destructive, is 2.44 times the planet's radius. Since all the rings lie within this distance, we may reasonably assume that their origin was a satellite that was torn to pieces by Saturn's tidal action.

The innermost ring is so faint that it was not discovered until 1850, when larger telescopes became available. An interesting phenomenon, however, is noted at the division between the two outer rings. This division is called *Cassini's division;* no particles are found in that space. This is believed to be the result of a resonance effect, for any particles that were located in it would have a period of revolution exactly half that of Mimas, Saturn's innermost moon (there are eight others). Thus every two revolutions, a particle would get an impulse from Mimas; these impulses tend to add as a mechanical resonance, pulling the particle out of its orbit.

In addition, it is believed that all nine moons together are responsible for the 6000-mile gap between the innermost edge of the rings and the planets.

There are other illustrations of the importance of resonance effects. One is provided by the effects of Jupiter on the orbits of the asteroids.

When a frequency analysis is made of the orbital periods of the asteroids, some periods are found to be missing; there are no asteroids in those orbits. These periods are exactly $0.33\frac{1}{3}$, 0.4, and 0.5 of Jupiter's sidereal period. These voids are called *Kirkwood's gaps*. It is believed that this is evidence for a resonance effect; an asteroid with one of these periods will cross the sun-Jupiter line regularly at the same heliocentric longitude. Thus the perturbations by Jupiter's mass will add, deflecting the asteroid from such an orbit.

At most, very little microwave radiation has been detected from Saturn. The presence of the rings may inhibit the development of zones of trapped radiation around this planet.

4.14   Uranus

Uranus is known chiefly for two things: it was discovered by accident (by William Herschel in 1781), and its axis of rotation is almost exactly in its orbital plane. The seasons on Uranus are unusual, to say the least; the axis is inclined only 8° to the ecliptic. It takes Uranus 84 years to complete its orbit around the sun; the polar night on Uranus is nearly a half-century long.

The five moons of Uranus (Figure 4.17) move almost exactly in the plane of the planet's equator. They therefore move almost perpendicular to the ecliptic, since the orbital inclination of Uranus to the ecliptic is very small ($\sim 0.75°$). The mean density of the planet appears to be somewhat higher than that found for Jupiter and Saturn; a mean value of ~ 1.5 gm cm^{-3} seems about right for Uranus.

4.15   Neptune

Neptune is too faint to be seen by the naked eye. It was discovered in 1846 through an analysis of the orbit of Uranus, which indicated perturbations, presumably caused by another planet.

The prediction of Neptune's existence beyond Uranus had to assume a mean distance from the sun for the hypothetical planet. The distance chosen was taken from *Bode's law*, which is not a law and was not discovered by Bode; J. D. Titius of Wittenberg noted in 1772 that there seemed to be a series of numbers that expressed, with rough accuracy, the distances from the sun of the six then-known planets. The "law" may be expressed as $d_n = 0.4 + 0.3(2^n)$, if the distances from the sun d_n of the planets (each assigned a number n) are expressed in astronomical units. In this scheme $n = -\infty$ for Mercury, $n = 0$ for Venus, $n = 1$ for Earth, $n = 2$ for Mars,

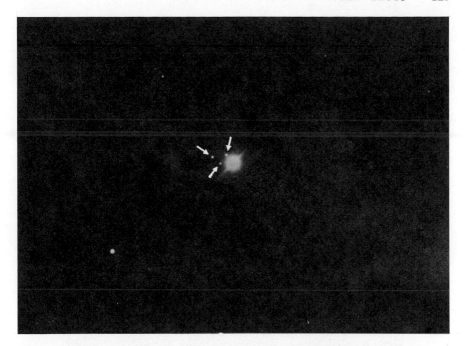

Figure 4.17   Uranus and the satellites photographed with the 120-inch reflector. Lick Observatory photograph.

$n = 3$ for the asteroids, etc. Bode organized a search for the "missing planet" at $n = 3$, and the law therefore acquired his name. Neptune ($n = 7$) should lie at $\sim d_7 = 39$ A.U.; the true mean distance is 30.1 A.U. and the disagreement becomes even worse at $n = 8$, since Pluto's mean predicted distance from the sun is only about 57% of the true mean distance.

The sidereal period of Neptune is 165 years; it has not yet completed a full revolution since its discovery. Neptune is extremely cold; the subsolar temperature is about $-200°C$. There are two moons. Triton, the larger, is a little bigger than our moon and has about twice the lunar mass. Like the other Jovian planets, Neptune's mean density is small; a value of 1.3 gm cm^{-3} has been recently estimated from the occultation of a relatively bright star by the Neptunian atmosphere. This observation of the planetary diameter also indicates the scale height high in the atmosphere (above the absorbing regions) to be ~ 40 km; a hydrogenous composition is indicated (Figure 4.18).

4.16   Pluto

Pluto is the outermost planet of the solar system; the orbit extends to an aphelion of 50 A.U. Perihelion occurs at 29.7 A.U., well within the orbit

Figure 4.18   Neptune and its satellites seen with the 120-inch reflector. Lick Observatory photograph.

of Neptune. This leads to an eccentricity of 0.25, easily the largest in the solar system. The orbital inclination of 17° to the ecliptic is also the largest in the solar system.

Since Pluto (Figure 4.19) is so far from the earth, little is known about it. The day appears to be about 1 week in duration. The size is very much in doubt; a diameter of 4700 miles has been measured, but this may well refer only to a central bright "spot" on the planet. If the diameter is in fact only 4700 miles, then the density comes out, from a recent mass determination of 0.18 earth mass, to be about 1.3 gm cm^{-3}, rather like the Jovian planets. Pluto may once have been a Neptunian satellite. It is not known whether this yellowish "star" itself has any "moons."

Figure 4.19   Two photographs of Pluto, showing motion of planet in 24 hours. Photographs were taken with the 200-inch Mount Palomar telescope. Courtesy Hale Observatories, California Institute of Technology.

Problems

4.1. (a) What would the synodic period of Jupiter be as seen from Saturn?
   (b) Of Mars as seen from Uranus?

4.2. Under what conditions will the synodic period of the moon, say, be shorter than its sidereal period?

4.3. Why cannot the moon's parallax be used to evaluate the astronomical unit?

4.4. Assume that Venus has a moon with a sidereal period of 40 days and that the sidereal period of Venus about the sun is 225 days. Find the synodic period of this moon, relative to an observer on Venus.

4.5. Show that Stefan's law follows from the radiation law developed by Planck.

4.6. If a planet moved so that its angular separation from the sun, as measured on earth, remained constant, what would be the distance of the planet from the sun in astronomical units? What would its period be in years?

4.7. If one assumes that the atmosphere and earth reflect incident solar radiation equally well and that the atmosphere acts as a single isothermal layer that absorbs all long-wave radiation falling on it, find the equilibrium temperatures T_e of the atmosphere and the earth's surface. (*Answer;* $T_e = 296°K$.)

4.8. How much greater is T_e in the preceding problem than the temperature predicted by black body radiation? (*Answer;* 48°K.)

4.9. If the average surface temperature of the earth is 288°K, find the "effective absorptivity" of the atmosphere for long-wave radiation.

4.10. If one takes the known distribution and luminosity of stars, one finds that the energy density u of starlight in interstellar space is nearly 1 eV per cm³; it turns out that $u = 7.7 \times 10^{-13}$ erg cm⁻³. Calculate the "temperature of interstellar space," the temperature that would be assumed by a black body in a perfect vacuum.

4.11. A greenhouse with two glass plates is used to boil water at normal atmospheric pressure. The reflectance of each glass plate for sunlight is 4%, and the absorption of the outer plate is 10%. The inner plate absorbs an additional 3% of the light transmitted by the outer plate, and the heater surface absorbs 95% of the incident energy. The emittance of the heater surface and the glass is 0.8. Approximate the heat available for boiling water for 1 square meter of surface.

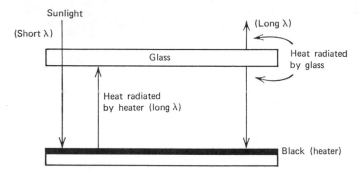

4.12. Consider a hypothetical planet that is always 2 A.U. from the sun. It has an albedo of 0.15 in the visual. The measured infrared temperature of the planet is 189°K. Is this value of the temperature consistent with that expected from the ideas of black body radiation?

4.13. There is an undetected satellite of Saturn that rotates very rapidly. Estimate the surface temperature of this satellite if it is in radiative equilibrium with the sun and has an albedo of 0.2.

4.14. Compute the Roche limit for Saturn. Do the same for Jupiter. How does the existence of a tensile limit (as is found in solid materials such as rock) affect this calculation?

4.15. The synodic period of Pluto is 367 days. How far has it moved in its orbit about the sun since it was discovered in 1931?

4.16. Eclipses on earth are of two kinds; one occurs when a third body is interposed between the light source (such as the sun) and the earth, while the second occurs when the second body moves into the earth's shadow.

(a) Can there be an eclipse of the half-moon?

(b) Explain why the greatest possible number of solar eclipses in a year is five.

4.17. (a) Suppose that the molecule HBr is present in a planetary atmosphere. At what wavelength, in microns, will the $J = 10$ pure-rotation line occur?

(b) Recalling that there are $(2J + 1)$ components of each state, what is the intensity of this $J = 10$ state (compared with that of the $J = 1$ state) if the temperature of the HBr is $200°K$? (Assume that both wavelengths can be detected.)

4.18. A rapidly rotating planet with extensive atmosphere is at a distance of 100 stellar radii from a star whose surface temperature is $10,000°K$. Calculate the change in equilibrium temperature for the planet if its albedo changes from 0.6 to 0.7.

4.19. The planet Jupiter has a sidereal period of 11.86 years and an orbital radius of 5.2 A.U. How long after opposition does quadrature occur?

Chapter 5

AIRGLOW

The "black" night sky is not completely dark: the stars, the planets, the zodiacal light, the gegenschein, and the moon all illuminate the surface of the earth. At high latitudes the aurora flares and flickers across the sky, and this phenomenon even occasionally occurs at midlatitudes. Even when allowance is made for all these effects (and for artificial illumination), however, the night sky of the earth is not completely dark: our planet's atmosphere itself glows; the process is called *airglow*. It seems likely that the importance of airglow research will increase as planetary exploration proceeds.

5.1 Airglow Observations

Airglow is present day and night on the earth, but is most easily demonstrated at night, when "nightglow" occurs. This is the radiation emitted by the atmosphere when it is not subject to direct sunlight. Nightglow may be detected by holding one's hand up against the sky on a dark, moonless, cloudless night. Your eyes must first be sufficiently dark-adapted; it is usually necessary to spend at least 5 minutes in the dark before the observation may be successfully attempted. You may then note that your hand appears darker than the surroundings. The intensity of the glow is roughly the same in all azimuths. (With sufficiently sensitive instruments it would be found that there is structure or patchiness in the airglow and that there is a zenith angle dependence.) The amount of light received is about equal to that cast by a candle ~100 meters away or, equivalently, the brightness of one star of magnitude +22 (near the visibility limit for the Palomar telescope) per square second of arc, under good seeing conditions. It does not depend significantly on the observer's latitude, although it is difficult to conduct high-latitude airglow observations because of the aurora. The characteristics of nightglow are summarized in Table 5.1.

5.2 The Planetary Origin of the Radiation

The observation discussed in Section 5.1 does not prove that the light is due to a terrestrial phenomenon as opposed to an extraterrestrial one. An observer on the surface, however, does note that the intensity depends on

128

TABLE 5.1. NIGHTGLOW CHARACTERISTICS

Parameter	Atomic oxygen		Sodium atoms	Hydroxyl radicals
		Atmospheric constituents		
Spectrum	Green line, $\lambda = 5577$ Å	Red line, $\lambda = 6300\text{–}6314$ Å	Yellow lines, $\lambda = 5800\text{–}5806$ Å	Bands, infrared
Zenith intensity (Rayleighs)	250	100	30–200	4.5×10^6
Altitudes (km)	95 ± 5	~300–400	90 ± 5	60–100
Secular variation	Time scale order of many minutes			
Diurnal variation	None clearly detected			
Seasonal variation	None	None	Maximum in winter	?
Solar-cycle variation	?	Possible increase when sun spot number is maximum	?	?
Spatial variation	Generally cellular with dimensions of ~100's (km)			
Polarization	Less than 1%			

the zenith angle; it is greater at greater zenith angles, at least up to zenith angles χ of 70 or 80°. This means that the light emanates from a shell nearly edge-on. We are then looking along the emitting shell, rather than through it, as in the case when we look vertically upward (Figure 5.1). An observer *outside* the shell (in a spacecraft, say) will see a ring of light around the planet. In the case of the earth the light is greenish; the brightest line in the visible part of the spectrum is at 5577 Å.

The airglow layer at ~100 km of altitude has been shown, through pictures taken from sounding rockets, to have a thickness of 30–40 km (see Figures 5.2 through 5.5). Satellite-borne telescopes have also been used; the spacecraft OSO-B2 conducted airglow observations during 1965. Figure 5.6 shows typical OSO data.

5.3 The Height of the Emitting Region

There are three methods that have been used to determine the altitudes from which nightglow radiation is emitted. These are the van Rhijn method, the triangulation method, and the sounding rocket method.

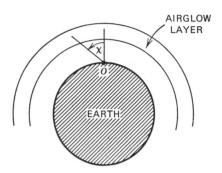

Figure 5.1 Illustration of the airglow layer that is deduced from the zenith angle (ψ) dependence of the airglow intensity, measured by an observer on the surface at O.

The van Rhijn technique, the oldest of the three, uses the fact that the intensity of the light from a homogeneous layer brightens as the zenith angle increases. In fact, the intensity emitted in any direction θ (measured from the normal to the layer) from an arbitrarily thin homogeneous layer at a finite height is just proportional to sec θ. If the height of the emitting layer is infinite, the intensity is independent of zenith angle.

If R_E is the radius of the earth and z is the altitude of the layer, (see Fig. 5.7),

$$\sin \theta = \frac{R_E}{R_E + z} \sin \chi. \tag{5.1}$$

Now the intensity I from a thin layer may be evaluated in terms of the intensity I_0, the light observed from the zenith; it is

$$I = I_0 v \left(\frac{z}{\chi} \right), \tag{5.2}$$

where $v(z/\chi)$ is called the van Rhijn function and is

$$v(z/\chi) = \left\{ \left[1 - \left(\frac{R_E}{R_E + z} \right)^2 \sin^2 \chi \right]^{1/2} \right\}^{-1}. \tag{5.3}$$

Table 5.2 presents values of v as a function of the observer's zenith angle χ and the altitude of the emitting layer z. If the layer is thick with a lower boundary at altitude z_1 and an upper edge at altitude z_2,

$$I = I_0 \left[\frac{R_E + z_2}{(z_2 - z_1) v(z_2/\chi)} - \frac{R_E + z_1}{(z_2 - z_1) v(z_1/\chi)} \right]. \tag{5.4}$$

The intensities I and I_0 may be measured, so that z may be determined.

TABLE 5.2. VAN RHIJN FUNCTIONS $v(z/\chi)$

z (km)

χ	60	80	100	125	150	175	200	250	300	350	400	1000
0°	1.0000	1.0000	1.0000	1.0000	1.0000	1.0000	1.000	1.000	1.000	1.000	1.000	1.000
5	1.0037	1.0037	1.0037	1.0036	1.0036	1.0036	1.004	1.004	1.003	1.003	1.003	1.003
10	1.0151	1.0150	1.0149	1.0148	1.0147	0.0145	1.014	1.014	1.014	1.014	1.014	1.011
15	1.0345	1.0343	1.0341	1.0338	1.0335	1.0333	1.033	1.033	1.032	1.032	1.031	1.026
20	1.0628	1.0624	1.0620	1.0615	1.0609	1.0604	1.060	1.059	1.058	1.057	1.056	1.047
25	1.1011	1.1004	1.0997	1.0988	1.0979	1.0971	1.096	1.095	1.093	1.090	1.090	1.074
30	1.1511	1.1499	1.1488	1.1474	1.1460	1.1446	1.143	1.141	1.138	1.136	1.133	1.109
35	1.2152	1.2134	1.2116	1.2095	1.2073	1.2052	1.203	1.199	1.195	1.192	1.188	1.152
40	1.2969	1.2942	1.2915	1.2882	1.2849	1.2818	1.279	1.273	1.267	1.261	1.256	1.203
45	1.4012	1.3971	1.3930	1.3880	1.3831	1.3783	1.374	1.365	1.356	1.348	1.340	1.263
50	1.5356	1.5291	1.5229	1.5152	1.5077	2.5005	1.493	1.480	1.467	1.454	1.443	1.334
55	1.7113	1.7012	1.6913	1.6793	1.6678	1.6566	1.646	1.625	1.606	1.587	1.570	1.416
60	1.9465	1.9299	1.9138	1.8946	1.8761	1.8583	1.841	1.809	1.779	1.751	1.725	1.508
65	2.2711	2.2425	2.2151	2.1826	2.1518	2.1226	2.095	2.004	1.997	1.954	1.915	1.609
70	2.7381	2.6846	2.6347	2.5765	2.5227	2.4726	2.426	2.341	2.267	2.200	2.141	1.714
75	3.4438	3.3336	3.2342	3.1227	3.0231	2.9335	2.852	2.710	2.590	2.487	2.398	1.816
80	4.5565	4.3009	4.0860	3.8599	3.6697	3.5068	3.365	3.130	2.943	2.789	2.660	1.905
85	6.1985	5.5852	5.1278	4.6921	4.3550	4.0842	3.861	3.511	3.247	3.039	2.870	1.966
90	7.3378	6.3695	5.7103	5.1222	4.6893	4.3539	4.084	3.672	3.372	3.040	2.953	1.988

From Chamberlain, Physics of the Aurora and Airglow.

This treatment ignores the fact that there is an atmosphere between the layer and the observer. The intervening atmosphere will in general scatter the emitted light, as well as absorb some of it. Light will be scattered out of the observer's line of sight to the airglow layer, and scattering in will also occur. Some wavelengths will be more strongly absorbed (e.g., ultraviolet by the ozone), and reflection from the ground will also occur. Thus the actual situation is quite complicated, and an accurate treatment should be along the lines of radiation transport, including all these effects and the correct cross-section with the proper dependence on wavelength. If properly done, the resulting equations will enable one to relate the intensity observed at sea level

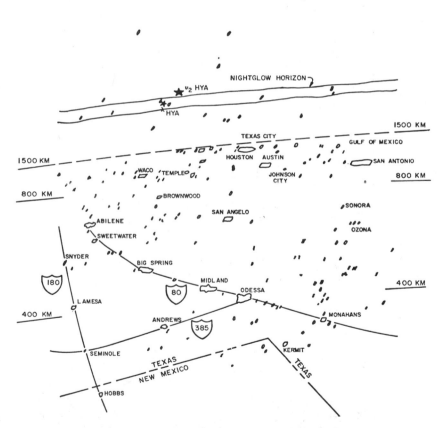

Figure 5.2 Map illustrating the location of light sources visible on the ground from following rocket photos. From J. Hennes and L. Dunkelman, Courtesy *Journal of Geophysical Research*, 1966.

Figure 5.3 The night airglow above western Texas, photographed from 184 km in a three-second exposure with an $f/1.4$ camera using Tri-X film. The earth's horizon is 1500 km away. The exposure was made at 2319 MST, 30 November 1964, at latitude 32.7° N, longitude 106.5° W. From J. Hennes and L. Dunkelman, *Journal of Geophysical Research*, 1966.

Figure 5.4   A 10-second exposure taken at 183 km. The Gulf of Mexico occupies the dark region at the upper right horizon. From J. Hennes and L. Dunkelman, *Journal of Geophysical Research*, 1966.

Figure 5.5   A three-second exposure taken at 93 km. The rocket is moving down at 1.3 km per second, but its attitude remained stationary with respect to the star background. From J. Hennes and L. Dunkelman, *Journal of Geophysical Research*, 1966.

to the flux of light emitted by the glowing layer. Such a proper treatment, for example, indicates that the intensity observed on the ground decreases with zenith angle for zenith angles in excess of $\sim75°$ (see Fig. 5.8).

5.4   Emission and Absorption of Radiation

In this section as well as the following three we shall briefly review, first microscopically and then macroscopically, how electromagnetic radiation is transported in planetary and stellar atmospheres. Since the wavelengths in

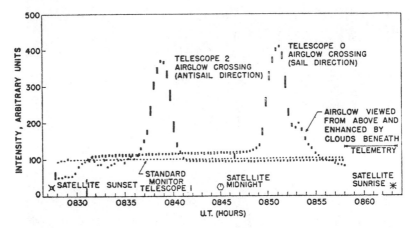

Figure 5.6 Intensities measured by telescopes on satellite 0S0-B2 during a typical satellite night. The spectral response of photometer 0 corresponded approximately to an average of astronomical blue and visual, whereas photometer 2 had a visual response, and the "monitor telescope" (1) looked parallel with 0 but viewed a c^{14} impregnated plastic scintillator for calibration purposes. From J. Sparrow, E. Ney, A. Burnett, and J. Stoddart, *Journal of Geophysical Research*, 1968.

which we are primarily interested for the purposes of this discussion are approximately those of visible light, we will first concern ourselves with the sources of such radiation, atoms and molecules.

Quantum mechanics describes the manner by which atoms and molecules emit and absorb energy. Emission and absorption are associated with transitions of atoms or molecules between quantum states of different energies.

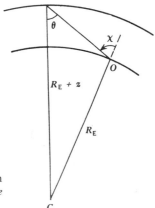

Figure 5.7 Illustration of the geometry of the van Rhijn method. From J. Chamberlain, *Physics of the Aurora and Airglow*, Academic Press, Inc., 1961.

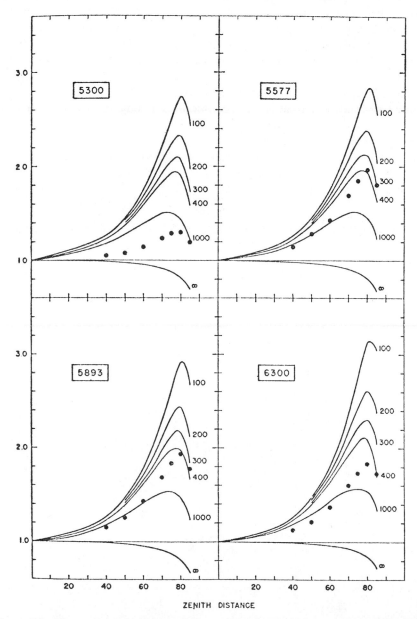

Figure 5.8 Theoretical calculations of the ratio of (I_x/I_0) versus zenith angle for four wavelengths. The various heights of the emitting layer are given in kilometers and the ground albedo is taken to be 25%. From F. Roach and A. Meinel, *The Astrophysical Journal*, University of Chicago Press, 1955.

Let all the quantum states for an atom be enumerated in a single series. When an atom is in state n there is a certain probability that during an interval of time dt it will jump spontaneously into another state j, with the emission of a photon of frequency $(E_n - E_j)/h$. This probability is denoted by $A_{nj}\,dt$. Out of a large number N of atoms in state n, NA_{nj} will "jump" per second into state j.

Consider the history of N_0 atoms that start in state n. If N of them are still in that state after a time t, during time dt a number $-dN$ will leave by spontaneous radiative transitions, where $dN = -N\gamma\,dt$ and $\gamma = \sum (j)A_{nj}$. Here $\sum (j)$ denotes a sum over all states into which a spontaneous jump can occur out of state n; that is, over all states having lower energy. In a particular case some A_{nj}'s may, of course, be zero. Integrating our expression for dN, we find that (see Fig. 5.9)

$$N = N_0 e^{-\gamma t}.$$

Obviously there is no absolute limit to the length of time that an individual atom may remain in a given quantum state. The *average* time τ_n spent by the atoms in state n is called the *mean life* of an atom in that state. To calculate τ_n, consider that during each interval of time dt, $N\gamma\,dt$ of the N_0 atoms initially in state n will leave the state. Let us say that they spent a time t in the state. Therefore, since $N = N_0 e^{-\gamma t}$, we may write

$$\tau_n = \frac{1}{N_0}\int_0^\infty t(N_0 e^{-\gamma t}\gamma\,dt) = \gamma\int_0^\infty t e^{-\gamma t}\,dt$$

$$= -t e^{-\gamma t}\int_0^\infty dt + \int_0^\infty e^{-\gamma t}\,dt,$$

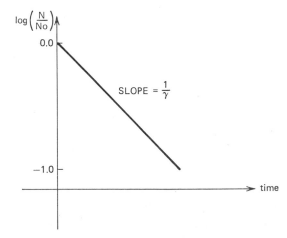

Figure 5.9　The exponential decrease in the number N of atoms in a given initial state as they decay through radiative transitions.

after integrating by parts. But $te^{-\gamma t} \to 0$ as $t \to \infty$; and $\int_0^\infty e^{-\gamma t}\, dt = 1/\gamma$. Hence

$$\tau_n = \frac{1}{\gamma} = [\textstyle\sum (j)A_{nj}]^{-1}.$$

The mean life of an excited state, as expected, is inversely proportional to the transition probability A_{nj}.

For atomic states involved in visible or ultraviolet emission, τ_n is found to be of the order of 10^{-8} seconds for many optical states. As we shall see in Section 5.6, however, some of the most interesting lines in the airglow spectrum involve much longer times. The mean life is usually much shorter than 10^{-8} second for emission of x-ray photons and is correspondingly longer in the infrared.

Absorption may also be treated in terms of the same coefficient A_{nj}. As indicated above, these coefficients are strongly frequency-dependent. To evaluate the frequency dependence, consider the number of atoms N_n in state n (the higher-energy state) if N_j are in a lower-energy state j. This relation is summarized in the quantum-mechanical version of Boltzmann's law; it may be written as

$$N_n = N_j \exp \left(-h\nu_{nj}/kT\right)$$

for a system of atoms in a thermodynamic equilibrium at temperature T. Let $\rho(\nu)\, d\gamma$ be the energy density of isotropic radiation with frequencies in the range ν_{nj} to $(\nu_{nj} + d\nu)$. Then the number of atoms transferred from state j to state n per unit time by the absorption of radiation will be proportional to $\rho(\nu)N_j$, provided that the absorption line is so narrow that $\rho(\nu)$ is nearly constant for the spectral range for which there is appreciable absorption. Call the coefficient of proportionality B_{jn}. In a unit of time a certain fraction of the atoms in state n will be transferred to state j by spontaneous emission. These transitions are not dependent on the presence of the radiation. Their number is proportional to N_n; the constant of proportionality is just A_{nj}. We must also consider that in addition to the spontaneous emissions, which are independent of the presence of the radiation, "stimulated" emissions may occur. The number of stimulated emissions is proportional to the radiation density; the coefficient of proportionality is called B_{nj}. We then have

$$\rho(\nu)B_{jn}N_j = A_{jn}N_n + \rho(\nu)B_{nj}N_n.$$

Using Boltzmann's law, we therefore find that

$$\rho(\nu) = \frac{A_{nj}}{B_{jn} \exp \left(h\nu_{jn}/kT\right) - B_{nj}}.$$

But, Planck's law (see Section 4.2) also gives the energy density as a function of frequency, since the flux of energy crossing a unit area per unit time is related to the energy density by the factor c. A comparison of the two expressions for $\rho(\nu)$ shows that both are satisfied if and only if

$$B_{nj} = B_{jn},$$

and the transition probability (by emission), A_{nj}, is given by

$$A_{nj} = \frac{8\pi h\nu_{nj}{}^3}{c^3} B_{nj}.$$

The coefficients A_{nj}, B_{nj}, and B_{jn} are called the Einstein coefficients.

5.5   Calculation of Transition Probabilities

Texts on quantum mechanics show how the transition probabilities A_{nj} may be calculated. The probability that an atom will radiate a quantum $h\nu_{nj}$ depends not only on what the atom does in the initial quantum state before it radiates, but also on what it is going to do in the final quantum state after it radiates. When the atom radiates, it actually is in neither the initial state or the final state, but rather in some linear combination of the two.

The transition probabilities A_{nj} can be shown to be approximated by (in centimeters-gram-second units)

$$A_{nj} = \frac{64\pi^4 e^2 \nu_{nj}{}^3}{3hc^3} (|x_{nj}|^2 + |y_{nj}|^2 + |z_{nj}|^2).$$

The transition rate for spontaneous emission is proportional to the cube of the frequency of the emitted quanta. This expression is strictly true if the wavelength of the radiation is so long, compared to the dimensions of the atom, that any variation of the electric field over the atom may be neglected. This expression is consistent with the short lifetimes of x-ray states compared with the lives of those states involved in visible radiation.

In this expression for A_{nj} the quantities x, y, and z are the Cartesian coordinates of the jumping electron, and x_{nj}, y_{nj}, and z_{nj} denote the electric dipole matrix components between the two quantum states. If ψ_n and ψ_j denote the wave functions for the two states, then, in the case of a 1-electron atom,

$$x_{nj} = \iiint \psi_n{}^* x \psi_j \, dx \, dy \, dz,$$

with similar integrals for y_{nj} and z_{nj}.

5.6   Forbidden Transitions and Collisional De-Excitation

If the dipole terms in A_{nj} all vanish, this particular transition cannot occur by dipole emission. The transition is said to be "forbidden" (as far as dipole emission is concerned). Observationally these are transitions for which the spectral lines are found to be absent or extremely weak.

The transition rates are predicted to be zero in these cases, because the integrals involved in the electric dipole matrix elements vanish identically. In the language of the quantum mechanics *selection rules* exist; these rules are a set of conditions on the quantum numbers of the wave functions of the initial and final states such that the electric dipole matrix elements are zero when calculated with a pair of wave functions whose quantum numbers violate these conditions.

We noted, however, that faint radiation may be observed at the frequency in question; forbidden transitions may in fact occur in spite of the selection rules. If the transition cannot take place by the normal means of an interaction involving the electric dipole moments and the electric fields (called the *electric dipole* interaction) there is a very small probability that it can take place through an interaction of the magnetic dipole moments and the magnetic fields that are also present in electromagnetic radiation. In addition there exist small probabilities of radiation because of the omitted "quadrupole" or higher-order terms in A_{nj}. When the higher approximation is calculated, it is found that the transition probability for a typical electric quadrupole radiation is less than that for a typical dipole in the ratio $(a/\lambda)^2$, where a is the atomic radius and λ is the wavelength of light. Thus if typical dipole transition probabilities are of the order of 10^7–10^9 per second, the quadrupole probabilities (which involve products of the type $[ex^2]$ instead of $[ex]$) are of the order of 0.1–100 per second.

We have seen that airglow is predominantly at the 5577-Å line of atomic oxygen. It is interesting that this transition (between the $^1S_0$ and $^1D_2$ states of the oxygen atom) is forbidden; the most prominent line in the airglow spectrum is not allowed by the usual dipole interaction. (Calculation shows that it is an electric quadrupole line.) Figures 5.10 and 5.11 illustrate the energy levels of the ground configuration for monatomic oxygen and nitrogen (OI and NI, respectively), as well as their singly-ionized counterparts, OII and NII.

The reason why a forbidden line is the brightest that is observed is explained as follows. Suppose that *all* the downward radiative transitions have extremely low probability. Then the mean life (against radiation) of the state will be very long. If atoms are somehow transferred to this state (by chemical reactions or by slow-electron bombardment, for example), they remain in this state for relatively long periods. In the case of the $^1S_0$ oxygen state this time is 0.74 second, about three-quarters of a second. Such a long-lived state is said to be *metastable*.

At ordinary pressures atoms in a metastable state lose their energy by collision with other atoms. The energy of excitation may appear as an increase

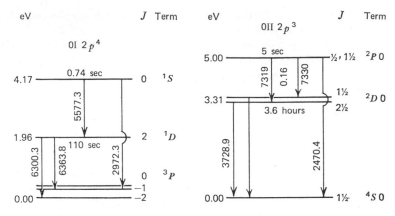

Figure 5.10   Level structure and multiplets of the ground configuration of the oxygen atom. All transitions shown are dipole-forbidden. Wavelengths shown are in angstroms. From J. Chamberlain, *Physics of the Aurora and Airglow*, Academic Press, Inc., 1961.

of the kinetic energy of the two colliding atoms. In the upper atmosphere, however, pressures are very low and the mean time between collisions is very long, so that the atoms have time to radiate even though the radiative transition probability is very low. Thus the intensity and spectrum of airglow from a given region of an atmosphere depend on the mean free path between collisions at that altitude; they depend on the composition, altitude, and scale height of that atmosphere.

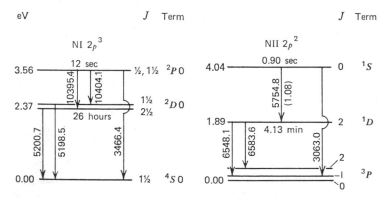

Figure 5.11   Level structure and multiplets of the ground configuration of the nitrogen atom. All transitions shown are dipole-forbidden. Since the terms shown are less than 5 eV, they are fairly readily excited by chemical reactions. From J. Chamberlain, *Physics of the Aurora and Airglow*, Academic Press, Inc., 1961.

The study of the energy levels of atoms is, of course, part of atomic physics, Experimental and theoretical work on the atomic reactions and excitation de-excitation mechanisms relevant to atmospheres comprise parts of what some people call "laboratory astrophysics." *Aeronomy* is the science of the (upper) regions of atmospheres, those regions where dissociation and ionization are present.

5.7 Radiative Transfer

The differential equation of radiative transfer is not too difficult to formulate, although an exact solution of it is rather a different matter. Consider Fig. 5.12 and assume for the sake of simplicity that monochromatic light of frequency v is incident on the top of the atmosphere with a specified directional dependence. The optical thickness of the atmosphere is t_v (for this frequency), where $t_v \equiv \alpha_v \int_0^z N(z') \, dz'$ and α_v is the cross-section for either absorption or scattering, and N is the number density of particles capable of scattering light at an altitude z.

We wish to determine the radiation lost and gained by a beam in a small cylinder; the net change of intensity over distance ds is

$$dI_v(t_v \mid \mu, \phi) = -I_v(t_v \mid \mu, \phi) \cdot N\alpha_v \, ds$$

$$+ \frac{1}{4\pi} \int_0^{2\pi} \int_{-1}^{+1} I_v(t_v \mid \mu', \phi') p(\mu, \phi; \mu', \phi') \, d\mu' \, d\phi' \cdot N\alpha_v \, ds. \quad (5.5)$$

Here $I_v(t_v \mid \mu, \phi)$ is the differential energy flux (ergs per square centimeter per

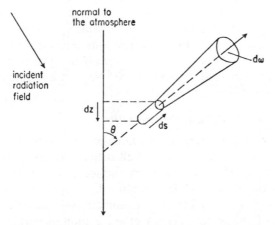

Figure 5.12   The geometry of radiative transfer. The beam passes through a small cylinder of length ds into an element of solid angle dw. From J. W. Chamberlain, *Physics of the Aurora and Airglow*, Academic Press, Inc., 1961.

second per steradian) at depth t_v and toward direction $\mu(= \cos \theta)$ and ϕ. The integral term, called the *source function*, gives the intensity scattered into a unit solid angle in direction μ, ϕ. The quantity p is the phase function; it determines the character of the scattering. If the scattering is isotropic, p is a constant called the albedo; if p is molecular (Rayleigh scattering), $p = \frac{3}{4}(1 + \cos^2 \theta)$.

This equation of transfer (5.5) may be rewritten as

$$\mu \frac{dI_v}{dt_v} = I_v - S_v, \tag{5.5a}$$

where S_v is the source function and $dt_v = N\alpha_v\, dz$. Also $dz = -\mu\, ds$. Equation 5.5a has formal solutions; these may be written as

$$I_v(\tau_v \mid -\mu, \phi) = I(0 \mid -\mu, \phi)e^{-\tau_v/\mu}$$
$$+ \int_0^\tau S_v(t_v \mid -\mu, \phi) \exp\left[(-\tau_v - t_v)/\mu\right]\frac{dt_v}{\mu}, \qquad 0 \leqslant \mu \leqslant 1, \tag{5.6}$$

and also as

$$I_v(0 \mid +\mu, \phi) = I_v(\tau_v \mid +\mu, \phi) \exp\left(-\tau_v/\mu\right)$$
$$+ \int_0^{\tau_v} S_v(t_v \mid +\mu, \phi) \exp\left(-t_v/\mu\right)\frac{dt_v}{\mu}, \qquad 0 \leqslant \mu \leqslant 1, \tag{5.7}$$

where (5.6) refers to the intensity of the downward radiation at the bottom of the atmosphere ($t = \tau$). The relation (5.7) describes the radiation diffusely reflected at the top of the scattering region of the atmosphere. The function $I_v(0 \mid -\mu, \phi)$ is the incident radiation field; according to our assumptions it is above the atmospheric region of interest for scattering. It may, for example, be described in terms of the van Rhijn functions. The factor $I_v(t_v \mid +\mu, \phi)$ is the outward intensity at the ground; it is zero if the ground albedo is zero, but is nonzero if the ground is snow-covered, say. At any rate it is clear that exact solutions are difficult, to say the least.

5.8   Other Means of Altitude Determination

The second method of altitude determination relies on the patchiness of the emitting layer. If the structure is well defined, two observatories some tens of kilometers apart on the planetary surface may view the same feature the coordinates of which may be established in terms of the celestial coordinates of nearby stars. Spherical trigonometry will then result in a value for the altitude of the feature. This method is more accurate than the van Rhijn method, because the latter neglects the patchiness in addition to the already noted scattering effects.

Finally, one may directly measure the emitted photons with a rocket-borne photometer aimed toward the zenith. As the rocket passes upward

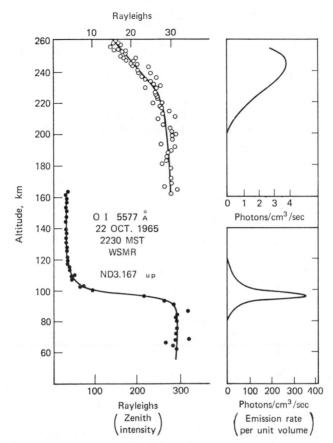

Figure 5.13   (a) Sounding rocket measurement of airglow at 5577-Å wavelength. The data were obtained at night from photometers as the rocket ascended; (b) (see page 146) same as a, except that the filter in front of the photometer transmitted 6300 Å. From Gulledge, Packer, Tilford, and Vanderslice, *Journal of Geophysical Research*, 1968.

through the emitting layer a considerable decrease in the output will be noted. Differentiation of the curves of intensity versus altitude yields a curve of the emission rate versus altitude (see Fig. 5.13). There is a fourth method that may be used to determine the altitude of the twilight glow; it will be discussed in Section 5.9.

5.9   Dayglow

Figure 5.14 illustrates the differences between dayglow, twilight glow, and nightglow. *Dayglow* is that light emitted by a layer of the atmosphere that is

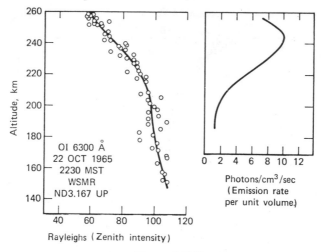

Figure 5-13(*b*)

illuminated from above by the sun. *Twilight glow* is that radiation emitted by the layer when it is illuminated from *below* by the sun. Thus both effects represent photoexcitation of atmospheric constituents; resonant scattering and fluorescence both occur. Faint emissions from potassium, calcium, and lithium have been detected.

The most interesting feature of the earth's atmosphere to be found through airglow studies is the *geocorona*, an envelope of hydrogen surrounding the earth at very high altitudes ($1000 \simeq 100{,}000$ km and more). There is only about $10 \simeq 20$ rayleighs of visible light from hydrogen at the earth's surface in the airglow; it is found at the wavelength of Hα emission (6563 Å).

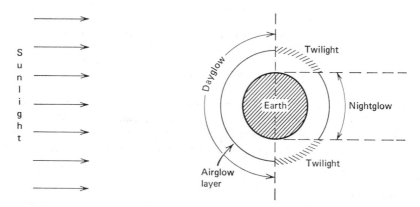

Figure 5.14   Illustrating the differences between dayglow, twilight-glow, and nightglow. All three originate in the airglow layer at the same altitudes.

Rocket-borne ultraviolet photometers, however, have detected Lyman alpha (Lα) radiation at altitudes in excess of 80 km from the zenith. Approximately 3×10^{-3} erg per square centimeter per second per steradian are detected, or about one *kilorayleigh*, below $\sim$200 km. Along the earth-sun line, an Lα experiment on the Mariner 5 spacecraft detected $\sim$3 kr at 20,000 km; the hydrogen density of atoms appeared to follow an $R^{-3.07}$ power law, where R is the distance from the center of the earth. Such a distribution is consistent with simple evaporative models of the exosphere, in which no satellite-orbit particles are allowed for. The precision of the experimental results, however, is not sufficiently great that the presence of satellite (i.e., gravitationally trapped) hydrogen atoms in the geocorona may definitely be ruled out. In any event it does seem clear that the geocorona has been detected out to $\sim$15 R_E, in the earth-sun direction.

The Lα radiation is weakest in the antisolar direction. Referring to Figure 5.14, this means that resonant scattering of solar Lyman alpha from hydrogen surrounding the earth has been observed. This hydrogen radiation thus should be included in dayglow, even though it was first detected at night.

It is also possible to find the temperature of the hydrogen atoms comprising the geocorona from dayglow measurements. Daytime spectral measurements of Lα from the sun show an absorption "dip" in the center of the emission line. The emission comes from the hot sun. The dip, however, is due to absorption by the cold ($\sim$2000°K) neutral hydrogen of the geocorona. The depth of the dip implies that there are $\sim$2 $\times$ 10^{12} hydrogen atoms per square centimeter between 145 km of altitude and the sun.

An interesting technique may be used to measure the altitude of the emitting layer from the planet's surface, through observations of the twilight glow. For example, just after sunset on the surface, the earth's shadow slowly rises higher and higher through the atmosphere above the observer, who measures the light from a sodium line, say. There is $\sim$30 kr of NaD radiation in the dayglow and about 5 kr in the twilight glow. If there is a thin emitting fluorescent layer at an altitude z, this will glow with a constant intensity until the shadow reaches $\sim z$, when the emissions will cease. Thus the time dependence of the intensity may be used to determine the altitude profile of the emitting region, to an accuracy governed by the sharpness of the shadow and by partial absorption of the solar radiation by the layer on the other side of the earth.

5.10 Causes of Airglow

As we noted previously, dayglow and twilight glow may be explained in terms of direct solar illumination of the upper atmosphere. Nightglow is more complicated and requires more explanation.

There is an immense reservoir of chemical energy stored in the upper atmosphere as a consequence of the work done on it by absorbed sunlight during the day. For example, solar ultraviolet dissociates molecular oxygen into atoms during the day; the atoms subsequently recombine, liberating 5.11 eV for each such recombination. (The analogous figure for nitrogen is 9.76 eV.)

Quantitatively, about 10^{12} oxygen molecules per square centimeter per second are dissociated during the day. In an equilibrium situation, therefore, about 10^{12} recombinations per square centimeter per second occur, so that $\sim 5 \times 10^{12}$ eV per square centimeter per second is made available. The total power radiated in nightglow is $\sim 5 \times 10^{12}$ eV per square centimeter per second (OH infrared radiation), or some 2000 times more power than observed in the optical.

The fact that nightglow is observed at the equator indicates that photochemistry, rather than bombardment of the upper atmosphere by charged particles, is responsible for the nightglow, since low-energy charged particles do not impinge on the terrestrial atmosphere at the equator. The details of the photochemical reactions, however, are still in some doubt.

Oxygen may recombine through a three-body reaction (known as the Chapman mechanism),

$$O + O + O \rightarrow O_2 + O^*, \tag{5.8a}$$

$$O^* \rightarrow O + h\nu. \tag{5.8b}$$

The cross-section is not sufficiently large, however, to produce the observed intensity of 5577-Å radiation, given the (known) concentrations of monatomic oxygen at ~ 100 km.

The excitation mechanism for the intense OH emission is most likely

$$H + O_3 \rightarrow OH^* + O_2^*, \tag{5.9a}$$

$$OH^* + O \rightarrow H + O_2. \tag{5.9b}$$

The sodium light may be produced by the following chain of reactions:

$$Na + O_2 + X \rightarrow NaO_2 + X, \tag{5.10a}$$

$$NaO_2 + H \rightarrow NaH + O_2, \tag{5.10b}$$

$$NaH + H \rightarrow Na^* + H_2, \tag{5.10c}$$

where the symbol X represents another atmospheric constituent.

The temperature of the electrons, T_e, found at ionospheric altitudes may also affect the spectrum of the nightglow; various states may be excited in monatomic oxygen, for example, through collisions. Tidal oscillations of the upper atmosphere, which change the pressure and hence the recombination rate at a given altitude, may be responsible for the temporal variations in the nightglow, and atmospheric turbulence may give rise to the observed patchiness of the intensity.

Problems

5.1. Suppose that a source of airglow excitation deposited an equal energy density per second at all altitudes between 60 km and 500 km. What would be the altitude dependence of the intensity of 5577 Å and 6300 Å? Why?

5.2. (a) If the green line in airglow was produced by the Chapman mechanism, how would its intensity depend on the pressure at a given (high) altitude?

(b) The daytime height of the E-layer seems to vary by ~ 1 km because of lunar tidal oscillations. If the scale height is 8 km, how much fluctuation in the intensity of the green line is to be expected because of lunar tides?

5.3. Although excitation of atomic oxygen by photoelectrons seems adequate to explain $\lambda = 5577$ Å dayglow above 130 km of altitude, a very similar distribution of intensity with height can result from dissociative recombination of O_2^+. If the rate coefficient for the latter reaction is 5×10^{-8} cm^3 per second (see Problem 6.4), how efficient must the process be to account for the 1-kr intensity found at ~ 150 km?

5.4. (a) Suppose that 10^{10} oxygen atoms in a 1-cm^2 column are suddenly excited into the $^1S_0$ state. Assume that the $^1S_0$ state decays only by emission of $\lambda = 5577$ Å; if the $^1D_2$ state were to decay only by the emission of 6300 Å energy, calculate the time dependence in rayleighs of the $\lambda = 6300$ Å and $\lambda = 5577$ Å lines, if there is no further excitation of the $^1S_0$ state.

(b) What is the intensity of this "two-wavelength airglow" 1.5 seconds after the initial population takes place?

(c) If the solar constant is 2.00 calories per square centimeter per minute and the sun behaved as a 6000°K black body, compare the glow calculated in (b) with the light received from the full moon at the top of the atmosphere. Take the optical albedo of the moon to be 0.073 and assume the terrestrial albedo to be zero.

Chapter 6

THE AURORA

At high latitudes on the earth the night sky is sometimes brilliantly lit by moving forms called aurorae. They are in themselves transparent: stars may be observed through them. The light is almost bright enough to read by and is usually, although not always, yellowish green. The *aurora borealis* (northern lights) and the *aurora australis* (its southern counterpart) may be seen on almost any clear night, although the intensity of the light varies, depending on a number of parameters. The Milky Way may be "washed out" by a bright aurora. There is also evidence that aurorae occur during the daytime, so that light is always emitted by the high-latitude atmosphere. The presence of aurorae has been recognized for many centuries, if not millenia, but many questions still remain concerning them. It was at one time thought that aurorae were sunlight reflected from the icecaps. Another "theory" held that the light was emitted by the gods. Present-day ideas are more prosaic, but it is still not clear where the acceleration of the responsible charged particles occurs.

6.1 Auroral Research

The fact that the frequency of aurorae is controlled by latitude (in particular by *geomagnetic latitude*; see Chapter 9) tells us that charged particles are involved in the phenomenon, since charged particle trajectories are controlled by the earth's magnetic field. Particles such as electrons and protons follow the lines of force of the earth's magnetic field; they impinge on the planet's atmosphere at high latitudes. We therefore define an aurora to be the emission of light by the upper atmosphere when it is bombarded by charged particles. Research has shown that electrons, rather than protons, are most often responsible.

Auroral research may be divided into three main types: (a) the study of aurorae themselves, (b) studies of the charged radiation causing the aurorae, and (c) investigations of phenomena related to aurorae. Type (a) research has resulted in classifications of the different and yet reproducible auroral shapes, as well as data on the spectra of the light emitted. It has also been concerned with the altitude and geographical distribution of aurorae and the

150

Figure 6.1   A rayed band aurora. The color is due to atomic oxygen radiation at 5577 Å and is typical. Courtesy B. J. O'Brien.

Figure 6.2 The great red aurora of 12 February 1958. Star trails in this time exposure may be seen through the display. Courtesy B. J. O'Brien.

variation of these factors with time. It still remains to be definitely proved that conventional aurorae are identical at opposite ends of a magnetic field line, the so-called *conjugate points*. This would show whether the aurorae source lies on closed, orderly field lines. Research of type (b) seeks to identify the particle types, fluxes, energy, and pitch angle distributions, and also the distribution of the particles in space. The primary problem of auroral research today is to learn where and how the particles responsible for the aurorae are accelerated to the energies observed; solar wind energies are far too low and there are far too few particles available in the Van Allen radiation belt. Type (c) research consists of studies of phenomena such as geomagnetic storms and atmospheric heating.

6.2 Geographical Distribution and Morphology

Aurorae are seen most frequently in two rings or zones around the two magnetic poles of the earth (Figure 6.3). The *auroral zones* are the regions between magnetic latitudes λ of $\sim$65–70° where aurorae are most often observed; they occur there sometime during every clear night during magnetically quiet times. The latitudes of the auroral zones change if the geomagnetic field is disturbed. The zones advance toward the equator and retreat from the pole as magnetic activity increases. There is always an aurora present in the zones, although it sometimes may not be discernible to the naked eye (Figure 6.4).

In addition to the visible aurorae, *radio* aurorae also occur; radio aurorae include radio-frequency emissions from aurorae as well as the structures detected by radar, and these emissions therefore refer to ionization in association with aurorae. They may be observed conveniently at frequencies ranging from 30 to 3000 megahertz.

Radar echoes are detected in the same general regions as the optical aurora, although either can occur without the other. The radar echoes are the results of weak scattering from irregularities in the aurorally associated ionization; they may also result from total reflection, at least at the lower frequencies. Altitude measurements show that echoes originate in the range 90–120 km, about the same range as the lower borders of the optical aurorae. Radar returns are most easily detected when the line of sight is perpendicular to the magnetic field at the reflection point. It therefore appears that the source ionized regions are in the forms of strongly field-aligned bundles. Unlike the optical, radar aurorae are most often detected near dusk and dawn.

Sporadic VHF and UHF radio emissions have been reported. They are picturesquely termed *auroral hiss*, and may be due directly to the precipitated electrons.

The frequency of occurrence of aurorae is diminished at latitudes removed from the auroral zones. At magnetic latitudes removed from the auroral

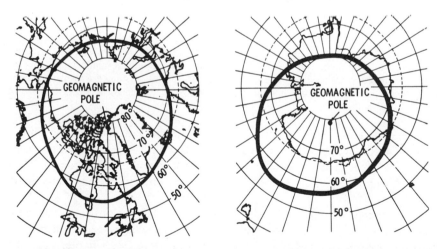

Figure 6.3 The locations, in geographic coordinates, of the zones of maximum auroral frequency of occurrence. Fewer aurorae occur both poleward and equatorward of the two ovals. From W. N. Hess, Ed., *Introduction to Space Science*. New York: Gordon & Breach, 1965, p. 208.

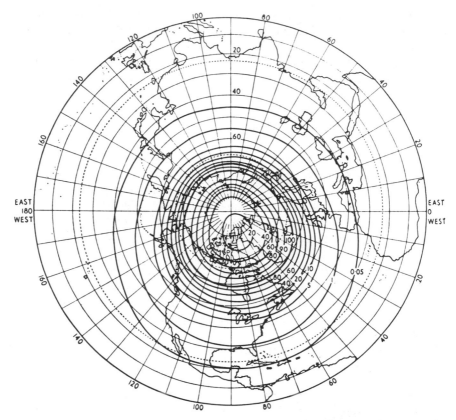

Figure 6.4   The frequency of aurorae observed anywhere in the sky (corrected for visibility variations) in geographic coordinates. The numbers are the relative number of nights on which an aurora might be seen at some time during the dark hours and are for the northern hemisphere. From *Journal of Geophysical Research*, E. H. Vestine, *Terr. Magn.* **49**, 77–102 (1944).

zones. At magnetic latitudes of ~40° (geographical latitudes of ~30° in the United States, for example) aurorae may be seen only once or twice each solar cycle (~11 years), and then only during intense magnetic disturbances. They are seen very rarely in places such as Singapore, which is on the geomagnetic equator.

The occurrence frequency of visible aurorae is also low at the geomagnetic poles themselves, although a special category known as *polar-glow* aurora is occasionally seen there, following a large solar flare (see Chapter 12). The protons emitted by the sun during such a flare are excluded from the earth by the geomagnetic field except near the poles; the Lorentz force deflecting incoming charged particles is small there.

In any event, the fact that an aurora is always present in certain zones proves that there is always particle precipitation in those regions. Thus the particle source must be always active.

Auroral shapes may be classified as having either horizontal or vertical patterns (or combinations of the two), where "vertical" is here taken to mean *magnetically vertical*; that is, a vertical shape such as a "ray" (see Fig. 6.5) is aligned parallel to the geomagnetic field vector **B**. When one views a set of rays in the magnetic zenith, it is apparent that they exist over a great range of altitudes, for a perspective effect called *corona* is observed.

Horizontal shapes include patterns such as *arcs*, *bands*, and so forth. Horizontal structures are generally brightest at an altitude between 100 and

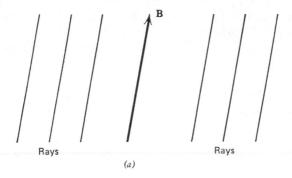

Rays   Rays

(*a*)

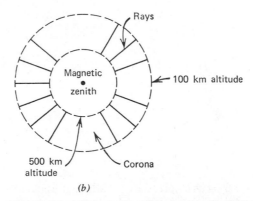

(*b*)

Figure 6.5   (*a*) Rays aligned along the geomagnetic-field vector **B** and viewed from the side; (*b*) same as (*a*) but viewed from directly "below" the magnetic zenith. The *corona* observed is evidently an effect of perspective.

120 km. They are characteristically ~10 km high and about 1–10 km thick in latitude, and they may be ~5000 km long in longitude. They therefore may be visualized as a long, thin ribbon wound edge-on around an appreciable portion of the earth. When viewed transversely they resemble searchlight beams along the geomagnetic lines of force. If viewed from underneath (i.e., in the magnetic zenith) they appear not to be separate entities, but merely "folds" in the auroral sheet (see Fig. 6.6).

There is another kind of auroral display, called the *auroral mantle*. This is an aurora without discrete form, so that it resembles enhanced airglow but has the auroral spectrum. It is the mantle aurora that makes the night sky so bright in the auroral zones that the Milky Way is "washed out" by the display.

All the aurorae so far discussed are of the conventional type, except for the polar glow aurorae previously mentioned. There are also two other unconventional auroral types, the *red auroral arcs* and the *discrete polar cap aurora*. The latter type, however, may well be an "ordinary" polar aurora, since it resembles the conventional aurora in structure and spectrum. The only difference is that the frequency of occurrence of polar-cap aurorae decreases with increasing magnetic activity, but this may merely be a consequence of the fact that the auroral zones retreat from the poles during disturbed times.

The red auroral arcs are, as their name suggests, horizontal structures that emit in the red, at $\lambda = 6300$ Å. Their most puzzling characteristic is the virtual

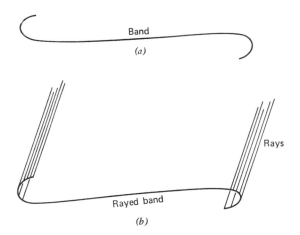

Figure 6.6 (*a*) Auroral horizontal band, viewed from below; (*b*) an auroral horizontal rayed band, viewed transversely, but is the same "structure" (seen differently) as in (*a*). The rays are parallel to the local geomagnetic field vector.

absence of the oxygen green line at 5577 Å. Only about 2 eV is required to excite the parent state of 6300 Å, and only an additional ~2 eV is required to excite the parent state of 5577 Å. Charged particles that possess many kilo-electron volts of energy are now known to cause the "ordinary" polar aurorae; such particles can easily supply both excitation states, so that it would seem unlikely that the red auroral arcs are caused by particle bombardment. But they are subject to magnetic control: they may extend for tens of degrees along a parallel of magnetic latitude, and they are seen in the auroral zones and midlatitudes, where they often occur on the equatorial boundary of active aurorae. Their frequency of occurrence depends on the solar cycle. The red arcs may occur several times each year around the maximum of solar activity, and they are very rare in the years of the quiet sun. Thus they exhibit features that remind one of charged particle bombardment.

6.3   Time Variations in Aurorae

There are large-scale time variations in a given auroral display that have time scales of the order of many minutes, while smaller scale changes may occur in the aurora within time periods of a few seconds. Investigations have shown that, statistically speaking, there are several features common to the displays. Thus the auroral patterns may be regarded as fixed with respect to the earth-sun line and the large-scale temporal variations that are observed are caused by the motion of an observer turning with the rotation of the earth. There is another explanation of the large-scale variations, however: they may better be explained as a consequence of the development and decay of individual aurorae, each of which may take ~30 minutes to evolve and an hour or two to decay.

On the shorter time scale individual rays may appear to move with velocities of ~100 km per second; typical horizontal motions of ~1 km per second are seen. Also, aurorae are occasionally observed to "flame": the brightest region seems to move upward. This last type of change may be due to the later arrival of lower-energy electrons that are not able to penetrate down to normal auroral altitudes. (It may also be due to the later arrival of trapped electrons that have larger pitch angles and consequently higher mirror altitudes; see Chapter 7.)

In general the biggest aurorae (i.e., the most extensive geographically) occur more often when solar activity is high. They occur about a day after particularly large solar flares. In the "great aurorae" that occur several times each year at sunspot maximum, displays may be seen near the magnetic equator and, more commonly, at midlatitude sites such as Washington, D. C. But it is important to recall that there is always an aurora in the auroral zone, even if it is too weak or diffuse to be seen.

6.4 Auroral Brightness

Aurorae are classified by brightness with international brightness coefficients; these are specified in Table 6.1, where the intensity is described in terms of the intensity of 5577-Å radiation.

TABLE 6.1. CLASSIFICATION OF AURORAL BRIGHTNESS

Classification	Brightness similar to	Emission of 5577 Å
IBC I	The Milky Way	10^9 photons cm^{-2} sec^{-1}
IBC II	Thin moonlit cirrus clouds	10^{10} photons cm^{-2} sec^{-1}
IBC III	Moonlit cumulus clouds	10^{11} photons cm^{-2} sec^{-1}
IBC IV	Ground illumination equal to that provided by full moon	10^{12} photons cm^{-2} sec^{-1}

6.5 Auroral Spectra

As in the case of airglow, the brightest visible emissions are from atomic oxygen and from ionized molecular nitrogen. The relative intensities of the two emissions in auroral displays, however, are quite different from those found from airglow. While the ratio of 5577 Å to 3914 Å (atomic oxygen and the band "head" of ionized molecular nitrogen, respectively) is $\geqslant 80:1$ in airglow it is only ~2:1 in aurorae.

A good energy balance for auroral phenomena has not yet been made, but it appears that an energy deposition of ~1 erg cm^{-2} second^{-1} by ionizing particles results in the emission of only ~200 rayleighs (200×10^6 photons cm^{-2} second^{-1}) of 3914 Å and only ~0.5% of the deposited energy is converted to visible light. It appears likely that ~20% of the deposited energy is in the little-studied vacuum ultraviolet portion of the electromagnetic spectrum emitted by the atmosphere. Perhaps ~30% of the particle energy goes into heating the atmosphere.

One of the most interesting spectral features of aurorae is that of Doppler-shifted Balmer radiation. This is indicative of protons spiraling downward along the geomagnetic field lines and radiating as they come in.

As a proton spirals downward into the denser atmosphere, it may capture an electron and become a hydrogen atom in an excited state. The atom may then return to the ground state by radiating a photon of an energy equal to the energy difference between the excited and ground states. If the initial energy of the incoming proton is of the order of 100 keV, it may go through this process of charge neutralization and radiation hundreds of times as it plunges into the atmosphere, before finally being absorbed and assuming an identity as an atom of hydrogen in the upper atmosphere. Most of the 100 keV will be used in ionizing atmospheric atoms, but the proton will emit many protons of the Balmer series, the visible radiation from hydrogen. Detailed estimates are that such a particle will radiate ~60 photons of Hα, the red line nominally at 6563 Å.

However, the hydrogen atoms are moving with respect to the observer on the ground when they emit this light. Hence their wavelength λ is Doppler-shifted by an amount $\Delta\lambda$, where $\Delta\lambda/\lambda = v/c$ and where v is the velocity of the atom relative to the observer along the line of sight. Wavelength shifts of several angstroms are observed, as shown in Figure 6.7.

The profiles of Figure 6.7 are consistent with the above description. The light comes from atoms that on the average spiral around and down along the geomagnetic field lines. The neutral hydrogen atoms are not guided by the field, but when each loses its electron—as it does repeatedly in atomic collisions in the atmosphere—the remaining proton is then guided by the Lorentz force that acts on any charged particle moving in a magnetic field. When one views the magnetic horizon, one looks edge-on at the spiraling particles, so that equal numbers are moving away from and toward the observer. The line profile is simply broadened by the Doppler mechanism. However, when one looks toward the magnetic zenith most of the particles

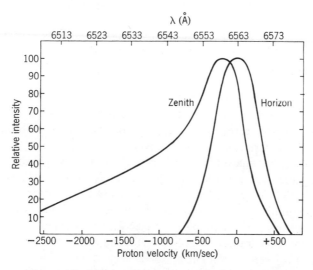

Figure 6.7 Results of many combined measurements of 6563-Å line profiles, toward and perpendicular to the magnetic zenith. The blue-shift of the radiation toward the zenith indicates downward-moving hydrogen atoms. The width of the line viewed toward the horizon is due to protons gyrating about the magnetic field lines; at any instant some protons are moving toward the observer and others are moving away.

are moving toward the observer on the surface, so that the Doppler profile is shifted toward shorter wavelengths.

These 1950 observations provided the first conclusive proof of particle bombardment in aurorae. Most auroral emissions, however, are not sustained by proton bombardment, as shown by the following analysis of photometric observations.

The intensity of the line from ionized molecular nitrogen is proportional to the total energy dissipated by the charged particles, while the intensity of a line in the hydrogen series is roughly proportional to the flux of incident protons. Now, since 1 erg cm^{-2} second^{-1} deposited by charged particles will produce 200 rayleighs of 3914 Å at 100 km and since the intensity of the 4861 Å line of the Balmer series (known as Hβ) will be 150 rayleighs for a proton energy flux of 1 erg cm^{-2} second^{-1}, we would expect that the 3914 Å and 4861 Å lines would be nearly equally bright when excited by protons. The observation, however, is that the 3914 Å emission is usually 10 or more times brighter than the Hβ brilliance.

6.6 Particle Bombardment

The spectral measurements conducted from the surface and described in Section 6.5 showed that protons do bombard the atmosphere during aurorae, but they also showed that protons do not sustain aurorae. Direct rocket flights into aurorae, however, have shown that electrons with energies of the order of $\sim$1 to $\sim$10 keV are the primary agency that excites aurorae. The spectra are quite soft (exponential differential spectra with e-folding energies of $\sim$5 keV are frequently found) and vary rapidly with time. The pitch angle distributions also fluctuate in time (the pitch angle is the angle between the instantaneous velocity vector of the particle and the geomagnetic field line). The causes of these variations are unknown.

It is interesting that the intensity of trapped radiation (see the following chapter) at altitudes between 100 and 2000 km does not appreciably decrease during an aurora. Indeed, it actually increases somewhat during an aurora. Thus it appears unlikely that the trapped radiation (otherwise known as the Van Allen radiation) is the source of the electrons that excite aurorae.

That the Van Allen radiation seems inadequate as the auroral electron source may also be seen by comparing the energy required to supply a bright auroral display with the energy stored in the Van Allen belt. The total energy stored in electrons in the radiation belt is less than 10^{20} ergs. About 10^{17} ergs per second are required for a bright aurora. Thus it appears that the entire Van Allen radiation belt would be expended in about 15 minutes in producing a bright aurora.

It is also important that the particles more or less steadily emitted by the sun—the so-called "solar wind" (see Chapter 10)—have energies that are typically 0.5–1 keV. Thus the energies of solar wind electrons are too low for auroral excitation, even though the flux of solar wind particles is ample. Also, the energies of solar wind radiation have been measured to be about the same, all the way from the orbit of Venus out to the boundary of the ordered geomagnetic field (~10 earth radii from the terrestrial center). Therefore, if the solar wind particles are indeed responsible for the aurora, they must be *locally accelerated*. That is, the solar wind electrons would have to have their energies increased somewhere in the neighborhood of the earth.

There is some evidence that the solar wind particles do (somehow) become auroral particles. Direct measurements of the ratio of the flux of helium nuclei (alpha particles) to the flux of protons in auroral radiation appear to agree with the composition of the solar wind as measured in interplanetary space; the α/p ratio is ~2% in both locations. Thus the problem of understanding auroral excitation becomes one of understanding the local acceleration mechanism. This problem is being actively attacked both theoretically and experimentally, since the acceleration mechanism may well be a general feature of plasma physics. Thus the theoretical consequences of the interaction of the solar wind with the earth's magnetosphere are being explored, while experiments are attempting to locate the acceleration region and the motion of the particles from interplanetary space down to auroral altitudes.

6.7 Particle Trajectories

It is relevant at this point to examine some characteristics of the motion of charged particles in the geomagnetic field, since it is such particles that excite aurorae. The motion of a charge q moving with a velocity $\mathbf{v}$ at a pitch angle α with respect to the geomagnetic field $\mathbf{B}$ is governed by the Lorentz force $\mathbf{F}$, where

$$\mathbf{F} = q(\mathbf{v} \times \mathbf{B}) \tag{6.1}$$

and where it is assumed that no electric fields $\mathbf{E}$ are present. Equation 6.1 may be expanded to read

$$F = qvB \sin \alpha \tag{6.1a}$$

in scalar notation.

In Chapter 7 (on Van Allen radiation) it is shown that the magnetic moment I of the particle is a constant, provided that $\mathbf{B}$ does not change during one spiral rotation around $\mathbf{B}$. The "magnetic moment" of a particle arises from the fact that its motion is in effect a current loop around $\mathbf{B}$; the spiraling charge may be thought of as a magnetic dipole. It can be shown that the value of this magnetic moment is

$$I = \frac{\frac{1}{2}mv^2 \sin^2 \alpha}{B}. \tag{6.2}$$

If the value of $\mathbf{B}$ does not change with time, the magnetic field does no work on the particle; the kinetic energy of the free particle, $\frac{1}{2}mv^2$, is a constant. We may therefore write that

$$\frac{\sin^2 \alpha}{B} = \text{constant} = \frac{1}{B_m}, \tag{6.3}$$

where B_m is the field strength at the place where $\alpha = 90°$. A particle with $\alpha = 90°$ moves perpendicular to the lines of force; it is said to *mirror* at such a location, for it will travel backward along the field line if $\alpha > 90°$.

Now particles that cause aurorae must hit the atmosphere in order to do so. Since aurorae occur at altitudes of ~100 km and neglecting scattering and absorption, the mirror points of auroral particles must therefore be at or below 100 km; $B_m \gtrsim B_{100 \text{ km}}$ follows from this, since the field strength B increases as the particle nears the earth. The symbol $B_{100 \text{ km}}$ refers to the field strength at 100 km of altitude.

We may define a *loss cone* of pitch angles bounded at any altitude by a value α_D, where

$$\frac{\sin^2 \alpha_D}{B} = \frac{1}{B_{100 \text{ km}}}. \tag{6.4}$$

Particles with $\alpha \leqslant \alpha_D$ are in the loss cone and they are *precipitated* into the atmosphere. The value of α_D decreases with altitude (or B). The boundaries of the loss cone should not be regarded as perfectly sharp; atmospheric scattering makes them somewhat fuzzy. Since the amount of scattering depends on the altitude, the loss cone is somewhat diffuse and the degree to which it is diffuse depends on the altitude.

Neglecting the effects of scattering, in the equatorial plane for field lines that pass through the auroral zone $\alpha_D \sim 2\text{--}3°$. Thus particles within $2\text{--}3°$ of the geomagnetic field lines may excite aurorae.

6.8 Related Phenomena

The atmosphere is heated during aurorae. An IBC III aurora, for example, deposits ~400 ergs cm^{-2} second^{-1} in the atmosphere, of which perhaps 20% will go into the form of random kinetic energy of atmospheric constituents. There are ~2×10^{18} atoms and molecules in a one square centimeter column above an altitude of ~110 km. If they share equally in this kinetic energy, each such atmospheric particle receives ~2×10^{-5} eV per second. Taking their normal temperature as ~300°K, their normal kinetic energy is $kT \sim 3 \times 10^{-2}$ eV. Thus they will be heated to twice their original temperature in ~1000 seconds.

Auroral heating has been observed. A heated atmosphere rises, increasing the drag on artificial satellites. The heated atmosphere cools through the processes discussed in Chapter 3: radiation and downward conduction. *Gravity waves*—the bulk motion of the atmosphere in the form of waves that arise because of the "restoring force" provided by the earth's gravitational field—also provide a means of cooling the atmosphere.

Another phenomenon observed to occur when and where aurorae occur is audible sound. Such sounds (a crackling) could not propagate through the rarefied atmosphere that exists at auroral altitudes. Hence they must be created indirectly somewhere near the planetary surface. One possibility may be electric discharges in surface objects. These could be caused by an electromotive force induced in conductors by the intense currents postulated to be flowing along auroral arcs. Some 10^5 amperes are thought to flow in these heavily ionized regions; these currents may account for the fluctuations in the geomagnetic field strength observed on the ground during aurorae.

6.9   Auroral Particle Configurations

Observations of particle precipitation during aurorae have resulted in several models of the configuration in space of auroral particles. (These models are presently being investigated experimentally.)

Figure 6.8 shows the relationship observed between particle fluxes and

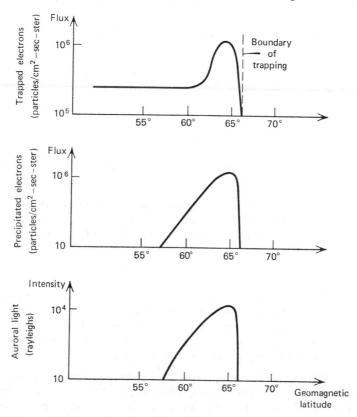

Figure 6.8   Approximate relationship between particle fluxes and auroral brightness, as a function of geomagnetic latitude.

auroral light. The curve labeled "trapped electrons" refers to electrons with pitch angles greater than those angles contained in the loss cone.

The regions of proton precipitation tend to appear on the equatorial side of quiescent electron-produced aurorae before local midnight. The reverse holds true after midnight, and the reason for this is as yet unknown.

The various models of auroral particle acceleration invoke extraction of energy from the solar wind at its interface with the magnetosphere. At this writing it seems clear that the "tail" of the magnetosphere (see Chapter 10) plays an important role in the acceleration; the tail extends at least as far as the moon (60 earth radii) in the antisolar direction and may be $\sim 10^3$ earth radii in extent. It may be that the acceleration process occurs all along the tail, as well as in front of the magnetosphere (i.e., in the solar direction). This field of research is so active at the present time that there are reasonable grounds for hoping that this problem will be solved shortly.

Problems

6.1. Derive expressions for the variation of the size of the loss cone with altitude and latitude.

6.2. Calculate the energy flux of energetic charged particles (units of ergs per square centimeter per second) required to produce an IBC I aurora. Why did you use the brightness of 3914 Å in your calculation rather than the brightness of 5577 Å by which the IBC I is defined?

6.3. Prove analytically that the brightness of an IBC I aurora will always be measured as the same if the aurora fills the field of view of a vertically pointing, narrow-angle photometer. Show that this is so for an aurora at 100 km of altitude and at 1000 km of altitude.

6.4. Any collisional process, such as those that excite or ionize atmospheric atoms, may be described in terms of its cross-section Q_{12} (i.e., the system changes from state 1 to state 2) or by a *rate coefficient* S_{12}, which is defined by

$$S_{12} = \int_{v_{12}}^{\infty} Q_{12}(v) v f(v) \, dv.$$

In this integral v_{12} is the minimum velocity capable of producing the reaction. If $Q = Q_0/v$, where Q_0 is constant, and if $f(v)$ is the Maxwellian distribution $f(v) \, dv = 2(m_e/kT)^{3/2} v^2 \exp[-m_e v^2/2kT_e] \, dv$, for electrons of mass m_e at a temperature T_e, find the rate coefficient that describes excitation by inelastic electron collisions of the type $X + e \rightarrow X^* + e$.

6.5. Suppose that in a faint proton aurora the total energy influx amounts to 10^{12} eV cm^{-2} second^{-1}. Take the atmospheric heating efficiency to be $\sim 50\%$ and find the rate of temperature rise in degrees Kelvin per minute. Assume that the heat is equally shared by the $\sim 2 \times 10^{17}$ atoms and molecules above 140 km.

6.6. An upper limit to the auroral reflection coefficient for radar waves may be obtained from the reflection coefficient R for a sharp surface, where $R = 4 \times 10^{10} N_e{}^2 v^{-4}$ and where N_e is in electrons per cubic centimeter and v is in megahertz. Find R at 72 MHz for a total proton flux of 5×10^7 cm^{-2} second^{-1} varying as $\exp(-E/20 \text{ keV})$ by first finding the maximum electron density.

6.7. (a) If the approximate radius of the cross-section of the magnetosphere perpendicular to the flow of the solar wind is ~10 earth radii, calculate the number of protons arriving per second at the magnetosphere, if the solar wind proton density is 5 cm^{-3}.

   (b) Consider the proton precipitation zone to extend uniformly between geomagnetic latitudes of 70° and 60°. If 5×10^7 protons cm^{-2} second^{-1} are precipitated in a proton aurora, express the rate of proton precipitation as a percentage of the total number of protons arriving at the magnetosphere.

6.8. If the number density of solar wind particles is 5 cm^{-3} and each particle has 500 eV of energy, express the energy in an IBC III aurora observed over the zone described in the preceding problem as a fraction of the solar wind energy brought into the magnetosphere.

6.9. A cloud of barium ions is released at a very low geomagnetic latitude where aurorae normally do not occur. Within minutes it forms a broad arc with ray structure parallel to the geomagnetic field lines, showing nearly perfect resemblance to a rayed auroral arc. In light of this experiment, discuss the relative importance of the ionospheric plasma dynamics and the spatial extent and structure of the auroral acceleration mechanism in determining the structure of a rayed auroral arc.

Chapter 7

VAN ALLEN RADIATION

The discovery in 1958 by Van Allen of an intense zone of charged radiation around the earth provided the first major surprise of research conducted in space. Most of the other results obtained with spacecraft have confirmed and/or refined ideas based on research conducted within the earth's atmosphere. The existence of geomagnetically trapped radiation was not entirely discounted in early theoretical discussions, but the pre-1958 treatments of particle populations in near-earth space did not predict the large fluxes actually observed The source(s) and loss mechanisms of the Van Allen radiation are still not understood in complete detail, but the existence of high fluxes of energetic electrons, protons, deuterons, and so on, as well as their distributions in time and space, are now well established and their continued study defines the problems of injection and acceleration of charged radiation in magnetic fields.

7.1 General Discussion

At great altitudes above the earth—wherever the atmospheric density is sufficiently low and the geomagnetic field sufficiently well ordered—there are nonthermal energetic particles that are *durably trapped*. They are mostly electrons and protons. The energies of the former extend up to a few million-electron volts; proton energies up to 700 MeV have been observed.

In their normal motion the particles are continually subjected to a Lorentz force that is perpendicular both to the local geomagnetic field vector and to their instantaneous velocity vector. They therefore gyrate around the lines of the local magnetic field. The center of the circle of gyration drifts along the field lines. This guiding center drifts in latitude along a field line until the pitch angle of the spiral increases to 90° (the pitch angle α is the angle between **B** and the particle velocity **v**). At this point the particles *mirror*; they start spiraling back along the field line, so that the direction of drift is reversed. The particle thus "bounces" between the northern and southern hemispheres and we speak of a trapped particle. The trapped particles all have pitch angles, of course, that are outside the loss cone (see Section 6.7).

At the same time the particle drifts in longitude. It moves westward if positively charged and eastward if negative. Thus a cloud of energetic charged particles surrounds the earth.

7.2   Motion of Charged Particles in Magnetic Fields

The equation of motion of a particle with momentum $\mathbf{p}$ and charge ze is

$$\frac{d}{dt}\left(\frac{pc}{ze}\right) = \mathbf{v} \times \mathbf{B}. \tag{7.1}$$

If the magnetic field $\mathbf{B}$ is approximately uniform, the particle rotates around the field very much as it does in a cyclotron; the angular frequency of rotation ω is

$$\omega = \frac{zeB}{\gamma m_0} = 1.76 \times 10^7 \frac{B}{\gamma} \frac{\text{radians}}{\text{second}} \tag{7.2}$$

for electrons, where

$$\gamma = \left(1 - \frac{v^2}{c^2}\right)^{-\frac{1}{2}}.$$

The radius of the circle ρ is just

$$\rho = \frac{p_\perp c}{300 z B} \text{ centimeters}, \tag{7.3}$$

where $p_\perp = \gamma m v_\perp$ and $v_\perp$ is the component of velocity perpendicular to the magnetic field. The center of the circle drifts along the field line with a speed $v_\parallel$, the longitudinal component of the velocity vector.

The above statements will be true if the field varies slowly in space and time so that $|\nabla B/B| \ll 1$ over a distance ρ and/or a time $\tau = 2\pi/\omega$. In this case it is possible to decompose the particle's motion into motion of the guiding center plus rotation of the particle around the center. In this motion the quantity I_1, called the *magnetic dipole moment* of the particle, is conserved, where

$$I_1 = \frac{m v_\perp^2}{2B}. \tag{7.4}$$

The kinetic energy $T = mv^2/2$ is also a constant of the motion if $\mathbf{B}$ is constant in time, but T is not conserved if $\mathbf{B}$ is not constant, since a time-varying magnetic field is equivalent to an electric field.

The electromotive force ϵ around one loop of the orbit is

$$\epsilon = \oint \mathbf{E} \cdot d\mathbf{r} = -\frac{1}{c}\frac{\partial}{\partial t}\int_{\text{area}} \mathbf{B} \cdot d\mathbf{A} = -\frac{1}{c}\frac{\partial}{\partial t}(\pi \rho^2 B). \tag{7.5}$$

Inserting the value (7.3) for ρ, ϵ becomes

$$\epsilon = -\frac{1}{(300)^2 c}\frac{\partial}{\partial t}\left(\frac{\pi m^2 c^2 v_\perp^2}{q^2 B}\right) \tag{7.6}$$

for a particle of charge q. If **B** does not vary in time, therefore, ϵ is zero and consequently T is zero, since $\epsilon = 2\pi/\omega \, dT/dt$. Also, since B is a constant, $mv_\perp{}^2/2B$ is a constant; I_1 is conserved.

When B varies, the expression (7.6) may be written

$$\epsilon = -\frac{1}{c}\frac{\partial}{\partial t}\left(\frac{\pi m c^2}{q^2}\frac{2T_\perp}{B}\right)\frac{1}{(300)^2}, \qquad (7.6a)$$

or

$$\epsilon = \frac{2\pi m c}{(300)^2 q^2}\left(\frac{1}{B}\frac{\partial T_\perp}{\partial t} - \frac{T_\perp}{B^2}\frac{\partial B}{\partial t}\right) = \frac{2\pi}{\omega}\frac{dT}{dt}, \qquad (7.7)$$

where $T_\perp \equiv mv_\perp{}^2/z$. If the field is uniform the partial derivatives in (7.7) may be replaced with the total derivatives, so that

$$\text{constant} \times \frac{dT_\perp}{dt} = \frac{T_\perp}{B}\frac{dB}{dt}. \qquad (7.8)$$

This result tells us that $(T_\perp/B)$ is a constant. But $T_\perp/B = I_1$, so that the magnetic dipole moment of the particle is conserved in a uniform field, even when the field varies in time.

This result may be expressed in terms of the pitch angle α. We have that $v_\parallel = v \cos \alpha$ and $v_\perp = v \sin \alpha$. If T is the kinetic energy of the particle,

$$I_1 = \frac{T}{B} \sin^2 \alpha. \qquad (7.9)$$

An important feature of the motion is that

$$\frac{I_1}{B} = \frac{\sin^2 \alpha}{B}, \qquad (7.10)$$

so that the magnetic moment in an inhomogeneous magnetic field is an *adiabatic invariant*.

Adiabatic invariants appear in systems that have periodic motion. They are connected with slow perturbations of the Hamiltonian, which maintain the basic periodic character of the motion and do not resonate with it. The number of adiabatic invariants is less than or equal to the number of degrees of freedom of the system. They are "approximately constant"; they are constant only for very slow changes of the variables involved.

The guiding center drifts along the field lines until a value of $B = B_m$ is reached where $\alpha = 90°$. The particle is said to "mirror" at this point, and (7.10) may be rewritten as

$$\frac{\sin^2 \alpha}{B} = \frac{1}{B_m}, \qquad (7.11)$$

where B_m is the maximum value of the field strength to which the orbit may penetrate. Thus if the particle is in a weak field bounded on either side by stronger fields of strength B_m, it will be trapped in the weak field and bounce back and forth between the two "boundaries."

There are other types of field geometries of interest here. They include the cases when $|B|$ changes in a direction perpendicular to **B** and where the field B is curved. In the first case the guiding center drifts across the field lines in a calculable way; we speak of the "gradient drift" of the guiding center. The velocity $\mathbf{v}_g$ of the gradient drift of the guiding center is given by

$$\mathbf{v}_g = \frac{v_\perp^2}{2\omega} \frac{\mathbf{B} \times \nabla B}{B^2}. \tag{7.12}$$

The second case gives rise to a "curvature drift" velocity $\mathbf{v}_c$, which may be expressed as

$$\mathbf{v}_c = \frac{v_\parallel^2}{\omega R^2} \frac{\mathbf{R} \times \mathbf{B}}{B^2},$$

where **R** is the radius vector from the center of curvature of the field lines to the field line. If $\nabla \times \mathbf{B} = 0$, then

$$\frac{\mathbf{R}}{R^2} = -\frac{\nabla_\perp B}{B}, \tag{7.13}$$

so that if curvature drift occurs, the guiding center moves with a speed $v_\parallel$ along the field line and stays on the same field line as seen in the planes containing **R**. It drifts, however, in the $\mathbf{R} \times \mathbf{B}$ direction.

Curvature drift occurs because in the moving frame of reference there is an apparent electric field $\mathbf{E}'$ in the $(-\mathbf{R})$ direction of strength,

$$\mathbf{E}' = \gamma \left(\frac{\mathbf{v}_c}{c} \times \mathbf{B} \right),$$

that is just sufficient to keep the particle on the field line. In other words,

$$q\mathbf{E}' = -\frac{\mathbf{R}}{R} \frac{mv_\parallel^2}{R}. \tag{7.14}$$

The relation (7.14) expresses the force balance on the particle in curved fields.

Indeed, gradient drift is but a special case of curvature drift. The gradient drift is due to variations in the instantaneous radius of curvature of the orbit; the radius of curvature of the orbit will not be a constant all along one loop of the orbit if a gradient of the field exists in a direction perpendicular to **B**.

7.3 Motion in a Dipole Field

The trapped radiation around the earth is found where the magnetic field lines are sensibly closed. It is therefore pertinent to discuss trapping in a dipole field.

We saw in Section 7.2 that a charged particle in a low field region bounded by two stronger fields of strength B_m will be trapped in the low field region. Thus a particle will bounce back and forth between the northern and southern hemispheres of the earth. The guiding moves down in altitude in the high-latitude regions in each hemisphere until it reaches the altitude where B becomes as strong as B_m:

$$\frac{1}{B_m} = \frac{\sin^2 \alpha_1}{B_1},$$

where α_1 is the pitch angle at some point in the trajectory at which the field strength is B_1.

In Section 9.4 it is shown that the field strength of the equivalent (geomagnetic) dipolar field varies with geocentric radius r and colatitude θ as

$$|B| = \frac{B_0}{r^3} [3 \cos^2 \theta + 1]^{\frac{1}{2}},$$

where B_0 is the value of the field at the equator on the planetary surface ($B_0 = 0.311$ gauss, for the earth). The point here is that the dipolar field is strongly altitude-dependent, the field falling off inversely with the cube of the geocentric distance.

7.4 Motion in the Real Geomagnetic Field

The motion of a charged particle in the real field is similar to that discussed above for the dipolar field, even though the real field deviates somewhat from that of dipole, even for small distances from the earth. In the dipolar approximation the guiding center moves on a closed shell of field lines; the shell is a surface of rotation of a single field line. (The particles drift westward if positively charged.)

In the real field the shell is no longer a figure of rotation. The motion may be described with the use of three "adiabatic invariants." The first such invariant is I_1, as given in (7.4). The second invariant is called the *longitudinal invariant I_2.*

The second invariant is derived from the fact that in nearly cyclic motion the quantity J_i is a constant of the motion, where

$$J_i = \oint p_i \, dq_i \tag{7.15}$$

for the ith particle, which has momentum component p_i canonically conjugate to coordinate q_i. This is true in the adiabatic case, which applies when the equations of motion change but little over a cycle. If $p_\| = p \cos \alpha$, we set $J = \oint p_\| \, ds$, where ds is an element of distance along the field line and $p_\|$ is the component of momentum parallel to **B**. The integral is taken along a field line, from one mirror point to the other and back. Since $I_1 = $ constant,

$$p_\| = p \left(1 - \frac{B}{B_m} \right)^{1\!\!/\!\!2} \tag{7.16}$$

and

$$J = 2p \oint_A^{A'} \left(1 - \frac{B}{B_m} \right)^{1\!\!/\!\!2} ds = \text{constant.} \tag{7.17}$$

The limits A and A' in (7.17) are the mirror points in the opposite hemispheres. Since p is a constant of the motion where B is a constant, the quantity I_2, which is defined by

$$I_2 \equiv \frac{J}{2p}, \tag{7.18}$$

is also an adiabatic invariant.

The longitudinal invariant I_2 is a property of the mirror points A and A' only. Equations 7.10 and 7.18 tell us that as a particle drifts in longitude every mirror point has the same value of B_m and I. Referring to Figure 7.1 (following a discussion due to B. J. O'Brien), suppose that we have a trapped particle that mirrors at P and P^*. It therefore has a value of the longitudinal invariant I_2 given by I_0. Also, the value of the mirror field strength B_m is given by B_p. We wish to know where it will be when it drifts in longitude, say to the right-hand side. It could mirror at Q and Q^* and satisfy $I_2 = I_0$, but then $B_x \neq B_p$; the second but not the first invariant would be conserved. The situation is reversed if it mirrored at A_1 and A_1^*, or A_3 and A_3^*. Only if it mirrors at A_2 and A_2^* will both invariants be conserved. Hence it will mirror at A_2 and A_2^*.

There is a third invariant of the motion, called the *flux invariant* I_3. It is defined by

$$I_3 = \int_{\text{surface}} \mathbf{B} \cdot d\mathbf{A}, \tag{7.19}$$

where the surface of integration is any surface over the polar regions bounded by the locus of mirror points in one hemisphere. Clearly I_3 will be constant if I_1 and I_2 are constant and the magnetic field is static. However, if the field slowly expands or contracts (because of a change in the planetary magnetic moment or in any external currents), the fact that I_3 remains constant assures us that the shell on which a particle drifts will expand or contract with the field. In this case the kinetic energy and momentum of the particle may change, but I_1, I_2, and I_3 will not.

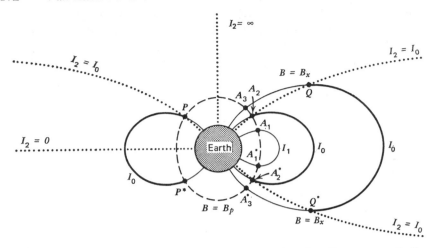

Figure 7.1   Illustration of the various parameters used to describe geomagnetically trapped radiation. A particle that mirrors at P and P^* will also mirror at A_2 and A_2^*. Reprinted from A. J. Dessler and B. J. O'Brien, Penetrating Particle Radiation, in Francis J. Johnson, Ed., *Satellite Environment Handbook*, second edition, with the permission of the publishers, Stanford University Press. Copyright 1965 by the Board of Trustees of The Leland Stanford Junior University.

7.5   The Three Characteristic Times of the Motion

There is first the time required to complete one orbit about a field line, τ_1. It is given by

$$\tau_1 = \frac{2\pi}{\omega} = \frac{2\pi mc}{zeB}\,\gamma \text{ seconds.} \tag{7.20}$$

There is also the characteristic time τ_2 required for one complete bounce from one mirror point to the other and back; τ_2 may be calculated from

$$\tau_2 = 2\int_A^{A'} \frac{ds}{v_\parallel} = \frac{4}{v}\int_0^{A'} \frac{ds}{(1 - B/B_m)^{1/2}}. \tag{7.21}$$

It is frequently more convenient to express τ_2 in terms of the pitch angle α_0 at the equator. This dependence on α_0 is approximately given by

$$\tau_2 = \frac{4R_0}{v}[1.3 - 0.56 \sin\alpha_0] \tag{7.21a}$$

seconds, where R_0 is the distance from the center of the earth (at the equator) of the field line.

Finally, a characteristic time τ_3 is required for the particle to drift once completely around the earth in longitude. This time is somewhat more

TABLE 7.1. CHARACTERISTIC TIMES AT 2000 KM ALTITUDE NEAR THE EQUATOR

Energy	Type	ρ (cm)	τ_1 (sec)	τ_2 (sec)	τ_3 (min)
50 keV	$e^{\pm}$	5×10^3	2×10^{-6}	0.25	690
1 MeV	$e^{\pm}$	3×10^4	7×10^{-6}	0.10	53
1 MeV	P	1×10^6	4×10^{-3}	2.2	32
10 MeV	P	3×10^6	4×10^{-3}	0.65	3
500 MeV	P	3×10^7	6×10^{-3}	0.11	0.08

complicated to calculate; the result is

$$\tau_3 = 172.4 \left(\frac{1 + \delta}{2 + \delta}\right) \frac{G}{F} \text{ minutes}, \qquad (7.22)$$

where $\delta = \gamma - 1$ and where the function G/F ranges between 1.0 and 1.5, depending on the mirroring latitude.

Typical values of τ_1, τ_2, and τ_3 are given in Table 7.1 for particles mirroring near the equator at an altitude of ~2000 km. It is seen that <1 second is required in most cases for a complete bounce to be made and that these particles usually drift completely around the globe in less than an hour.

The considerations discussed above are valid only if the field changes are small during a characteristic time. That is, the field changes must be small in these periods for the corresponding "adiabatic invariant" actually to be invariant.

7.6 The Maximum Trapped Flux

There is one other simple principle that applies to trapped particles. It is that the magnetic field due to the superposition of the fields from the dipole moments of the individual particles is antiparallel to the trapping field and must not exceed ~10% of the trapping field. This stems from the principle that the kinetic energy density of the particles must be small compared with $B^2/2\mu_0$, the magnetic energy density of the field. This limits the flux of particles typically found around the earth at $R_0 = 1.5$ Re (in the equatorial plane) as follows:

	100-keV proton	40-keV electron	1-MeV electron
Omnidirectional flux	9×10^{10} cm^{-2} second^{-1}	6×10^{12} cm^{-2} second^{-1}	6×10^{11} cm^{-2} second^{-1}
Density	200 cm^{-3}	500 cm^{-3}	200 cm^{-3}

7.7   Conservation of Unidirectional Flux

Liouville's theorem of classical mechanics states that if each state of a number of identical systems is represented by a point in phase space (a separate point for each system) the density of points in phase space is constant. If the "systems" are charged point masses, the density of points in the six-dimensional phase space (three momentum dimensions, three conjugate coordinate dimensions) will remain constant if the particles move in a static magnetic field **B**. Liouville's theorem then implies that the unidirectional flux of particles is constant along the trajectory in ordinary space; the unidirectional flux **j** measured at one point on the field line is equal to the flux to be found at any other point along that line. The meaning of **j** may be defined with the aid of Figure 7.2.

If dN_1 is the number of particles (of a given type and energy) that cross an element of area dA_1 in unit time and within the element of solid angle $d\Omega_1$, then

$$dN_1 = j\, dA_1 \cos\, \varphi_1\, d\Omega_1 \qquad (7.23)$$

defines j at dA_1 in the direction $d\Omega_1$. Now Liouville's theorem tells us that

$$dN_2 = j\, dA_2 \cos\, \varphi_2\, d\Omega_2. \qquad (7.24)$$

Consider the motion of trapped particles as they bounce in latitude. The constancy of j means that if j is measured as a function of α at some point on a particular line of force (where $|B|$ is known), then j is known for every α and $|B|$ along the field line in the direction of increasing $|B|$; for example, if

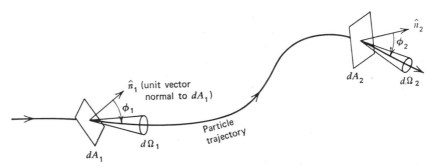

Figure 7.2   Illustration of Liouville's theorem, one consequence of which is that the intensity of particles in any allowed direction is the same as it is at their starting point. If the distribution at infinity of cosmic particles is isotropic, their intensity near earth is the same in all allowed directions for any given energy; to calculate the intensity, therefore merely multiply the subtended solid angle by the intensity in any other (allowed) direction.

α_0 is the pitch angle at the equator, where the field is B_0,

$$j(\sin \alpha, B) = j_0(\sin \alpha_0, B_0) = \text{flux}\left[\frac{\text{particles}}{\text{seconds-cm}^2 \text{ ster}}\right]$$

at α_0 and B_0, where $(\sin^2 \alpha)/B = (\sin^2 \alpha_0)/B_0$. If we assume cylindrical symmetry about the field line B, we can also determine the *omnidirectional flux* J at every point along the field line, for

$$J(B) = 2\pi \int_0^\pi d\alpha \sin \alpha_0 \cdot j(\sin \alpha, B)\left[\frac{\text{particles}}{\text{seconds-cm}^2}\right]. \tag{7.25}$$

7.8   The B-L Coordinate System

The invariant I_2 is difficult to visualize, and for this reason a different invariant parameter, called the L-parameter, is used. In a pure dipole field the L-value of a particular point is just the distance from the center of the earth at which the field line through that point crosses the equatorial plane. The unit of L is an earth radius, usually taken as 6374 km. The coordinate L is computed as follows.

In a dipole field we can compute $|B_m|$ and I_2 for any point (r, θ), using the equations

$$|B| = \frac{B_0}{R_0^3 \sin^6 \theta}[3 \cos^2 \theta + 1]^{1/2} \tag{7.26}$$

and (7.18). Since B is a function of R_0 and θ only, it is possible to show that $I = R_0 f_1(\theta)$, where R_0 is calculated from $r = R_0 \sin^2 \theta$ (see Section 9.4). From (7.26) we obtain

$$\theta = f_2\left[\frac{BR_0^3}{B_0}\right],$$

which may be substituted into f_1 to yield

$$\frac{I_2^3 B_m}{B_0} = \frac{R_0^3 B_m}{B_0} f_3^3\left[\frac{R_0^3 B_m}{B_0}\right]. \tag{7.27}$$

Inversion of (7.27) gives

$$\frac{R_0^3 B_m}{B_0} = F\left[\frac{I_2^3 B_m}{B_0}\right],$$

or

$$R_0 = f(I, B_m, B_0). \tag{7.28}$$

In a pure dipolar field $R_0 = L$. Numerically, therefore, L is such that if the geomagnetic field were that of a pure dipole, the value of L would represent (in units of earth radii) the equatorial radial distance from the dipole center to the magnetic shell. The statement in (7.28) tells that we may indeed transform from (I, B_m) to (L, B_m) and that L is constant along a field line.

The significance of the (B, L) coordinate system is that with certain important exceptions the trapped radiation is identical at all points in three-dimensional space that have the same value of these two parameters B and L.

Matters are not quite so simple in the real geomagnetic field; $f_1(\theta)$ is different for different field lines. In other words, I is a different function of B along different field lines. Hence there is no universal function $f(I, B_m, H_0) = L$ that is constant along a line of force. A satisfactory procedure for calculating an L-value that is nearly constant is as follows:

1. For every point calculate the "real" geomagnetic field vector, B, from an expansion such as the international geomagnetic reference field (see Section 9.6).
2. Use this "real" B to calculate I from (7.18).
3. Calculate an L-value from (7.28), using the dipole function f.

It is clear that a computer is most useful for this task.

A particular particle always mirrors at the same values of L and B_m, but L is not a constant of the motion along a field line, so that L varies along the line. Experimentally, however, the value of $j_\perp$ is constant to within $\sim 4\%$ at all points having the same (B_m, L). Let L_0 be the distance from the center of the geomagnetic equivalent dipole to the point along the field line where $|B|$ is a minimum. We find that particles mirroring at the same L but different B will move on lines with different L_0 as they drift in longitude, but the variation of L_0 is only about 1% around the earth, if $L_0 \lesssim 8R_E$. This is about the same as the variations of L along a particular field line. Thus the (B, L) coordinate system is still quite useful; plots of the trapped flux in (B, L) space display a high degree of ordering, while they seem quite disorderly in geographical space (see Figures 7.3 and 7.4).

Sometimes it is useful, while thinking about the trapped radiation, to try to use some more easily visualized coordinates, such as the actual physical geometry. This may be accomplished by transforming B and L to polar coordinates, using the relationships

$$ B = \frac{M}{R^3}\left(4 - \frac{3R}{L}\right)^{1/2} ; \qquad R = L\cos^2\lambda $$

(where M is the magnetic dipole moment of the earth). Thus a radial distance R and a "latitude" λ may be computed. Care should be used, however, in interpreting these parameters; the irregular variation of the magnetic field results in longitude-dependent "latitudes" λ. Figure 7.5 illustrates the mapping of the polar coordinates onto the (B, L) plane.

7.9   Observed Distribution of Radiation

Rather thorough explorations of the space over the equatorial regions of the earth (latitude $\lesssim 30°$) have now been made with satellites, up to altitudes of ~ 10 earth radii, R_E. There is thus a rather considerable body of data on

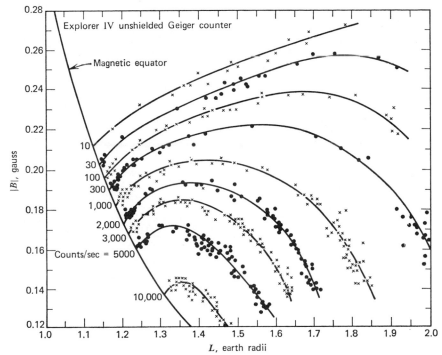

Figure 7.3   A plot of the distribution in space of protons with energies greater than 30 MeV. This plot, due to McIlwain, was the first to show the ordering of the trapped flux in *B*, *L* space. The data are for the period 26 July–26 August 1958. The geometric factor of the counter is 0.54 cm²; the threshold for protons is 31 MeV. From C. E. McIlwain, *Journal of Geophysical Research*, 1961.

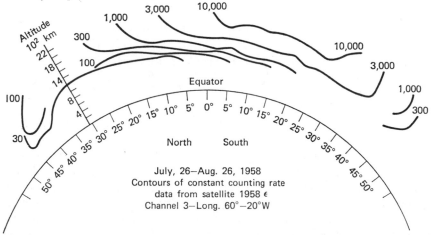

Figure 7.4   The same data given in Figure 7.3 but for one longitude range only. It is clear that this presentation in geographic space is much less orderly than that displayed in the preceding figure. From W. N. Hess, *The Radiation Belt and Magnetosphere*, Blaisdell Publishing Co., 1968.

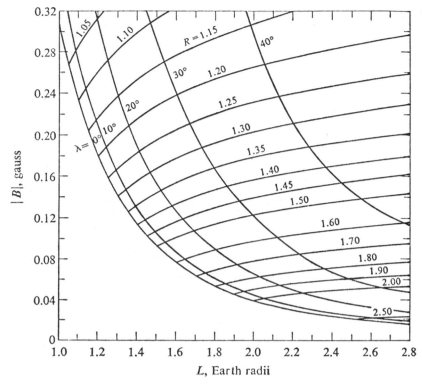

Figure 7.5   The mapping of the polar coordinates R and λ on the B-L plane according to the transformation

$$B = \frac{M}{R^3}\left(4 - \frac{3R}{L}\right)^{\frac{1}{2}}, \qquad R = L\cos^2\lambda.$$

The polar coordinates correspond approximately to distance from the earth's center (in earth radii) and to latitude but are longitude-dependent and should be used only as intuition aids. From C. E. McIlwain, *Journal of Geophysical Research*, 1961.

the trapped radiation at low latitudes; the data become less regular at L-values much in excess of ~ 2.

It was thought in the early days of this exploration that the trapped radiation was divided into inner and outer "belts." It is now known, however, that the particle population is continuously distributed throughout the distorted-dipole trapping region and that the apparent division into zones merely reflects an altitude-dependent energy spectrum, rather than two separate trapping regions. Particles are not, of course, trapped outside the magnetosphere (see Section 9.11), so that trapping does not occur beyond $\sim 10R_E$ on the sunlit side of the planet, and trapped radiation is not even at this high an L-value on the night side; $\sim 8R_E$ appears to be the limit there.

7.10 Composition of the Trapped Radiation

Electrons and nuclei comprise the trapped particles, with protons constituting the vast majority of the latter. Protons account for 99% of the nuclei with energies greater than 35 MeV. Deuterons ($E \geqslant 50$ MeV) are only $\sim 0.5\%$ of the trapped nuclei, as are tritons ($E \geqslant 60$ MeV). Alpha particles with energies of a few million-electron volts per nucleon have also been detected, but the alpha to proton ratio at a given energy appears to be only $\sim 2 \times 10^{-4}$.

Plasma is also found (see Section 3.16) within the magnetosphere of the earth. The energies here ($E \leqslant 50$ eV) are considerably lower than what is normally ascribed to the Van Allen radiation ($E \geqslant 40$ keV), but these numerous protons and electrons may provide significant clues as to the origin of the trapped radiation.

7.11 Trapped Protons

Protons are distributed continuously throughout the region $L = 1.2$–8 (earth radii). The peak flux at given L occurs near the equatorial plane, where the minimum value of the field strength, $|B|$, occurs. Figures 7.6 and 7.7 show the approximate distribution in space of protons with various energies. Proton energy spectra on the equator are shown in Figure 7.8.

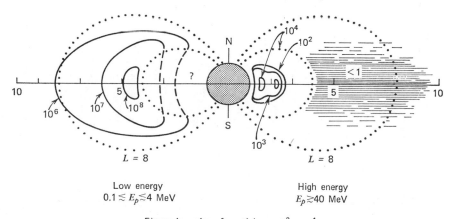

Fluxes in units of particles cm^{-2} sec^{-1}

Figure 7.6 The approximate distribution in space of geomagnetically trapped protons. One earth radius is the unit of distance (schematic representation). Reprinted from A. J. Dessler and B. J. O'Brien, *Penetrating Particle Radiation* in Francis S. Johnson, Ed., *Satellite Environment Handbook*, second edition, with the permission of the publishers, Stanford University Press. Copyright 1965 by The Board of Trustees of the Leland Stanford Junior University.

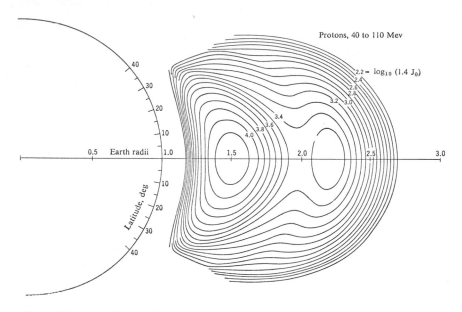

Figure 7.7 An *R-λ* plot of the data obtained for protons with the spacecraft Explorer 15. C. E. McIlwain, *Science*, 1966. Copyright 1966 by The American Association for the Advancement of Science.

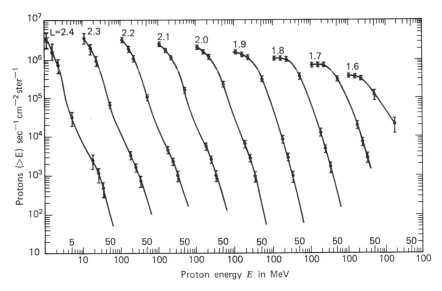

Figure 7.8 Various measurements, as summarized by Fillius and McIlwain, of the proton energy spectrum in the equatorial plane for values of 1 to 2.4. R. W. Fillius and C. E. McIlwain, *Phys. Rev. Letters*, 1964.

The four general characteristics of the proton fluxes are given below. These results have been obtained in many experiments with rockets and satellites.

1. The energy spectra get softer with increasing L; the average energy of the trapped protons decreases with increasing L. Thus the empirical energy spectrum for the trapped protons near the equator (that is, for the unidirectional flux of protons perpendicular to **B** near the equator) is

$$I(E)\, dE = Ke^{(-E/E_0)} \frac{\text{protons}}{\text{cm}^2\text{-second-ster-Mev}} \tag{7.29}$$

for energies between 0.1 and 100 MeV. The value of the e-folding energy, E_0, is dependent on L, however. For values of L between 1.2 and 8, it is found empirically that

$$E_o = (306 \pm 28)\, L^{-(5.2)\pm 0.2}\ \text{Mev}, \tag{7.30}$$

approximately an L^{-5} dependence (for E_0). The values of the constant K in (7.29) may be inferred from Figures 7.3 and 7.4.

2. Trapped protons of lower energies reach a peak flux at greater L-values than do trapped protons that have higher energies.

3. At small L the flux of trapped protons with energies less than ~ 1 MeV is greater than that computed from (7.29).

4. The distribution in pitch angle of 1-MeV protons on the equator is proportional to $\sin^3 \alpha$.

7.12 Trapped Electrons

High-energy electrons are the most numerous particles in the high-energy, geomagnetically trapped radiation. Figures 7.9 and 7.10 show the distribution in space of trapped electron fluxes for two energies that are easily studied experimentally (40 keV and 1.6 MeV). (This predominance does not, however, lead to a negative charge for the near-earth space; charge neutrality is preserved by the presence of [positive] thermal ions.)

Four general characteristics of the electron fluxes are as follows:

1. The energy spectra become softer with increasing L.

2. The trapped electron flux exceeds the flux of protons that have energies greater than 0.1 MeV in the region $L = 1.2$ to $L = 3.5$. The electron and proton fluxes are comparable at greater distances.

3. The electron flux is quite variable in time for $L \geqslant 3$.

4. The electron pitch angle distribution varies in time; at low altitudes for $L = 5$–10 the flux of downward-moving electrons may be isotropic.

The data on both trapped protons (Section 7.11) and trapped electrons, when combined, show how the confusion over 'inner'' and "outer'' belts of

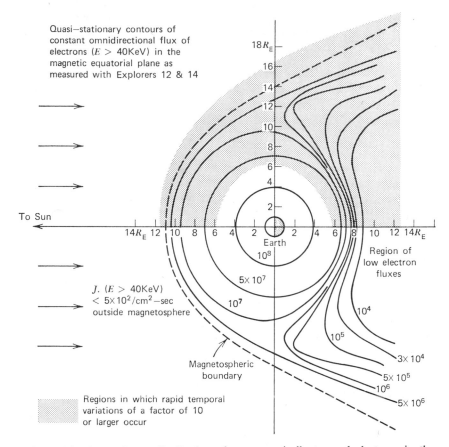

Figure 7.9 Approximate distribution of geomagnetically trapped electrons in the magnetic-equatorial plane. Fluxes shown are electrons per square centimeter per second. L. A. Frank, J. A. Van Allen, and E. Macagno, *Journal of Geophysical Research*, 1963.

trapped radiation originally arose. The distribution of particles is such as to cause two maxima to appear in the counting rate of a lightly shielded Geiger counter as it is removed farther from the earth.

7.13 Artificial Electrons

The detonation at high altitudes of thermonuclear test devices will liberate large quantities of high-energy fission electrons from the beta decay process. The isotopes may decay in the detonation region or in the magnetically conjugate region. The equilibrium spectrum of fission electrons in units of beta particles per fission has the form $N(E) \, dE = 3.88 \, \exp(-0.575E - 0.055E^2) \, dE$, where E is the electron energy in million-electron volts and only

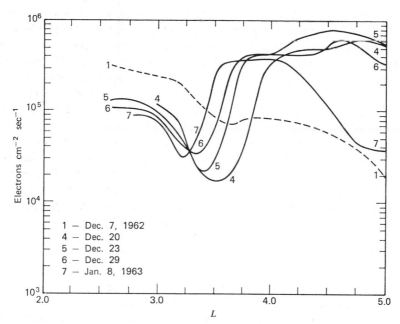

Figure 7.10   Electrons with energies greater than 1.6 MeV, measured by Frank et al. in late 1962 on the Explorer 14 satellite. There is evidence here for radial diffusion of the electrons inward after a period of geomagnetic disturbance that took place on 17–20 December 1962. L. A. Frank, J. A. Van Allen, and H. K. Hills, *Journal of Geophysical Research*, 1964.

electrons between $E = 1$ and $E = 7$ are considered. There are about six such electrons per fission fragment, and about 10^{23} fission fragments are emitted per kiloton equivalent of TNT; approximately one-sixth of the electrons are emitted in the first second and the remainder at a rate $\sim(t - 1)^{1.2}$, where t is in seconds, so that fission is an important source hours after the detonation.

The Starfish explosion of 9 July 1962 over Johnston Island in the Pacific apparently was of the order of 1500 kilotons; $\sim10^{27}$ electrons were liberated at ~400 km of altitude. About 1.2% of these electrons were still trapped 100 days later, and the characteristic $(1/e)$ time for decay of the electron flux for $E \geqslant 0.5$ MeV is nearly a year; artificial electrons will apparently be detectable above background for another decade or so.

7.14   Radiation in the Distant Magnetosphere

If one probes outward from the earth toward the magnetopause (see Figure 9.11), the magnetic field and also the charged particle distribution become increasingly asymmetric about the dipole axis. Finally the boundary is reached at $L \sim 10$ on the sunward side and $L \sim 8$ on the midnight side of the earth.

This asymmetry is responsible for a low-altitude asymmetry; the high-latitude limit is trapped radiation is found at greater latitudes on the sunlit side of the planet than at night. At latitudes higher than these limits particles are no longer trapped in the sense that they execute the periodic motion in latitude and longitude described above. The charged radiation is still, however, constrained to spiral about and to drift along the local magnetic field. Consequently there are permanent but not constant features of the radiation in these distant portions of the magnetosphere.

In these portions the plasma flux (i.e., electrons and protons with $E \leqslant 1$ keV) is more or less constant when averaged over a few minutes. The more energetic particles appear sporadically with high intensity; their frequency of occurrence and the peak flux attained decrease with increasing energy. The plasma observations are discussed in Section 7.16; the high-energy reports are outlined in Section 7.17.

7.15   Local Acceleration Mechanisms

We saw in Chapter 6 that particle acceleration to the energies observed in the aurorae and in the trapped radiation must occur close to the earth; the particle energies in the interplanetary medium are far too small.

The energetic charged particles observed within the magnetosphere are almost surely accelerated by fluctuations and waves in the magnetic field. This acceleration may occur either locally or farther out in the geomagnetic tail; the field lines in the tail connect with the boundaries of trapping discussed in Section 7.14.

The geomagnetic field is subject to at least three types of variation. As the solar wind pressure or the interplanetary magnetic field changes, the magnetopause is pushed in or expanded outward to balance the wind and field pressure. For example, the subsolar magnetopause has been reported as close as $6R_E$ during magnetic storms. This changing size of the magnetosphere constitutes a fluctuation in the strength of the magnetic field.

Second, as the earth executes its diurnal rotation and its annual revolution about the sun, the angle between the magnetic dipole axis and the incoming "solar velocity vector" changes. The "solar velocity vector" is $\sim 5°$ east of the sun in the ecliptic plane, as seen from the earth during its ~ 30 km per second revolution about the sun. The angle changes between extremes given by the $\sim 23°$ obliquity of the ecliptic and the $\sim 11°$ tilt of the dipole with respect to the rotational axis; the extremes are $90° \pm (23° + 11°) = 56°$ and $124°$.

Finally, there is good reason to think that even a rather steady solar wind produces fluctuations ("flutterings") in the magnetopause. These fluctuations propagate into the magnetosphere in various hydromagnetic wave modes (see Section 9.12).

7.16   Magnetospheric Plasma

Figure 7.11 summarizes the available data on particles with plasma energies ($E \leqslant 1$ keV) in the geomagnetosphere. The solar wind that flows radially outward from the sun is deflected by the magnetosphere. The solar wind flow is supersonic, and so a standing shockwave is formed upstream of the magnetopause (the boundary surface between the geomagnetic and interplanetary magnetic fields). Some of the energy of directed flow is converted to transverse velocity when the wind passes through the shock. Hence the temperature and density of the plasma increase and the velocity distribution becomes more isotropic. The plasma flows with subsonic speed parallel to the magnetopause, gradually becoming supersonic again as it passes the earth.

Although the solar wind does not penetrate into the magnetosphere with solar wind velocity, the magnetosphere is filled with plasma, most of it with energy $\leqslant 50$ eV. (This plasma may be thought of as a very high-altitude part of the ionosphere.) This plasma does appear to execute bulk flow (see Fig. 7.12, due to Freeman and his co-workers). For example, higher-energy plasma has been observed in the neutral sheet; it appears to carry a large electric current from east to west. The plasma motions are under active investigation at the present.

7.17   High-Energy Radiation in the Distant Magnetosphere

The electron energy threshold that has been the most studied experimentally is $\sim$45 keV; a summary by Anderson is given in Figure 7.13. Outside the trapping region the most stable flux of these electrons is in the tail, followed by the dawn and dusk regions just outside the trapping zone.

The "islands" of high electron flux are observed in the tail at least as far back as $30R_E$, and also above and below the neutral sheet. The number density of islands declines with distance from the earth and with distance from the neutral sheet. These islands are sporadic in time and are not fixed in position. Electron islands have been observed in the magnetosheath and even sunward of the shock.

Not many proton observations have yet been made. However, the studies that have been made of protons with energies $\geqslant 125$ keV indicate that occasional islands of protons are to be found in the tail; typical fluxes in these islands are 10^3–10^4 cm^{-2} second^{-1}.

At low altitudes and high latitudes (above the boundary of trapping) "spikes" of electrons ($E \geqslant 40$ keV) are occasionally seen. The unidirectional intensity of these electrons is much greater than $1/4\pi$ of the omnidirectional intensity of the islands in the tail. These spikes occur most frequently at latitudes just above the trapping boundary and near local midnight. Thus they appear on field lines that extend into the tail and are thought to be those

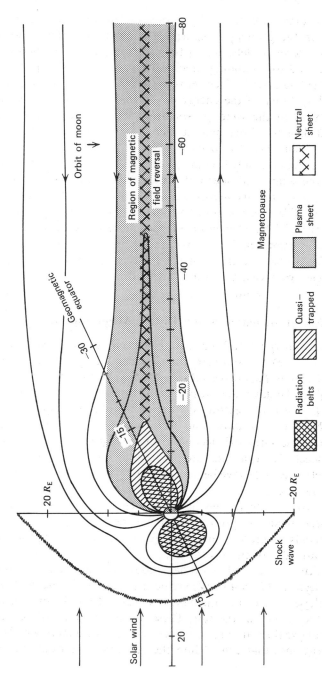

Figure 7.11 A summary of the configuration of the magnetosphere in the noon-midnight plane. N. F. Ness, *Reviews of Geophysics*, 1969.

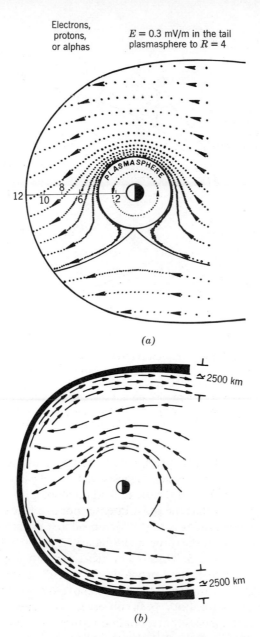

Figure 7.12 (a) Calculated drift paths in the equatorial plane; the plasmasphere is represented as a conducting sphere of radius $4R_E$. The positions of the magnetopause and of the forbidden region are indicated by solid lines. The distance between successive data is the distance a particle drifts in 10 minutes. (b) Schematic diagram of the plasma flow pattern in the magnetosphere. L. D. Kavanagh, Jr., J. W. Freeman, Jr., and A. J. Chen, *Journal of Geophysical Research*, 1968.

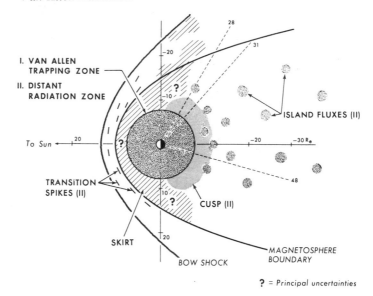

I. VAN ALLEN
   TRAPPING ZONE

II. DISTANT
    RADIATION ZONE

ISLAND FLUXES (II)

To Sun

TRANSITION
SPIKES (II)

CUSP (II)

SKIRT

BOW SHOCK

MAGNETOSPHERE
BOUNDARY

? = Principal uncertainties

Figure 7.13 Schematic representation of particle populations in the magnetosphere (ecliptic plane view). The particles in the "skirt" do not drift completely around the earth; they are not organized in (B, L) coordinates. "Cusp" electrons also do not execute complete drifts. They are confined near the ecliptic plane and have a pitch-angle distribution that favors small angles at the equator. K. A. Anderson, *Journal of Geophysical Research*, 1965.

portions of the islands that are precipitated into the atmosphere. A comparison of the spike and island intensities suggests that in the islands the directional flux is peaked along the local magnetic field direction.

It should be recalled that the most intense sporadic flux of both electrons and protons at low altitudes is that which produces the aurora. The aurora occurs most frequently at the trapping boundary, $L = 8$–10, but is sometimes observed as low as $L = 2$–3. Thus the particles certainly appear on "closed" (i.e., dipole-like) field lines, and probably on "open" lines as well. Since the flux of precipitating (auroral) and the flux of above-the-atmosphere-mirroring (trapped) electrons have both been observed to increase simultaneously during an aurora, it appears that a common source supplies both radiations. This source "simply" injects particles at both trapping and dumping pitch angles; the Van Allen radiation is not a reservoir of auroral radiation.

7.18 Lifetimes of Trapped Particles

By the "lifetime" of trapped particles, we mean the time required to reduce their flux to e^{-1} of the original value. At very low L-values, the atmosphere

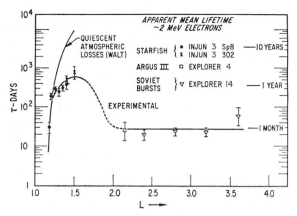

Figure 7.14   The apparent mean lifetime of 2-MeV electrons. From J. A. Van Allen, *Nature*, 1964.

absorbs the flux of electrons, with the lifetime increasing with mirror-altitude. The situation changes above $\sim L = 1.5$ (see figure 7.14, due to J. A. Van Allen).

It is found that the lifetimes are greatest at low L-values and become progressively shorter proceeding outward. Thus at $L = 1.2$–1.5 the lifetimes of electrons with energies ≥ 1 MeV is about a year, judging from the decay of the Starfish radiation. These same electrons, however, have lifetimes of days to months, from $L = 1.5$ to 2.5 (as do protons with $E \geq 1$ MeV). From $L = 2.5$ to the boundary of the trapping region the lifetime varies from weeks to hours, judging by the frequency with which the particle population is disturbed by magnetic storms. When world-wide magnetic storms occur, the flux of electrons ($E > 40$ keV) decreases at the beginning of the storm, then increases above the prestorm value, and finally returns to the prestorm value. The proton flux ($E \geq 100$ keV), on the other hand, typically increases temporarily and then recovers.

Beyond the boundaries of the trapping region the lifetimes are quite short. Times of minutes or less are inferred from the data.

7.19   Loss Mechanisms for Trapped Radiation

The loss mechanisms depend on L. In the region $L = 1.2$–1.5, the interaction with the atmosphere along the orbit of the trapped particle controls the lifetimes. The lifetimes are inversely proportional to the mass of atmosphere traversed per unit time. Ionization loss, or "Coulomb collisions," erodes the energy of the particles, as does the nuclear interactions of those protons that have energies in excess of ~ 500 MeV. Another phenomenon that occurs in this low L-value region is charge exchange of low-energy (≤ 100-keV) protons; this scattering process usually lowers the mirroring altitude, so that the particle is lost into the atmosphere on the next latitudinal bounce. Electron scattering by the atmosphere has the same affect on the (electron) trajectory; the mirror altitude is reduced. In particular, particles in this region may be lost over the south Atlantic magnetic anomaly; the field is weak there, and B_m may lie below the top of the atmosphere (100 km). There seems to be a constant loss of particles into this region, which therefore requires a constant source.

There are two main routes for a loss of particles at a given high L-value ($L = 1.5$–10). First, a gross change in the shape of the magnetosphere, such as those that occur during magnetic storms, result in a violation of the third adiabatic invariant, although the first two invariants remain valid. This violation may move the particles to higher L-values (out of the magnetosphere) or to lower L-values (where they remain trapped).

Second, a mechanism that is especially active at $L \geqslant 8$ is the action of small rapid variations in the field, such as waves. These can scatter trapped particles, so that the second and especially the first adiabatic invariants are altered. This results in a dumping of the particles into the atmosphere when their mirroring value of $|B|$ is increased sufficiently.

It must be noted, however, that the detailed mechanisms operative in either route are still obscure. We are not able at present to calculate correctly the rates of particle loss for the region $L = 1.5$–8 or 10; this is principally a result of inadequate knowledge of the geomagnetic field's behavior in this region.

7.20   Sources of the Van Allen Radiation

After the discovery of the trapped radiation around the earth it was early realized that the decay in flight of albedo neutrons represented a possible source mechanism for the radiation. The problem is to find a means of getting charged particles of the observed energies through the geomagnetic field and into the trapping region; the in-flight decay of neutral particles (whose paths are not affected by magnetic fields) provides such a means.

Neutrons have been observed for many years in the earth's atmosphere. They are the result of cosmic ray interactions with atmospheric nuclei. About 1 neutron cm^{-2} second^{-1} is scattered upward out of the atmosphere; the outward-moving flux is (poorly) termed the neutron *albedo* of the earth (Figure 7.15). The differential energy spectrum of the albedo is proportional to $E^{-1.3}$ below $E = 10$ MeV and steepens to $\sim E^{-2}$ at higher energies. Free neutrons decay; the decay scheme is $n \rightarrow p + e^- + \bar{\nu}_e$. The proton, according to the kinematics of the decay, carries most of the neutron's kinetic energy; the maximum electron energy from the beta decay of the neutron is 770 keV. The decay time in the rest frame is 10^3 seconds.

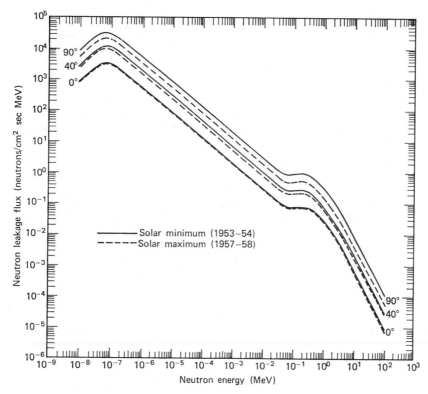

Figure 7.15   Calculated spectrum of the earth's neutron albedo at the top of the atmosphere for three different geomagnetic latitudes. From R. E. Lingenfelter, *Journal of Geophysical Research*, 1963.

Measurements have shown, however, that the neutron albedo can explain only a very small fraction of all the trapped particles. It appears that the albedo is adequate to explain the flux of protons with energies in excess of 30 MeV in the region $L = 1.2–1.5$, since the proton lifetime there is sufficiently long that this weak source can compensate for the loss.

From $L = 3$ to 7 protons with energies between 100 keV and ~60 MeV may be accounted for by inward diffusion and acceleration. We saw previously that there are perturbations of the magnetosphere that conserve I_1 and J but alter the flux invariant I_3 (and also the kinetic energy).

Now

$$I_1 = \frac{E \sin^2 \alpha_1}{B_1}, \tag{7.31}$$

where α_1 and B_1 are the equatorial values of the pitch angle and magnetic

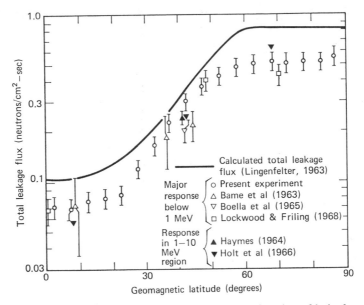

Figure 7.16   Measured neutron albedo flux as a function of latitude. From R. W. Jenkins, J. A. Lockwood, S. O. Ifedili, and E. L. Chupp, *Journal of Geophysical Research*, 1970.

field, respectively. The relation (7.31) may be rewritten as

$$I_1 = \frac{EL^3 \sin^2 \alpha_1}{0.311}. \tag{7.31a}$$

We also have that

$$J = m \oint v \cos \alpha \, ds = ELF(\alpha_1); \tag{7.32}$$

(7.32) provides a definition of the function $F(\alpha)$. The combination of (7.31a) and (7.32) tells us that

$$L \cdot \frac{\sin^2 \alpha_1}{F^2(\alpha_1)} = \text{constant}; \tag{7.33}$$

those particles that are moved to smaller L-values by the magnetic perturbation are also moved to larger values of $\sin \alpha_1$ (i.e., toward the equatorial plane). But since I_1 is conserved, the energy increases as L decreases. In the equatorial plane, where $\alpha_1 = 90°$, the energy E is proportional to $1/L^3$.

It is found that if the flux of 100–400-keV protons observed at $L = 7$ in the equatorial plane is transformed by these equations to $L = 2.4$, the transformed spectrum fits the observed proton flux at $L = 2.4$ for energies between

10 and 65 MeV. The data at points off the equator are fitted less well, but it is thought that this acceleration mechanism must operate, even though the detailed diffusion mechanism is not known.

It also appears that this diffusion acceleration may contribute to the energization of relativistic electrons in the magnetosphere. These electrons are observed to "diffuse" inward with a velocity that is proportional to L^8. Unfortunately, diffusion cannot explain the distribution of nonrelativistic electrons or of lower-energy protons ($E \leqslant 10$ MeV) near the earth.

Finally, it should be noted that the source of all the high-energy trapped particles (except for the neutron albedo protons) is probably the solar wind. The one observation difficult to understand in this context is the low alpha to proton ratio in the trapped radiation (the analogous ratio in the auroral radiation is like that of the solar wind). The wind also provides sufficient energy to "operate" the magnetospheric system. About 10^{18} ergs per second is required to supply the particle energy loss, and the solar wind brings $\sim 10^{20}$ ergs per second onto the magnetosheath.

Problems

7.1. Assume that the geomagnetic field may be described by the centered dipole approximation (see Section 9.1). If the surface field at the equator is B_0 and a geomagnetically trapped particle crosses the equator at a point where the field strength $B = B_0/8$, with a pitch angle of $45°$, find an equation depending only on R_0 for the radial distance at which the particle mirrors. (It is not necessary to solve this equation.)

7.2. What if any is the physical significance of L-values in excess of 10?

7.3. Express the albedo neutron decay density, $dn/d\tau$ (neutron decays per cubic centimeter-million-electron volts-second) in terms of the distance from the earth's center (in R_E), neutron speed v, lifetime L, and flux $\varphi(E)$, where $\varphi(E)$ is the differential neutron flux with energies between E and $E + dE$.

7.4. What is the flux of trapped protons with energies between 10 and 100 MeV at an L-value of 5.0 over the equator?

7.5. A 10-MeV electron has a pitch angle of $45°$ at four earth radii over the equator. Assuming the geomagnetic field to be purely dipolar, at what altitude will the electron mirror?

7.6. What is the maximum flux of 10-MeV protons that may be trapped 1.5 earth radii above the earth's equator?

7.7. Assume that a 100-kiloton nuclear device is detonated at $L = 2$. How does the artificial electron flux at $E = 1.6$ MeV compare with the natural 1.6-MeV flux normally found there?

7.8. A 10-MeV proton diffuses toward the earth from a place at $L = 7$ where the pitch angle is 45°. Assuming that the first invariant of the motion is not violated, find the energy of this proton at $L = 3$.

7.9. (a) Find an expression for the energy spectrum of the trapped protons with energies between 10 and 65 MeV at $L = 2.4$.
   (b) Assuming that the above spectrum was caused by the inward diffusion of 100–400-keV protons at $L = 7$, what was the spectrum of the latter at $L = 7$?

7.10. Show that the kinetic energy of the proton that results from the decay in flight of a neutron is comparable with the kinetic energy of the neutron.

7.11. Assume that all particles penetrating to 100 km of altitude are lost in the atmosphere, and that all particles mirroring above 100 km are durably trapped. What is the maximum pitch angle (loss cone) at the equator for geomagnetically trapped particles to be precipitated at a latitude of 60° N?

7.12. A satellite in a perfectly circular polar orbit at a geocentric distance of $4R_E$ measures a flux of energetic electrons with pitch angle of 45° having energy of 100 keV. A tracking station fixes the satellite latitude at 30° N at the time of the measurement. Assume that the earth's surface magnetic field at the equator is 0.3 gauss.
   (a) What are the radial and horizontal magnetic field components at the point of measurement, and what is the total field magnitude?
   (b) What is the field magnitude where the particles mirror, and where is the mirror point (radius and latitude)?
   (c) What is the pitch angle of the particles at the equator, and at what radial distance do the particles cross the equator?
   (d) Will another satellite in circular polar orbit at a geocentric distance of $3R_E$ cross through the particles' trajectory? If so, where (latitude and radius)? If not, why not?
   (e) At the point of the original measurement, what is the cyclotron frequency of the electrons, and what is the radius of gyration. (Mass electron = 9.1×10^{-31} kg.)

Chapter 8

PLANETARY INTERIORS

We have so far discussed the atmospheres of the various planets, including the neutral and ionized gases, the energetic radiation found around some of them, and phenomena such as aurorae and airglow. We return now to the solid objects themselves and what we know concerning their interiors. At this writing we have direct information on pressures, temperatures, etc., inside only one planet—the earth—but we know far less about the interior of our planet than we do about the space surrounding it. The study of geophysics, the science that bridges the gap between physics and geology, is therefore a most challenging undertaking. We shall not even attempt an introduction to the field here, but only indicate a few of the more prominent problems.

8.1 The Earth's Surface

The "crust" of the earth—the planet's outermost layer—is in itself not very well known, since 70% of the surface is covered by water. Oceanographic expeditions, however, are now beginning to fill in some of the gaps in our knowledge of the crust.

Thus the sea bottoms are apparently just as rugged as the exposed land on the continents; there are steep mountains and valleys. Among the latter, we may list the Marianas Trench, which is some 35,000 feet deep below the surface of the Pacific (and has recently been "visited" in a deep-submergence vehicle). Among the former we include Mauna Kea on the island of Hawaii, which is a "mere" 13,800 feet above sea level. But the Pacific is about 16,000 feet deep there, making Mauna Kea one of the world's tallest peaks.

As may be seen from Figure 8.1, the mid-Atlantic ridge runs submerged from Iceland south towards Antarctica, but it bifurcates so that a spur trails off eastward around the Cape of Good Hope, into the Indian ocean. The ridge thus splits the Atlantic into two basins. The only places where it appears above sea level are islands such as Ascension and the Azores.

In general the ridge appears to parallel the continental shelves of North America and of Europe; it is also approximately parallel to the shelf lines of South America and Africa in the southern hemisphere. Interesting patterns of magnetic anomalies have been detected in the ridge; the patterns support the ideas of geomagnetic polarity reversal and also a hypothesis called "sea floor spreading."

A glance at a map of the earth shows how well the continents fit together. One example is South America and Africa; it is just as though the two continents once touched and have since drifted apart. One of the original land masses, which apparently contained all of today's southern continents, has been named "*Gondwanaland*," for a key geological province in India. The northern land is known as *Laurasia*. Both may have started splitting some 200–300 million years ago.

The mechanism of continental drift has always been difficult to explain. A modern hypothesis is that the ocean basins grow by virtue of the tensile stresses associated with the formation of new material at the crests of structures such as the mid-Atlantic ridge or its Pacific ocean counterparts

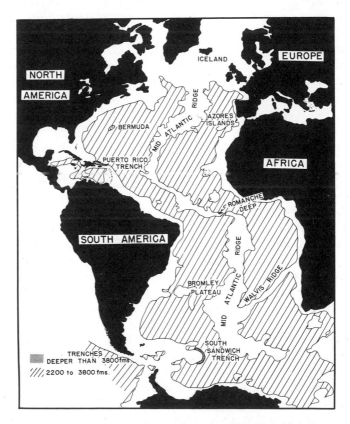

Figure 8.1   The principal relief features of the Atlantic Ocean. Trenches are slightly exaggerated in scale. Francis P. Shephard, *The Earth Beneath the Sea*, copyright Johns Hopkins Press, 1968; used by permission of the Press.

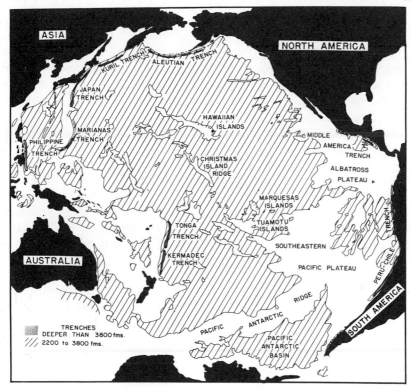

Figure 8.2 The principal relief features of the Pacific Ocean. Trenches are slightly exaggerated in scale. Francis P. Shephard, *The Earth Beneath the Sea*, copyright Johns Hopkins Press, 1968; used by permission of the Press.

(see Figure 8.2). Sea floor spreading is the name for this idea, which is that new crustal material is injected convectively from below (that is, from a layer called the mantle; see Section 8.4) and that this material is forced to "flow" horizontally (rather than vertically) at the surface, by the presence of a ridge of dense material at the site in question. The continents then move along on the convective flow; they float on top of and along with the mantle material. The speed in the Atlantic is 1–2 cm annually.

The mid-Atlantic ridge appears to satisfy the requirement of this hypothesis for a dense material barrier in the Atlantic. The other ocean floors apparently contain similar barriers or ridges that are parallel with continental coastlines. The East Pacific Rise provides an example.

8.2   Seismology

Most of our knowledge about the interior of our planet comes from seismology, the science that studies seismic waves as they propagate through the planet. (There are, to be sure, wave types that propagate along the planetary surface as well. They are called Love and Rayleigh waves, and have longish periods of 10^2 seconds; we neglect them here because they are easily distinguished by their slow propagation speed and may readily be sorted out.) Two types of such "earthquake waves" are possible: longitudinal and transverse. Seismologists call these waves *P*- ("push") waves and *S*- ("shake") waves, respectively.

They are also called "primary" and "secondary" because of their order of arrival. The time interval between the arrival of the *P*- and *S*-waves increases with the distance between the observer and the origin point of the shock.

The velocities of propagation of *P*- and *S*-waves may be measured with seismographs. A three-axis instrument may therefore yield data on both the distance to and direction of the shock center, but more accurate results are obtained when the data from several observatories are used, because of local irregularities in the density of the rock and hence the propagation vectors.

A one-axis seismograph records the local acceleration or the earth; it may consist of a mass suspended by a spring from a point securely attached to the earth. The mass then moves with respect to the earth whenever a seismic wave occurs, and the motion of the mass is recorded. A very sensitive instrument may be constructed with the use of laser interferometers, for such interferometers can detect extremely minute earth motions. "Richter magnitudes" are assigned on the basis of the amplitude of earth motions, as recorded by a particular type of seismograph.

The Richter scale of earthquake magnitudes attempts to assign a number to an event, a number that characterizes the amount of energy radiated from the earthquake source in the form of elastic waves. In analogy with stellar absolute magnitudes (Section 12.2), a logarithmic relationship is employed; if E is the radiated energy and M the magnitude, a reasonable approximation is $\log E = 9.9 + 1.9M - 0.024M^2$.

The use of the energy relation and earthquake observations indicates that $\sim 9 \times 10^{24}$ ergs per year is released in the form of seismic energy; this is about 10^{-3} of the annual flow of heat from the interior through the surface of the earth. Table 8.1 illustrates some features of the scale.

It is perhaps interesting that the largest recorded earthquake was rated 8.9 on the Richter scale; any larger quakes would release so much energy that the crustal rocks would yield under the strain, causing extraordinary catastrophes. Most of the recorded quakes are small (roughly two-thirds have $M \leqslant 2.9$), but 90% of the released annual seismic energy is accounted for by the few ($\sim 1\%$) quakes where $M \geqslant 7.0$.

The two types of body seismic waves may be distinguished on three-axis instruments. From the theory of elasticity, the velocity of propagation of a

TABLE 8.1. SOME FEATURES OF THE RICHTER SCALE

M	Energy (ergs)	Equivalent Amount of TNT	Remarks
0	8×10^9	0.01 oz	
2	4×10^{13}	2.5 lb	
2.5			Smallest felt shocks
6.5	8×10^{20}	20 kilotons	"Nominal atomic bomb"
8.9			Biggest known earthquake (Colombia, 1906)

P-wave, V_P, is

$$V_p = \left(\frac{k + \frac{4}{3}\mu}{\rho}\right)^{1/2}. \tag{8.1}$$

The corresponding propagation velocity, V_S, for a transverse wave is

$$V_S = \left(\frac{\mu}{\rho}\right)^{1/2}. \tag{8.2}$$

Usually the longitudinal waves travel twice as fast as the transverse; as we indicated above, P-waves reach a seismograph from an earthquake before the transverse oscillations do. In (8.1) and (8.2) ρ is the density, k is the bulk modulus or "incompressibility," and μ is the rigidity. The rigidity may be defined as the ratio of a shear component of stress to the corresponding shear. Now an "ideal" rigid body would be characterized by $\mu = \infty$; a "perfect fluid" would have $\mu = 0$. In general the rigidity is a function of ρ.

It is possible to learn about the earth's interior by measuring V_P and V_S as the energy spreads out from the "epicenter," the point on the surface directly above the origin or "focus" of the earthquake. By studying time delays between various seismic observatories, we may also learn something about the depth of the focus and also about the reflection and refraction of S- and P-waves; this information is of considerable help in constructing models of the planetary interior (see Figure 8.3). We should also note that "controlled earthquakes" (nuclear explosions conducted in known terrain types at given times) are of help, by calibrating the seismographs.

8.3 Density Variation

We may infer the density variation of the earth's interior from measurements of V_P and V_S. Combining (8.1) and (8.2), we have that the adiabatic

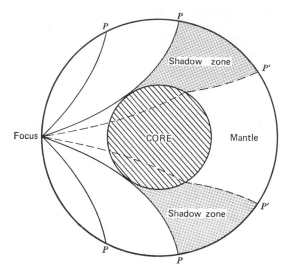

Figure 8.3   When *P*-waves cross the boundary between the mantle and the core, they are sharply bent. This creates a "shadow zone" (between *P* and *P'* in the schematic sketch) between 103 and 145° from the focus. The wave is indistinct on seismographs in the shadow zone.

bulk modulus k is

$$k = \rho\left(V_P^2 - \frac{4V_S^2}{3}\right). \tag{8.3}$$

The adiabatic bulk modulus k is defined to be

$$k = \rho\left(\frac{dp}{d\rho}\right). \tag{8.4}$$

We have assumed on the right-hand side of (8.4) that the pressure P depends only on density ρ. Thus, using (8.3) and (8.4), we have that

$$\frac{dp}{d\rho} = V_P^2 - \tfrac{4}{3}V_S^2 \equiv \Phi(r); \tag{8.5}$$

$\Phi(r)$ is usually called the *seismic parameter*.

The quantity $\Phi(r)$ may now be found from observations of V_P and V_S as a function of depth, an observation that requires data from many observatories. The resulting data may be summarized by plotting propagation velocity as a function of depth (see Figure 8.4).

Now we may write that

$$\frac{d\rho}{dr} = \frac{(dp/dr)}{(dp/d\rho)} = \frac{1}{\Phi(r)} \frac{dp}{dr} . \qquad (8.6)$$

For want of a better approximation to reality, let us again use the condition of hydrostatic equilibrium that we first met in Chapter 3. Here we assume that the weight of a parcel of rock is just balanced by the pressure, so that

$$\frac{dp}{dr} = -g\rho. \qquad (8.7)$$

We recognize that the local acceleration of gravity, g, is given by

$$g = \frac{GM(r)}{r^2} . \qquad (8.8)$$

Thus

$$\frac{dp}{dr} = - \frac{GM(r)}{r^2} \rho, \qquad (8.9)$$

and therefore we find that

$$\frac{d\rho}{dr} = - \frac{G\rho M(r)}{r^2} \frac{1}{\Phi(r)} , \qquad (8.10)$$

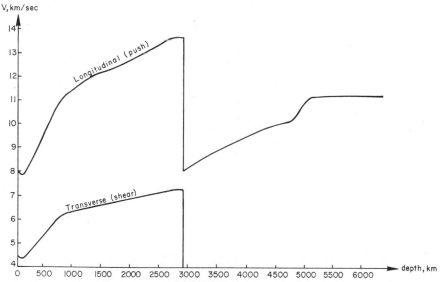

Figure 8.4 The dependence of propagation speeds on depth within the earth. The discontinuity at ~20 km is called the Mohorovičić discontinuity; that at 2900 km is known as the Gutenberg discontinuity.

for the desired result. We should remark also that some equations are common to many areas of space science; we shall see the relation (8.9) again when we study the structure of stars.

We may integrate (8.10) only when the seismic parameter varies smoothly; it does not apply at the depths where the discontinuities occur. This integration amounts to constructing a model of the planetary interior. An additional constraint and guide to the calculation is provided by the moment of inertia of the nonspherical earth. The moment contains information on the density distribution as a function of radius. Prior to the advent of artificial earth satellites, it was obtained by assuming the earth to be an ellipsoid, and by using the measured mass and measured polar and equatorial radii. As we saw in Sections 2.16 and 2.19, however, an analysis of satellite orbits gives a much more accurate portrayal of the moment.

It is interesting that the moment probably changes in time. One cause for the change is the melting of the Greenland ice caps, with a consequent redistribution of surface mass in the form of liquid water over the oceans.

8.4 Model Interiors

The results of calculations generally agree on the model shown schematically in Figure 8.5. It is seen that the planet is a well-differentiated structure.

The crust is much thicker at continental sites than under the oceans. The reason for this is that there appears to be some 20–25 km of granitic rock under the continents. The density of this rock is about $\rho = 2.7$ gm cm^{-3}.

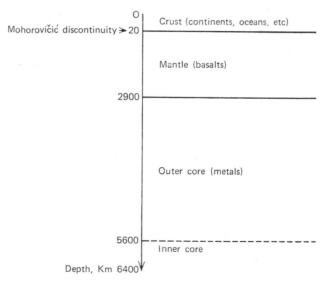

Figure 8.5 Schematic model of the earth's interior.

Now the oceans (which are perhaps some 5 km thick) are underlaid by about 1 km of sediment on top of a 5-km-thick layer of dark basaltic rock. (The density of the latter is about 3.0 gm cm^{-3}.) These same two layers also underly the continental granite, giving rise to the difference in crustal thicknesses mentioned above. It should be appreciated by the reader that the thicknesses and densities given above are only crude averages; there may be large departures locally from these figures.

Most shocks occur at relatively shallow depths. In fact, the deepest foci detected to date are only $\sim$700 km underneath the Pacific.

At the bottom of the basaltic layer the crust ends. The surface of termination is called the *Mohorovičić discontinuity*. It is at this discontinuity where V_S and V_P sharply increase (see Figure 8.4).

The region below the Mohorovičić is called the *mantle*; it is believed to be composed of ultrabasic rock with an average density of about 3.3 gm cm^{-3}. The much-discussed Project Mohole had the objective of drilling a hole from the oceanic floor down to the Mohorovičić discontinuity in order to ascertain conditions directly at those depths ($\sim$15 miles).

Observation indicates that V_S and V_P, after an initial decrease from the Mohorovičić discontinuity to about 100–150 km in depth, increase rather rapidly and irregularly with further increases of depth, down to $\sim$900 km in the mantle. The relatively high value of dV/dr, and the irregularities, suggests a gradual change in composition or phase, or both. Below 900 km V_P and V_S slowly increase with depth, suggesting that the lower mantle is fairly uniform.

The composition of the mantle, of course, is very much in doubt. One popular assumption is that is is the same as that found in the chondritic (stony) meteorites, a subject that we discuss in Chapter 10. It is possible that St. Paul's rocks, which are located some 80 km north of the equator on the axis of the mid-Atlantic ridge, are representative of the composition of the mantle at perhaps 30 km of depth.

There is another apparent discontinuity at about 2900 km of depth. There is a very sharp drop in V_P there; it jumps from 13.5 km per second down to 8.1 km per second. Even more interesting is the observation that V_S *suddenly goes to zero there*, so that *S-waves do not propagate at this and greater depths*. This has been interpreted as implying that $\mu = 0$, and the consequent inference is that the earth is composed of a "liquid-like" substance for nearly half the planetary radius.

Now we know that the mean density of the earth is 5.5 gm cm^{-3}. Therefore the average density of this "core" material must exceed 12 gm cm^{-3}. Indeed, some authors claim that the very innermost material is perhaps 14–17 gm cm^{-3}. These densities are higher than that of lead. Most calculations indicate that there is a discontinuous jump in ρ at 2900 km from about

6 gm cm^{-3} up to about 9 beyond the discontinuity. This suggests that the core material has a very different composition from the mantle.

It is thought that the core is composed of molten metals. One popular mixture is 92% iron and 8% nickel. If it is composed of molten metals, convective circulation is presumed to exist within the core. One school of thought holds that these convective currents may be likened to a dynamo and that they are the source of the geomagnetic field. We shall return to this idea in the next chapter.

To return to the model, there is some evidence that the innermost parts of the core are "solid-like" once more. A discontinuity in V_P has been detected at a depth of about 5200 km (about 800 km from the center). This has given rise to the concept that there exist "inner" and "outer" cores.

Over all, the earth behaves as a very rigid and highly elastic ball. That is, the elasticity and rigidity of the earth greatly exceed those of steel.

8.5   Temperature Profile

When we bore holes in the earth's crust and lower thermometers down the holes, we find that the temperature increases with depth. Measurements indicate a temperature coefficient of $+0.03°C$ per meter. This is equivalent to about 100°F per mile! We infer from this that the interior of our planet must be quite hot. Most models place the core temperature at something like 4000°C.

Now we know that the heat outflow must be quite small, since the temperature of the earth's surface is controlled by the sun, not by the interior. Thus the heat outflow must be small compared to the solar constant of approximately 2 calories per cm^2-minute. This is indeed the case; measurements made in the bore holes yield an average outflow of about 1×10^{-6} calorie per cm^2-second, less than 0.01% of the solar input. There are, as one might expect, variations in the heat flow over the planetary surface. These can be caused by (a) contrasts of thermal conductivity and the resulting refraction of heat, (b) contrasts in the sources of heat production, which are discussed below, (c) local temperature differences caused by intrusion of hot (or cold) material, and (d) convection of ground water.

Although the outflow is small compared with the sun, it is not negligible, and it is interesting to inquire into the source of this heat or energy. If the outflow were due to the burning of coal, a coal seam some 40 meters thick would have been consumed during the age of the earth.

Another possibility is that the outflow is due to "original heat." That is, one might suppose that in a previous epoch the earth was molten and has been slowly cooling ever since. This problem was attacked by Kelvin, who allowed for the fact that as the outermost rocks cooled they would solidify, insulating the outer layers from the molten interior regions, and the thickness of this insulator would steadily increase with time. The result of his calculation is that the outward flow of heat would be reduced below the observed flow in only 30 million years, assuming no other internal heat source. Thus the crust may have been molten only 30 million years ago. We know from other disciplines, such as paleontology, that this result is not allowed.

There is another possible source of heat, and that is the decay of radioactive isotopes. The elements uranium and thorium and the isotope K^{40} are observed to have concentrations in crustal granites of about 10^{-6} gm per gm, 10^{-5} gm per gm, and 10^{-2} gm per gm, respectively.

These concentrations are quite capable of supplying the required heat. In fact, it seems that there is too much radioactivity, if anything. For example, if the observed concentrations were to extend uniformly all the way down to the Mohorovičić discontinuity, we would obtain about twice the observed heat outflow! Therefore geologists believe that these radioactive substances must be concentrated near the planetary surface, rather than having a uniform density throughout the crust. It is also interesting that while the concentration of K^{40} is roughly that to be expected on the basis of the chondritic model referred to earlier, the observed concentrations of uranium and thorium are higher than expected. This supports the surface concentration idea.

A likely explanation for this surface concentration lies in diffusive separation. These substances are readily oxidized, and the resulting chemical compounds turn out to be rather light. Thus the oxides tend to "float" on top of the other rocks.

It is interesting that the heat outflow must have been greater in the past, because of the law of radioactive decay. This law says that

$$N(\tau) = N_0 e^{-\lambda t}, \tag{8.11}$$

where N_0 is the initial concentration of a given isotope and $N(\tau)$ is the concentration (number of nuclei per unit volume) at a time τ. In (8.11) the factor λ is the "decay constant" for a given species; it is related to the half-life T of that species through $T = 0.693/\lambda$. An analysis of the decay law tells us that the heat production 4 billion years ago must have been about $3\frac{1}{2}$ times the present figure. One concludes from this that the base of the crust must have been molten at that time.

8.6 Future Experiments

We can experimentally check some of these ideas concerning radioactivity in at least two rather different ways. One way—the most direct—is simply to dig a very deep hole (down to the Mohorovičić) and analyze the substances as a function of depth. This was one of the purposes of Project Mohole, a now-abandoned United States undertaking.

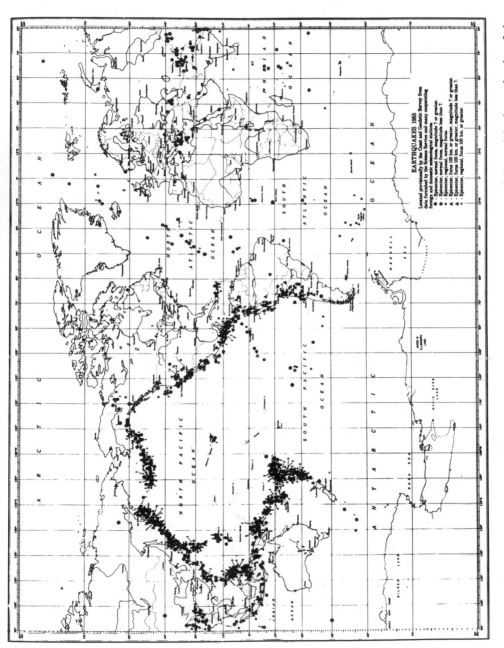

Figure 8.6 World-wide distribution of earthquakes for the year 1955. It is seen that most of the activity is on the rim of the

Another method might be to undertake measurements of the surface radioactivity of the moon. Some people believe that the moon came from the earth (perhaps where the Pacific now is). This belief is based on two observations. First, the moon is receding from the earth; if we extrapolate back at the present recessional speed, the two bodies were in contact some 1.5×10^9 years ago.

The day is presently lengthening at the rate of 15 microseconds per year, because of the recession of the moon; the spin angular momentum of the planet is being transferred through tidal friction to the orbital angular momentum of the satellite. Studies of fossils may provide information on the length of the day in the past, to see whether this linear extrapolation is valid.

Second, as seen in Figure 8.6, well over 90% of the earthquakes and active volcanoes on the earth are on the rim of the Pacific. This has suggested to some investigators that perhaps a violent upheaval once took place where the Pacific is now found.

8.7 The Moon

We should note at this point that it is by no means clear that the moon did come from the earth. If the moon came from the earth 1.5 billion years ago, there is no paleontological evidence for a catastrophic event at that time. Also, the earth must have been spinning rapidly at that time, since the moon's angular momentum should have contributed to accelerating the planetary rotation rate. There is no known evidence that the day was only ~ 1 hour long.

The moon can approach the earth's center to about 10,000 miles before tidal action tears it apart. This limiting distance is called the *Roche limit*, d.

To estimate d, picture the moon as composed of two equal liquid spheres, each one-half the size of the moon, as shown in Figure 8.7. These two spheres will remain together (that is, the moon will not be split in half) as long as their mutual gravitational attraction is greater than the destructive tidal action of the earth. If r is the lunar radius, the mutual attraction of the two spheres is

$$\frac{G(M_s/2)^2}{r^2},$$

where M_s is the mass of the moon. The tidal action of the earth tending to tear the moon apart is given approximately by

$$\frac{GM \cdot M_s}{2}\left[\left(\frac{1}{d-r}\right)^2 - \left(\frac{1}{d+r}\right)^2\right] \simeq \frac{2GM \cdot M_s r}{d^3},$$

where terms involving $(r/d)^2$ are neglected and where M is the mass of the earth. We can obtain the limiting value of d by equating these two expressions;

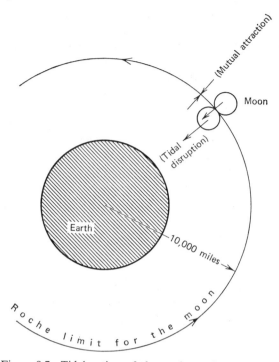

Figure 8.7   Tidal action of the earth on the moon.
The earth exerts a greater gravitational force on that
half of the moon that is closest to it; this "gravity-
gradient" force is counterbalanced by the mutual
gravitational cohesive force binding the two small
spheres representing the moon at distances from the
earth that are greater than the Roche limit. At
distances smaller than the Roche limit the moon
breaks up.

the result is

$$\frac{2GM \cdot M_s r}{d^3} = \frac{GM_s^2}{4r^2}, \tag{8.12}$$

so that

$$d^3 = \frac{8Mr^3}{M_s}. \tag{8.12a}$$

If the density of the moon is written as ρ_s, we finally obtain

$$d = \left(\frac{6M}{\pi \rho_s}\right)^{1/3} \tag{8.13}$$

for the value of the Roche limit d, in terms of the planetary mass M and
satellite density ρ_s. Since d is about 10,000 miles for the earth-moon system,

it is not clear how the moon could have existed as a single object ascending from the Pacific basin.

In another theory the moon was captured as a large body into the earth's orbit. The excess kinetic energy was absorbed through violent relocations of the matter in both the moon and the earth, such as volcanoes on the earth and temporary fragmentation of the moon itself.

A third main theory says that the moon originated from the coalescence of planetesimals. The point where they gravitationally came together is taken to be somewhere between the earth's surface and the present lunar distance. This is necessary because of the observed recession. This theory has won some favor, in that tremendous energy dissipation in the earth and moon is not required. There is uncertainty, however, regarding the number of planetesimals at the correct orbital inclination. The whole subject of the origin of the moon must be regarded as highly speculative.

Our first known look at lunar material occurred when samples of the lunar surface were returned to earth during 1969 by the Apollo 11 and Apollo 12 astronauts (Table 8.2). For comparison purposes, the analysis of a common

TABLE 8.2. CHEMICAL COMPOSITIONS, AS REPORTED BY ENGEL AND ENGEL, IN WEIGHT PERCENT, OF ONE DUST SAMPLE AND TWO ROCKS TAKEN FROM THE LUNAR SURFACE BY THE APOLLO 11 ASTRONAUTS

The difference in composition of the [lunar] known-to-be-primitive rocks in the planetary system indicates the complexities inherent in defining the solar abundances of elements and the initial compositions of the earth and moon.

Substance	Lunar Rock A	Lunar Rock B	Lunar Dust	Basalt
S_iO_2	42.01	39.79	41.50	48.01
T_iO_2	8.81	11.44	7.50	2.92
Al_2O_3	11.67	10.84	14.31	15.97
Fe_2O_3	0.00[a]	0.00?	0.06?	3.87
FeO	17.98	19.35	15.62	7.56
MnO	0.24	0.20	0.22	
MgO	6.25	7.65	7.95	5.26
CaO	12.18	10.08	11.84	9.04
Na_2O	0.48	0.54	0.48	3.73
K_2O	0.11	0.32	0.16	1.89
H_2O^+	0.00	0.00	0.00	1.33
H_2O^-	0.00	0.01	0.01	
P_2O_5	0.08	0.17	0.10	0.42
Total	99.81	100.39	99.75	100.00

[a] Actual value, -0.08.

terrestrial basalt is also shown in the table. The initial results of these studies also showed, from uranium-dating analysis, that some of the Sea of Tranquillity material returned by astronauts Armstrong, Aldrin, and Collins (Apollo 11) is over 4.6 billion years old. That portion of the moon, at least, is not significantly younger than the earth. It appears at this writing that some of the material returned from the Ocean of Storms during Apollo 12 by Conrad, Bean, and Gordon (Apollo 12) may be considerably younger; the two maria surfaces show different ages. Possibly this is evidence for some lunar activity billions of years ago that overturned some of the Ocean of Storms material.

It is interesting to compare the observed concentrations of radioisotopes in the crust with those found in meteorites and in tektites. Meteorites may be representative of the material generally found in the solar system (see Chapter 10), and some scientists believe that tektites originated in the lunar surface.

Tektites are opaque, glassy objects, usually black in color, that range in diameter from a few microns up to 10 cm or so. They appear to have been melted in their history. Tektites are found in only a few land areas on the earth, such as Australia, the region surrounding the South China Sea, and Texas.

The molten history (which is confirmed by their extremely low water content, as compared to continental, common rock) and the occurrence in strewn fields have suggested an extraterrestrial origin for tektites. This idea is strengthened by the occurrence in Philippine tektites of iron-nickel alloys found in iron meteorites but not in other natural material found on earth.

Two ideas for the origin of tektites are currently popular. One is that very large meteorites struck the earth, fusing the soil and throwing out molten material which hardened as it flew through the air to form tektites. The trouble with this suggestion is that the composition of tektites is unlike that of material known to be ejected by meteoritic impacts. The other possibility frequently discussed is that tektites come from the lunar surface, possibly again as a result of meteoritic impacts, but here as a result of impacts on the moon. It has been found that trajectories resulting in the known strewn fields and originating outside the atmosphere could only have come from the moon; it is the only known celestial body which could have been the source of objects having these trajectories.

The first two Apollo missions, however, have shown great differences in chemical composition between the tektites and maria material. If tektites do come from the moon, the most likely remaining places are the lunar upland regions.

8.8 The Other Planets

We really do not know anything at all about the interiors of the other planets, except for their average densities. But even this has shown that a peculiar situation exists. There appear to be at least two "kinds" of planets, the "inner" group and the "Jovian" planets. The first group (Mercury, Venus, Earth) have a relatively large average density—in excess of 5 gm cm^{-3}. The Jovian group (Jupiter, Saturn, Uranus, Neptune) appear to be largely composed of hydrogen, since their densities are approximately unity. Although the average density of Mars is only about 4.1 gm cm^{-3}, its size and character suggest that it belongs to the inner group. Too little is known about Pluto to permit comment.

Theories of the origin of the solar system must explain the differentiation of the planets into these groups. Additional observations are badly needed. For example, should we be successful in placing a spacecraft into orbit about another planet, we can observe the precession of the various orbital parameters. This will yield data on the distribution of matter within the planet. It will be most interesting to learn whether the earth's stratification (into crust, mantle, and core) is at all typical of the other planets, even of the inner group. Seismometers placed on the surfaces of the moon, the planets, and their satellites will shed light on this problem.

The Jovian planets are obviously different from our own; aside from the density difference, it is interesting that all the members of the Jovian group are much larger than any of the inner planets. Direct measurements of their interiors—which are presumably at extremely high pressures—may yield data on phenomena that are extremely difficult to observe in the laboratory.

Problems

8.1. A perfectly spherical planet has only a crust (of thickness d_1) and a liquid-like core (of radius d_2). The radius of the planet is $R = 1000$ km. As diagrammed, an explosive is detonated at A. A seismograph at A detects an S-wave reflected

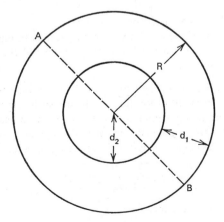

from the surface of the core $\tau_\beta = 620$ seconds after the explosion, and a similar seismograph at B (diametrically across the planet) detects a P-wave that travels along a diameter. The P-wave is detected $\tau_\alpha = 825$ seconds after the explosion. The known properties of the planetary interior are summarized below.

	Incompressibility (k)	Rigidity (μ)	Density (ρ)
Crust	5	4	3
Core	0	90	ρ_2

(The units are such that the propagation velocities are expressed in kilometers per second.)

(a) Find d_1 and d_2.

(b) Find the density of the core ρ_2.

8.2. Taking the thermal conductivity of crustal earth to be 2×10^{-4} calories second^{-1} cm^{-1} degree^{-1}, what would the temperature be at the base of the crust, if the surface is at a temperature $T_s = 273°K$ and the crust is 30 km thick? Assume that no heat is generated within the crust.

8.3. Considering the relative accelerations of the earth's center and of the sublunar point on the earth's surface, relate the tide-rising acceleration f to the mean earth-moon distance d. Compare f with g, the acceleration on the earth's surface due to gravity. Explain why maximum tides occur only when the moon is new or full (the so-called *spring tides*) and why the minimum tides (*neap tides*) occur at the first and last quarters of the moon.

8.4. Because of tidal friction (mainly in the shallow Bering sea), the earth's rotational rate is slowing, such that the day is lengthening by 1.5 seconds per 100,000 years.

(a) Where is the angular momentum going?

(b) What do you think will happen when the day becomes longer than the month?

8.5. (a) Compute the Roche limit for the moon.

(b) Using the data of the previous problem, and assuming that the slowdown rate has remained constant, how long ago could the moon have been at the Roche limit?

8.6. A lower limit to the central density ρ_c of the earth may be computed by considering that the moment of inertia I of the planet is larger than it would be if all the mass were at the center and the rest of the planetary material had a negligible effect. Taking a dense core of radius r_0 and density ρ_0 and using the equation for the moment of inertia of a rotating sphere,

$$I = \frac{8\pi}{3} \int_0^R \rho(r) r^4 \, dr,$$

find ρ_0 in terms of $\bar{\rho}$, the mean density.

8.7. Consider a terrestrial model composed of two components, the core and mantle, which have densities ρ_c and ρ_m, respectively.

(a) If αR_E is the boundary of the core, express ρ_c and ρ_m in terms of α and $\bar{\rho}$, the mean density of the earth.

(b) If $\alpha = 0.545$ and $\bar{\rho} = 5.52$ gm cm^{-3}, find ρ_c and ρ_m.

Chapter 9

PLANETARY MAGNETISM

We saw in the preceding chapter that we know disturbingly little about the interior of our planet and just about nothing concerning the internal structure of the other members of the solar system. The understanding of planetary magnetism is another source of frustration, for our understanding of even the earth's main field is very poor. In fact, about all that is in reasonably good shape is the description of the field; its origin is still most uncertain.

We concentrate our discussion again on the field of the earth, for the geomagnetic field is the only planetary field explored in detail. Fields have been indirectly detected around Jupiter (and some of the stars, including the sun) but their intensities are poorly known.

The fact that lodestones point approximately north-south has been known for many centuries. The concept of a planetary magnetic field to explain this pointing dates from the publication by Sir William Gilbert in 1600 of *De Magnete*. The pioneering work of Gauss should be mentioned in this connection, for he mapped the field with magnetometers, and the basic principles of these instruments have been utilized until very recently for magnetic measurements. In modern times the hyperfine splitting of atomic energy levels has provided a means of measuring the geomagnetic field accurately. The splitting is directly proportional to the ambient magnetic field. The gyromagnetic ratio of the proton has been measured in the laboratory with very great accuracy, and this atomic constant now provides the basis for modern magnetometers.

9.1 The Geomagnetic Coordinate System

The magnetic field of our planet is, to a first approximation, that of a dipole located near the center of the earth. Thus this approximation pictures the source of the field as a little bar magnet. The "equivalent dipole" would be offset 436 km from the center of the earth, displaced toward the Pacific ocean. It is tilted with respect to the earth's rotation axis by approximately 11° (see Figure 9.1).

The dipole axis intersects the surface of the earth at points far distant from the north and south poles. These intersection points are called the north

214

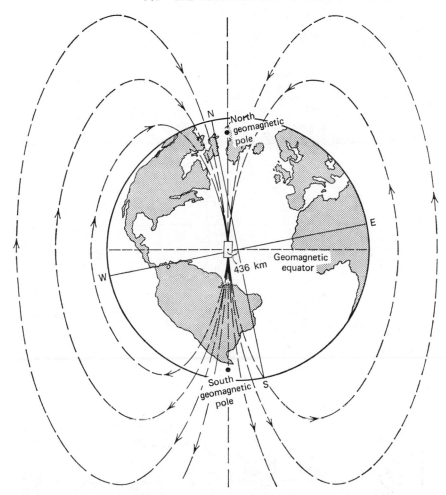

Figure 9.1   The eccentric-dipole model of the earth's magnetic field (schematic view). The equivalent dipole is ~436 km distant from the center of the planet and is closest to the surface in the hemisphere that contains the Pacific. Hence at a given altitude the field is stronger over the Pacific than it is over the Atlantic. The geomagnetic axis is tilted ~11.5° with respect to the earth's rotational axis (the N–S line in the figure).

and south "geomagnetic poles." The north geomagnetic pole is located in Greenland at 81.0° N, 84.7° W, in the geographic system of coordinates. The corresponding south geomagnetic pole lies in Antarctica, at 75.0° S, 120.4° E. (The figures given above apply for the year 1955.)

One may accordingly define a "geomagnetic equator"; this great circle is

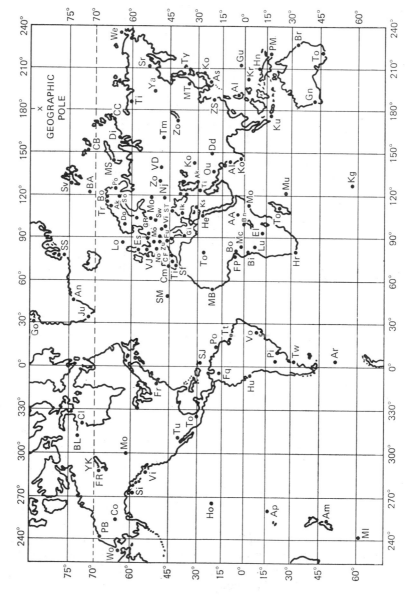

Figure 9.2 The world in geomagnetic coordinates, Mercator projection, showing magnetic observatories. From *Handbook of Geophysics*, United States Air Force Geophysics Research Directorate, The Macmillan Company, 1960.

90° away from either pole in "geomagnetic latitude." The geomagnetic equator passes through Singapore and is roughly 9° south of the geographic (or celestial) equator at the longitudes appropriate to the United States. Consequently a point in the United States that is at a geographic latitude of 32° N will be at a geomagnetic latitude of about 41° N. Figure 9.2 shows the world as it would appear in geomagnetic coordinates.

From the data for 1955 the geomagnetic latitude λ and longitude Λ (east) of a point with geographic latitude θ and longitude Φ is given by

$$\sin \lambda = \cos 78.3° \cos \theta \cos (\Phi - 291°) + \sin 78.3° \sin \theta. \qquad (9.1)$$

This expression (9.1) for the geomagnetic latitude is similar to that for the geomagnetic longitude; the latter relation is

$$\cos \Lambda = \frac{\sin 78.3° \cos \theta \cos (\Phi - 291°) - \cos 78.3° \sin \theta}{\cos \lambda}. \qquad (9.2)$$

In (9.1) and (9.2) Λ is positive for the eastern hemisphere and λ is positive for the northern hemisphere.

9.2 The Dip Poles

The offset of the equivalent dipole from the planetary center results in geomagnetic field lines that are not vertical where the dipole axis intersects the surface of the earth. Thus the field lines are inclined about 3.9° to the vertical at the geomagnetic poles.

The places where the field lines are vertical are known as the "dip poles." These locations are controlled both by the offset and by the substances in the crust.

Some observers believe the dip poles are located near 82.4° N, 137.3° W (Labrador), and at 67.9° S, 130.6° E (Antarctica). Other locations are shown in figure 1.12. The precise locations of the poles are difficult to establish accurately from on-site measurements. It is ironic that the dip coordinates— which should not be particularly representative of anything fundamental— seem to be a better coordinate system for discussion of the cosmic radiation than does the geomagnetic system of coordinates. The cosmic ray intensity as a function of latitude is discussed in Chapter 13.

9.3 Local Coordinates

Each geomagnetic observatory on the surface of the earth may measure the field in terms of local coordinates, which are shown in Figure 9.3. The total magnetic field vector is denoted by **F**. The inclination I is the smallest angle between **F** and the horizontal; it is sometimes called the magnetic dip. Another coordinate is called D, the declination. D is the angle between the projection of **F** on the horizontal and "true north." It is therefore the angle

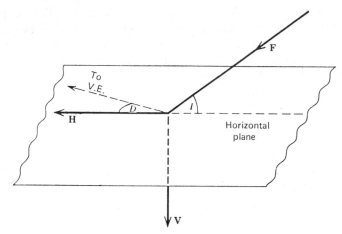

Figure 9.3   The local system of geomagnetic coordinates.

between north as indicated by a compass needle and the pole of rotation, and may be measured in degrees east or west. For example, a declination of 5° E means that a compass will point 5° to the east of true north.

A much-used quantity is H, the component of $|\mathbf{F}|$ in the horizontal plane. The corresponding vertical component is denoted by V. Since

$$\tan I = \frac{V}{H},\tag{9.3}$$

we should note that the dip latitude λ_D may be specified in terms of I. This may be done by resorting to the dipolar approximation.

9.4   Dipolar Relations

The equations for a static dipole field are

$$B_r = \frac{2H_0 \cos\theta}{r^3},\tag{9.4}$$

$$B_\theta = \frac{H_0 \sin\theta}{r^3},\tag{9.5}$$

and

$$B_\varphi = 0.\tag{9.6}$$

Referring to Figure 9.4, r is the distance from the dipole to the field point, θ is the colatitude, and φ is the azimuthal angle. The quantity H_0, as we shall see, is just the surface field at the geomagnetic equator, which may be measured directly.

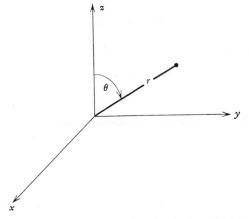

Figure 9.4 The geometry for a static dipolar field.

Now the equations of the field lines themselves may be obtained by setting

$$\frac{dr}{r\,d\theta} = \frac{B_r}{B_\theta}.$$ (9.7)

This tells us that

$$r = R_0 \sin^2 \theta.$$ (9.8)

Here R_0 is the geocentric distance of the equatorial crossing of the field lines (see Figure 9.5). Equation 9.8 may be rewritten in terms of the complement of θ, λ, where λ is the geomagnetic latitude. Thus the equation of the field

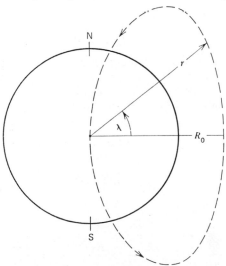

Figure 9.5 The geomagnetic latitude λ and the equatorial crossing distance of a field line R_0 are shown.

lines is

$$r = R_0 \cos^2 \lambda. \tag{9.9}$$

Since the magnitude of the field is given by

$$|B| = [B_r{}^2 + B_\theta{}^2 + B_\varphi{}^2]^{\frac{1}{2}}, \tag{9.10}$$

we see that

$$|B| = \frac{H_0}{r^3}(3\cos^2\theta + 1)^{\frac{1}{2}} \tag{9.11}$$

or

$$B = \frac{H_0}{r^3}(3\sin^2\lambda + 1)^{\frac{1}{2}}. \tag{9.12}$$

At the geomagnetic equator on the terrestrial surface $r = R_0 = R_E$; here $\lambda = 0$, and if we let $R_E = 1$, then

$$B = H_{0'}$$

as we previously indicated. The field lines are entirely horizontal at the equator, in this approximation. Measurements indicate that $H_0 = 0.311$ oersted = 0.311 gauss, since we are not dealing with magnetic material and the gauss is a more familiar unit. A useful unit for studies of the geomagnetic field is the *gamma*, where one gamma = 10^{-5} gauss.

The planetary magnetic moment M is another quantity that is of interest. It is the magnetic dipole moment of the equivalent dipole. We may rewrite (9.11) as

$$B = \frac{0.311}{r^3}\left(4 - 3\frac{r}{R_0}\right)^{\frac{1}{2}}. \tag{9.13}$$

The earth's magnetic moment is therefore $M_E = 0.311$ gauss-earth radii³, or, 8.06×10^{25} gauss-cm³.

9.5 The Dip Latitude

We may now return to our problem of expressing the dip latitude λ_D in terms of the inclination I of the field lines. Evaluating (9.7), we find that

$$\frac{B_r}{B_\theta} = 2\cos\theta. \tag{9.14}$$

But $\cos\theta = \tan\lambda_D$, and also

$$\tan I = \frac{B_r}{B_\theta}. \tag{9.15}$$

Thus

$$\lambda_D = \text{arc tan } [\tfrac{1}{2}\tan I], \tag{9.16}$$

so that a measurement of the angle that the field makes with the horizontal specifies the dip latitude.

9.6 Mathematical Descriptions of the Field

We have previously noted that, to a first approximation, the geomagnetic field is dipolar. It is possible to see this mathematically, and to also see what better approximations yield.

We start with the fourth Maxwell equation,

$$\mathbf{\nabla} \times \mathbf{B} = \left(\mathbf{j} + \frac{\partial \mathbf{D}}{\partial t}\right)\mu. \tag{9.17}$$

Now if there are no important currents and the time rate of change of electric fields around the earth is small,

$$\mathbf{\nabla} \times \mathbf{B} = 0. \tag{9.18}$$

We see, therefore, that we may express B as the gradient of a scalar "potential" V, so that $\mathbf{B} = -\mathbf{\nabla}V$. This indicates that we should seek to determine V everywhere, for $\mathbf{B}$ can be found from it. In fact, the north, east, and vertical components of the field are respectively

$$X = \frac{1}{r}\frac{\partial V}{\partial \theta}, \tag{9.19}$$

$$Y = -\frac{1}{r \sin \theta}\frac{\partial V}{\partial \varphi}, \tag{9.20}$$

and

$$Z = -\frac{\partial V}{\partial r}. \tag{9.21}$$

We may measure X, Y, and Z and deduce measured values of V from these.

Furthermore, if the amount of electric current flowing between the planet and space is small (a reasonable assumption), the potential V satisfies Laplace's equation, so that

$$\nabla^2 V = 0. \tag{9.22}$$

Laplace's equation may be readily solved; the potential is found in terms of a sum of spherical harmonic functions:

$$V = V^e + V^i = R_E\left\{\sum_{n=1}^{\infty} \left(\frac{r}{R_E}\right)^n T_n^{\ e} + \left(\frac{R_E}{r}\right)^{n+1} T_n^{\ i}\right\}. \tag{9.23}$$

Just as an arbitrary curve may be expressed in terms of a Fourier series, a function on a spherical surface may be represented by a spherical harmonic expansion.

TABLE 9.1. IGRF 1965.0 COEFFICIENTS

n	m	Main Field (gammas)		Secular Change (gammas/year)	
		$g_n^{\,m}$	$h_n^{\,m}$	$\dot{g}_n^{\,m}$	$\dot{h}_n^{\,m}$
1	0	−30339		15.3	
1	1	−2123	5758	8.7	−2.3
2	0	−1654		−24.4	
2	1	2994	−2006	0.3	−11.8
2	2	1567	130	−1.6	−16.7
3	0	1297		0.2	
3	1	−2036	−403	−10.8	4.2
3	2	1289	242	0.7	0.7
3	3	843	−176	−3.8	−7.7
4	0	958		−0.7	
4	1	805	149	0.2	−0.1
4	2	492	−280	−3.0	1.6
4	3	−392	8	−0.1	2.9
4	4	256	−265	−2.1	−4.2
5	0	−223		1.9	
5	1	357	16	1.1	2.3
5	2	246	125	2.9	1.7
5	3	−26	−123	0.6	−2.4
5	4	−161	−107	0.0	0.8
5	5	−51	77	1.3	−0.3
6	0	47		−0.1	
6	1	60	−14	−0.3	−0.9
6	2	4	106	1.1	−0.4
6	3	−229	68	1.9	2.0
6	4	3	−32	−0.4	−1.1
6	5	−4	−10	−0.4	0.1
6	6	−112	−13	−0.2	0.9
7	0	71		−0.5	
7	1	−54	−57	−0.3	−1.1
7	2	0	−27	−0.7	0.3
7	3	12	−8	−0.5	0.4
7	4	−25	9	0.3	0.2
7	5	−9	23	0.0	0.4
7	6	13	−19	−0.2	0.2
7	7	−2	−17	−0.6	0.3
8	0	10		0.1	
8	1	9	3	0.4	0.1
8	2	−3	−13	0.6	−0.2
8	3	−12	5	0.0	−0.3

TABLE 9.1. (CONTINUED)

n	m	Main Field (gammas)		Secular Change (gammas/year)	
		$g_n{}^m$	$h_n{}^m$	$\dot{g}_n{}^m$	$\dot{h}_n{}^m$
8	4	−4	−17	0.0	−0.2
8	5	7	4	−0.1	−0.3
8	6	−5	22	0.3	−0.4
8	7	12	−3	−0.3	−0.3
8	8	6	−16	−0.5	−0.3

The International Geophysical Reference Field for the period 1955.0–1972.0, as compiled by Commission 2, Working Group 4, of the International Association of Geomagnetism and Aeronomy (IAGA). A Fortran computer program to compute field values from the coefficients may be obtained from World Data Center A for Geomagnetism, U.S. Coast and Geodetic Survey–ESSA, Rockville, Maryland 20852.

In the expansion (9.23) the functions T_n are given by

$$T_n = \sum_{m=0}^{n} (g_n{}^m \cos m\varphi + h_n{}^m \sin m\varphi) P_n{}^m(\cos \theta). \qquad (9.24)$$

The functions $P_n{}^m$ (cos θ) are known to geophysicists as *Schmidt's functions*, and are related to the *associated Legendre polynomials* $P_{n,m}(\cos \theta)$ through the expression

$$P_n{}^m(\cos \theta) = \left\{ 2 \frac{(n-m)!}{(n+m)!} \right\}^{\frac{1}{2}} P_{n,m}(\cos \theta)$$

when $m > 0$. When $m = 0$ the two functions are equal. The associated Legendre polynomials are well known to science. They may be computed from a knowledge of the Legendre polynomials P_n themselves (see Section 2.17):

$$P_{n,m}(\cos \theta) = \sin^m \theta \, \frac{d^m P_n(\cos \theta)}{d(\cos \theta)^m}.$$

It will be noted that (9.23) is the sum of the two potentials, V^i and V^e. These are the potentials due to internal and external sources, respectively. Equation 9.23 may be differentiated with respect to r, θ, and φ and the derivatives set equal to their counterparts in (9.19), (9.20), and (9.21). It is the task of geomagneticians to deduce the values of the so-called *Gaussian coefficients* $g_n{}^m$ and $h_n{}^m$ from the surface magnetic data, where $r = R_E$, and calculate the field from these. Some analyses have used up to 80 coefficients, corresponding to terms through $m = n = 8$ (see Table 9.1). Table 9.1 lists the Gaussian coefficients of the International Geomagnetic Reference Field, epoch 1965.0. The main field (the field of internal origin) coefficients, as well as those for the secular variation, are listed. These apply for the period 1955.0–1972.0; any needed modifications will be made under the auspices of the International Union of Geodesy and Geophysics.

It is found from the field values computed from these coefficients that the external sources of the potential are minor; V^i describes over 99% of the surface geomagnetic field. Indeed, the available surface data are not really sufficiently accurate to deduce the presence of a V^e at all; satellite data have been much more useful in this connection, for the global coverage is vastly increased with satellites. The types of external sources that exist are a quiet-day ring current in the magnetosphere and the interplanetary field that is transported by the solar wind. It is rather remarkable that the 1% due to external sources is far better understood than the 99% of the geomagnetic field that (somehow) arises from internal sources.

9.7   The Secular Variation

One of the chief reasons why it has been so difficult to understand the origin of the geomagnetic field is that a secular variation exists. That is, the earth's main field varies in time. The change is slow compared with laboratory phenomena, but extremely rapid on a geological time scale. There are some regions on the earth where the field is changing by over 150 gammas per year; field descriptions at such points are out of date in 6 months, if 0.1% accuracy is desired. Another way of expressing this is to say that

$$\frac{dg_n{}^m}{dt} \quad \text{and} \quad \frac{dh_n{}^m}{dt}$$

are not equal to zero (see Table 9.1). To use the listed coefficients for any time other than 1 January 1965 (the epoch time), multiply $\dot{g}$ and $\dot{h}$ by $(t - t_0)$, where t_0 is the epoch time.

In general it appears that M, the earth's magnetic moment, is decreasing. The rate of decrease is sufficient to decrease the equatorial field by 50 gammas per year, or about 0.02% per year. In addition, the general field pattern is drifting westward at perhaps a quarter of a degree annually, and it is also slowly shifting northward. Another way of expressing this is that the offset distance of the equivalent dipole increased from 285 km in 1845 to 436 km in the epoch 1955, some 110 years later. This variation is sufficiently rapid to lead to a complete reversal of the field in a period of only 10^4 years or so. Indeed, there is some paleomagnetic evidence—the directions of magnetism as a function of depth in long deep cores (tens of meters deep)—that reversals have occurred in the past, although the result may also be complicated by continental drift. A typical time scale for these reversals is about 70,000–100,000 years; many such reversals appear to have occurred during the history of this planet.

9.8 Origin of the Main Field

We might suppose that the field is merely due to the presence of magnetic material distributed more or less uniformly throughout the sphere and that the specific magnetization A of this rock is such that $M = \frac{4}{3}\pi R_E^3 A$. But, if so, $A = 0.07$ gauss, which is in excess of the value for the material observed in the crust. Then, too, the temperature of the earth presumably exceeds the Curie point of iron (about $750°C$ at standard pressure) somewhere around the Mohorovičić discontinuity. This implies that it is unlikely that there is much magnetic material in the mantle.

In 1947 P. M. S. Blackett proposed that a massive rotating body such as the earth possesses a magnetic moment proportional to its angular momentum. This, however, is not in accord with the observed secular variation. We should also note that the Mariner 4 magnetometer, which set a low upper limit to the moment of Mars, does not lend support to this idea.

It is possible to explain the main field at least qualitatively on the basis of motions within the core. If, for simplicity, we assume the core to be an infinitely good conductor, any original magnetic field lines—say due to magnetic materials outside the core—will be dragged around by the currents within the core, just as though they were "frozen" into the core material. Now if the core rotates *nonuniformly* with depth, the field lines will become twisted around the axis of rotation, in such a direction as to oppose the initial field. The twisting action means that the lines of force become more tightly packed together, so that the field grows in intensity. It continues this growth until it neutralizes the initial field; a net reversal is even possible.

This model imposes some rather stringent requirements on the core material when we attempt to meet the boundary conditions that constitute the observed field. It requires a rather peculiar distribution of currents within the core, which (apparently) can be generated only if the center of the core is hotter than the remainder. This has led some investigators to suggest that the center of the core may be radioactive, but the differentiation process we discussed in Chapter 8 makes this hard to understand. Furthermore, the secular variation would have to be interpreted as a motion of the outer core with respect to the mantle. This might in fact be the case, but there is little other evidence for this. The only supporting observation is that the crust and mantle are nonspherical, while the core presumably is spherical. Thus we might expect the moon to exert a torque on the mantle that is not exerted on the core, so that relative motion of the two is plausible.

We should note that field amplification by the differential rotation of conductors has been eagerly grasped at by astrophysicists in an attempt to explain stellar magnetic fields. We shall return to this topic in Chapter 11.

Many investigators now believe that the important parameters in planetary magnetic fields are the size of the conducting core (if any) and its spin rate, as in a dynamo. The larger the product of these two parameters, the stronger the planetary field. Thus, if Mars has a small core (which might be consistent with the relatively small size of the planet), it would be expected to have a small magnetic field, even though it rotates as fast as the earth. Jupiter is thought to have a large conducting inner region (see Chapter 4) and it is a rapidly rotating planet; a strong field would be expected for Jupiter, a result consistent with the intense radio emissions that have been detected from the space around Jupiter.

9.9 Disturbance Indicators

In addition to the secular variation, there are several types of more rapid and temporary fluctuations in the geomagnetic field. There are several methods of describing the degree of disturbance of the field. The most widely employed is the so-called K-index. Each magnetic observatory assigns a figure, ranging from 0 to 9, every 3 hours, in terms of previously accepted criteria. These concern the departure from normal of H, the horizontal component. When $K = 0$ we have a situation of extreme quiet or "calm"; $K = 9$ corresponds to an extremely disturbed 3-hour interval. The K-indices are broken down into thirds, labeled $-$, o, and $+$, so that $K = 5+$ corresponds to $5\frac{1}{3}$, $K = 5$o corresponds to 5, and $K = 5-$ corresponds to $4\frac{2}{3}$. Twelve geomagnetic observatories—unfortunately not well distributed in latitude—combine their K-indices to form the *planetary K-index*, Kp. Kp is quasilogarithmic; a change of Kp from 1 to 2 represents only a small change in the amplitude of geomagnetic fluctuations, whereas a change from 8 to 9 indicates a violent increase in the planetary fluctuations.

There are eight such Kp indices in the 24-hour day, starting with the interval 0000–0300, U.T. Kp is published monthly by the Environmental Science Services Administration of the United States, ESSA.

Now, an attempt has been made to introduce a more linear index of geomagnetic activity. This index is called *ap* and is derived from the Kp values through the use of Fig. 9.6. The *ap* index ranges from 0 to 400. Thus we see that a violent geomagnetic disturbance is more apparent when described in

TABLE 9.2. GEOMAGNETIC OBSERVATORIES THAT COMBINE THEIR DATA TO FORM THE PLANETARY K-INDEX, Kp

Meanwood, Canada	Wingst, Germany
Sitka, Alaska	Witterveen, Netherlands
Lerwick, Shetlands	Hartland, England
Eskdalemuir, Scotland	Agincourt, Canada
Lovö, Sweden	Fredricksburg, U.S.A.
Rude Skov, Denmark	Amberly, New Zealand

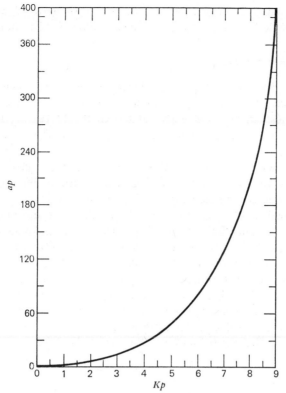

Figure 9.6 The relationship between Kp and ap. Reprinted from A. J. Dessler, Geomagnetism, in Francis S. Johnson, Ed., *Satellite Environment Handbook*, second edition, with the permission of the publishers, Stanford University Press. Copyright 1965 by the Board of Trustees of The Leland Stanford Junior University.

terms of ap, for a change from 300 to 400 is more dramatic than a change from 9 to 9+.

It happens that if ΔB is the range of the most disturbed magnetic element (in gammas) for the 3-hour period, then

$$\Delta B \approx 2ap.$$

For example, if $\Delta B = 90$ gammas, then $ap = 45$ (which corresponds to $Kp = 5$, a fairly disturbed interval). This linear relationship between the amplitude of field fluctuations and ap has encouraged the use of this index in some analyses of the field disturbances. Like Kp, values of ap appear monthly in the ESSA Research Laboratories series *IER-FB Solar Geophysical Data*. A related quantity, *Ap*, is the average of the *ap* values for 24 hours.

9.10   Geomagnetic Storms

Occasionally a sharp temporary change occurs in the geomagnetic field; these disturbances are called geomagnetic "storms." The frequency of occurrence of these storms is directly related to the 11-year cycle of solar activity. Whenever a large solar flare occurs, a geomagnetic storm starts a day or two later.

Figure 9.7 shows the variation of the horizontal component *H* of the geomagnetic field as seen at low and moderate latitudes during a typical storm. The "sudden commencement," S.C., is a sudden increase in *H*. It is presumably caused by a compression of the earth's magnetosphere by the solar wind and represents the first "gust" of the wind on the magnetosphere. The "initial phase," the period during which *H* remains above normal, usually lasts for several hours. It is thought to be the result of the continued compression of the magnetosphere by the plasma cloud ejected by the sun.

After the plasma cloud has passed the earth on its outward journey, we think the main phase begins. During the "main phase" *H* drops below the undisturbed, prestorm value for as long as a day or so; a negative "bay" forms. It is thought that the reduction in the surface field is due to the field of

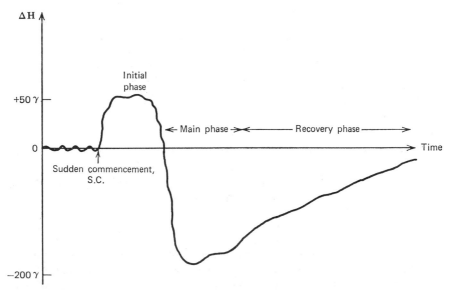

Figure 9.7   A typical midlatitude geomagnetic storm. One gamma $\equiv 10^{-5}$ gauss.

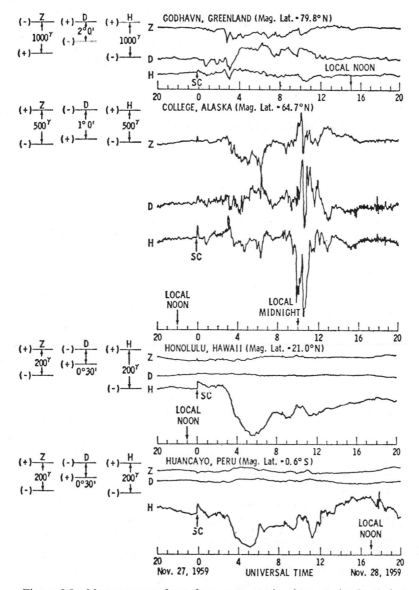

Figure 9.8 Magnetograms from four geomagnetic observatories located at various latitudes. The field is considerably more disturbed at College, in the auroral zone, than it is at Honolulu, where a classical geomagnetic storm is observed to be taking place. M. Sugiura and J. P. Heppner, *Introduction to Space Science*, W. N. Hess and G. D. Mead, Eds., Gordon and Breach, Inc., 1968.

a "ring current" at altitudes of several earth radii. The geomagnetic field is increased on the outward side of this ring current.

The ring current has only recently been detected directly. It has been located with satellites at geocentric distances of $3-5R_E$ and found to be carried mainly by kilovolt protons, although some electrons are also present.

The almost regular behavior described above for low latitudes is at complete variance with that observed in the auroral zones during a geomagnetic storm. Figure 9.8 shows magnetograms from four magnetic observatories during a storm of moderate intensity.

There are many irregular fluctuations and spikes in the College, Alaska, record and the amplitudes may be much greater than found at lesser latitudes. College is in the northern auroral zone.

One of the most notable general features of polar storms is the occurrence of *positive* bays, as well as the low-latitudelike negative departures from the prestorm values of the horizontal component. The longitudinal relationship of the negative and positive bays in polar storms suggests to some that they are caused by oppositely directed east-west ionospheric currents. The discussion of such currents is beyond the scope of this book.

9.11   The Magnetosphere and the Solar Wind

When seen from space, the magnetic field of the earth does *not* resemble that of the classical bar magnet that we have been discussing. Instead of reaching outward toward infinity, the field is sharply compressed into a finite volume on the daylit side of the planet by the solar wind.

The solar wind is a hot (about $10^5°K$) supersonic plasma streaming radially outward from the solar corona. It is always present, and the earth may therefore be considered to be immersed in the hot outer reaches of the solar atmosphere. The detailed chemical composition of the wind is somewhat uncertain, but it is composed mostly of equal numbers of protons and electrons (see Figure 9.9) and it is electrically neutral over distances greater than the *Debye length d.*

For a plasma of electron number density n at a temperature T, d is given by $(kT/4\pi ne^2)^{1/2}$; if e is in electrostatic units, d is expressed in centimeters. The Debye length is a measure of the distance over which the number density of the electrons can deviate appreciably from n_iZ (where n_i is the number density of the ions and Z their atomic number).

An aluminum foil was exposed by the Apollo 11 astronauts to the solar wind on the lunar surface on 21 July 1969. The exposure lasted for 77 minutes, and the foil was then returned to earth for analysis of the trapped solar wind particles. It could be determined (since the attitude of the foil was known) that the particles trapped on the foil were consistent with a directional flow coming from the unperturbed solar wind direction and with an

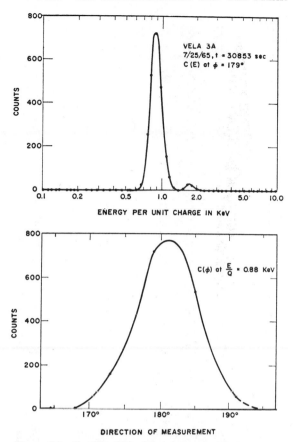

Figure 9.9 Satellite Vela 3A measurements of the solar wind conducted in July 1965. The energy spectrum (counts per unit charge as a function of energy) and direction are shown (180° corresponds to particles moving radially from the sun, viewed from the spacecraft). The small bump in the energy-spectrum curve may be due to He^{2+} ions. A. J. Hundhausen, J. R. Asbridge, S. J. Bame, and I. B. Strong, *Journal of Geophysical Research*, 1967.)

angular spread ($\sim \pm 5°$) corresponding to a temperature of $5 \times 10^5 °K$. A solar wind He^4 flux of $(6.3 \pm 1.2) \times 10^6$ cm^{-2} second^{-1} was determined; about 5% of the wind particles were He^4. Foil sections essentially shielded from the direct solar wind, but facing the lunar surface, showed a small He^4 content that could originate from a solar wind albedo of the lunar surface.

The existence of a continuous stream of corpuscular radiation from the sun was inferred by Biermann from observations of comet tails about 1950;

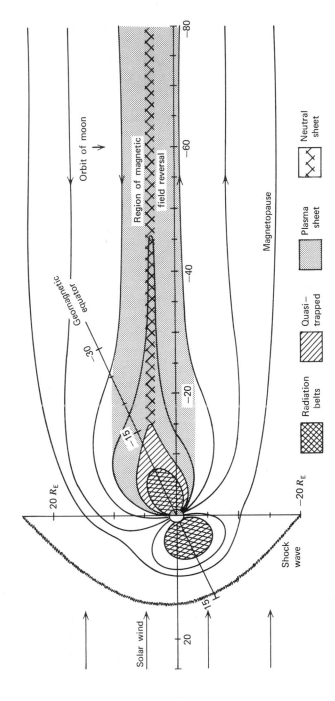

Figure 9.10 The magnetosphere of the earth, meridian plane view. The time is summer in the northern hemisphere.

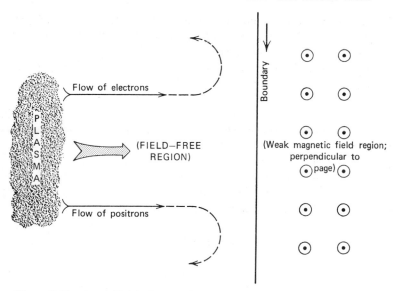

Figure 9.11   An artificial plasma of positrons and electrons pushes a weak magnetic field ahead of it. A surface current flows along the boundary between the two regions.

Parker later predicted the important characteristics of the wind theoretically, and direct measurements conducted with spacecraft have confirmed the phenomenon. A more detailed discussion of the wind will be deferred until the next chapter on the interplanetary medium.

When the hot plasma pushes against the obstacle created by the geomagnetic field, it deforms the field into the "magnetosphere" shown in Fig. 9.10. That a plasma will push aside a weak magnetic field may be seen by considering the following somewhat artificial situation. A plasma that consists of positrons and electrons streams with a velocity V into a region occupied by a uniform magnetic field of strength B (see Figure 9.11). (We use particles of equal mass to simplify the discussion.) The electrons will be deflected in one direction; the positrons execute circles in the opposite sense. (The radius of each circle is just the cyclotron radius mV/eB.) Thus a "current" flows that is perpendicular to the field. It is so directed that the magnetic field associated with the current opposes the original field on the plasma side of the current and adds to the field in front of the plasma. Thus the field is weakened in the plasma, and the plasma strengthens or compresses the field ahead of it. In other words, plasma has pushed the field ahead of it. The compressive process will continue until the magnetic energy density ahead of the plasma equals the kinetic energy density of the streaming plasma; the plasma stops streaming at this point.

We may estimate the "stand-off distance," the distance from the center of the earth, at which the boundary of the ordered geomagnetic field lines is found. The boundary surface is called the "magnetopause."

We may perform the calculation by equating the gas pressure of the wind to the magnetic pressure of the field, so that

$$\frac{B^2}{2\mu_0} = \frac{nmV^2}{2}.$$ (9.25)

In (9.25) V is the speed of streaming of the solar wind, n is the number density of ions (or electrons) in the wind, and m is the mass density of ions (or electrons) in the wind, and m is the mass of a representative ion, which we may take as the rest mass of the proton. Let us also assume that the geomagnetic field is dipolar, so that

$$B = B_0 \left(\frac{R_E}{r}\right)^3,$$ (9.26)

where $B_0 \approx 0.3$ gauss and $r \gg R_E$. Substitution of (9.26) into (9.25) yields

$$r = R_E \left(\frac{B_0^2}{\mu_0 nmV^2}\right)^{1/6}$$ (9.27)

for the stand-off distance, or that distance from the planetary center at which the two pressures balance each other, on the earth-sun line.

Measurements of the magnetopause location, as well as the shock wave that is sunward of it, have been made with spacecraft. Figure 9.12 shows some

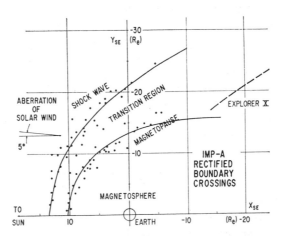

Figure 9.12 Measurements made with a rubidium vapor magnetometer, using the IMP-1 spacecraft. Some earlier Explorer 10 results are also shown. N. F. Ness, *Journal of Geophysical Research*, 1965.

of these as conducted with the Imp-1 (also called Explorer 18) satellite during 1963–1965. Experiments conducted on the moon, such as those left there by the Apollo 12 astronauts in 1969, may also shed considerable light on interplanetary phenomena. This is because the moon, at $60R_E$ from the earth, passes through the tail, the magnetopause, and bow shock, as well as interplanetary space.

The interaction of the solar wind with the atmospheres of the other planets will presumably be like that above, if they are shielded from direct bombardment by intrinsic magnetic moments. If they are not so shielded, the wind may interact with the outer portions of the atmospheres. While it is true that the ratio of the solar constant to the mean energy flux in the solar wind is 10^8 (see below), the efficiency for interaction with the rarefied gases found at great altitudes may be much higher for the particles than it is for electromagnetic radiation; the wind may be important for our understanding of the composition and scale heights of the ionospheres of planets such as Mars and Venus.

Direct measurements with interplanetary spacecraft have shown that typical numbers near the earth are $n = 5$ cm^{-3} and $V = 5 \times 10^5$ meters second^{-1}. Substitution into (9.27) reveals that the wind can penetrate to

$$r \approx 10R_E$$

from the center of the earth on the daylit side. At lesser distances than this and above the E-layer of the ionosphere, we have the "magnetosphere." The field is ordered within it. The magnetosphere contains energetic trapped particles as well as ionospheric plasma (see Chapters 3 and 7).

The "magnetosheath" has been named because of its similarity with the sheath of ionization surrounding a high-speed body immersed in an atmosphere. The magnetosheath is some $4R_E$ thick in the nose region; it contains the compressed, disorderly interplanetary magnetic field. Fields in the magnetosheath average some 20–25 gammas.

The magnetosheath is terminated on its outer surface by the "bow shock." This shock was postulated theoretically, by analogy with hydrodynamics, since the solar wind is supersonic. (By supersonic, we mean that the wind velocity exceeds the velocity of propagation of any applicable wave.) The magneto-acoustic wave velocity (the highest) is about 70 km per second near the earth, and the wind speed is about 400 km per second.

The shock thickness is approximately 10 km in extent, or the mean electron-proton cyclotron radius. It is also possible that there are really two shocks, one in front of the other, that merge into one near the earth-sun line. It appears that at least one shock has been detected experimentally, for "spikes" have been found in the energetic electron flux at about $r = 14R_E$.

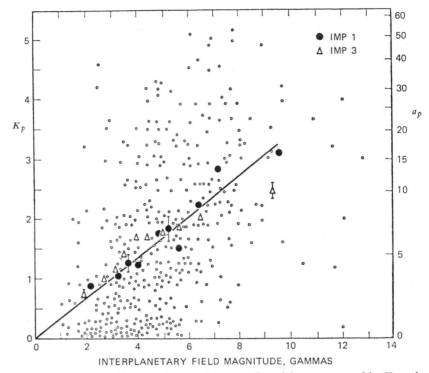

INTERPLANETARY FIELD MAGNITUDE, GAMMAS

Figure 9.13   Scatter diagram of terrestrial magnetic activity as measured by *Kp* and *ap* versus interplanetary magnetic field magnitude. Data are taken from IMP-1 and IMP-3, with the latter satellite's coverage extending throughout 1965. Special averaging techniques produced the large dots and open triangles and revealed an apparently linear relationship between *Kp* and field magnitude. J. M. Wilcox, K. H. Schatten, and N. F. Ness, *Journal of Geophysical Research*, 1967.

It has been found, from measurements conducted by Snyder and Neugebauer on the Mariner 2 spacecraft and since confirmed by others, that the speed of the solar wind is correlated with *Kp*. The curve that best fits the data is represented by

$$V = 8.44\Sigma Kp + 330, \tag{9.28}$$

where *V* is in kilometers per second and ΣKp denotes the sum over the day of the eight *Kp* values for that day. Thus the speed is typically some 400 km per second.

The magnitude of the interplanetary field is also correlated with *Kp*. The empirical relation is $Kp = 0.3B \pm 0.2$; *B* is the magnitude of the interplanetary field in gammas (see Figure 9.13).

The "tail" of the magnetosphere is the subject of much present research. The high-latitude field lines are dragged along by the solar wind, thus forming the geomagnetic tail, in the antisolar direction. The minimum latitude at which field lines are stretched out by the wind is called the "neutral point."

It is clear from spacecraft experiments that the tail extends out well past the moon ($60R_E$); some data indicate its detectability at $1000R_E$. The spacecraft Pioneer 7 reported "fluctuations" in the interplanetary field at $1000R_E$, but Mariner 4 data show no effects at $3300R_E$. It may be that the tail tends to break up at 1000–$2000R_E$.

If the wind blows as far as Pluto's orbit, then the solar wind meets the interstellar medium at around 10^2 A.U. At this distance, therefore, we should expect a shock to be set up that essentially destroys the ordered streaming of the solar wind. The region where the ordered streaming exists may be termed the *heliosphere*, in analogy with the earth's magnetosphere. The heliosphere may also be "comet-shaped" (like the magnetosphere) because of the motion of the sun with respect to the interstellar medium (see Section 10.28).

Experiments have not yet been fully able to delineate the configuration of the heliosphere, however. In any event, it appears reasonable to expect that a similar "magnetic cavity" should exist around any star with a stellar wind.

9.12 Rapid Fluctuations

There is set of fluctuations of the geomagnetic field called micropulsations; they are more or less continuous variations with frequencies ranging from about 0.001 Hz to perhaps 1 Hz. A suitable antenna may be used to detect them on the earth's surface, and they may then be recorded, say, on magnetic tape. If the tape is then played back at a higher speed, so that the recorded frequencies are translated up into the audio range, a really astounding collection of sounds is heard: groans, chirps, choruses, and so on. In other words, coherent wave trains are present. One such case is portrayed in Figure 9.14, where the frequency spectrum as a function of time is shown.

Those micropulsations near the lower end of the frequency scale may have

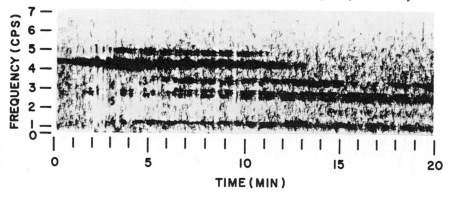

Figure 9.14 Example of the geomagnetic variations that occur at subaudio frequencies. Here a slowly descending frequency is apparent for 20 minutes.

amplitudes of perhaps 10^2 gammas, while the higher-frequency oscillations may only be 10^{-2} gamma in amplitude.

It is believed that micropulsations are the result of hydromagnetic waves, which are electromagnetic waves that propagate through a magnetized plasma with frequencies small compared to the cyclotron frequency ($eB/2\pi m$) of an ion. The magnetic and electric vectors of these waves are perpendicular to the field lines of the geomagnetic field.

The velocity of propagation of the waves along the field lines is quite slow. These waves, variously referred to as transverse HM waves or as Alfvén waves, propagate with a velocity V_A, where

$$V_A = \left(\frac{B}{4\pi\rho}\right)^{1/2} \text{(centimeters per second)} \qquad (9.29)$$

and where ρ is the plasma density in grams per cubic centimeter. The Alfvén speed V_A is typically less than 1 % of the speed of light in the magnetosphere, but depends upon altitude, because of the variation of **B** and ρ with height.

Alfvén waves may be visualized by thinking of the field lines as plucked strings. The disturbance may propagate along the field line to the planetary surface.

There is another hydromagnetic wave mode that travels faster than the Alfvén wave, and it may also propagate perpendicular to the field lines. Sometimes this latter type is called a "compression wave." It is useful to think of a local compression of the field, which in turn generates a variation in the magnetic pressure $H^2/8\pi$ that can propagate across the field lines like a sound pulse. The fact that compression waves are not guided by the field may explain why their amplitudes are smaller at the surface of the earth. The speed of propagation of the compressional waves is U, where

$$U^2 = V_A{}^2 + V_S{}^2, \qquad (9.30)$$

and V_S is the local speed of sound. It is these waves that are sometimes called "magnetoacoustic waves," the same type that we mentioned while discussing the supersonic solar wind.

It is believed that these waves are caused by plasma instabilities in the magnetosphere. The plasma interacts with geomagnetically trapped radiation to produce hydromagnetic waves. We may therefore say that the hydromagnetic waves recorded at sea level come from outer space with a velocity along the field lines whose magnitude is controlled by the ionization density along the path.

9.13 Daily Variations

The geomagnetic field at sea level is not completely constant, even when there is no solar activity and hydromagnetic waves are neglected. The quiet-time records of magnetic observatories have revealed that there are two types of daily variation; one is associated with the sun (*and called Sq, for solar quiet daily magnetic variation*) (see Figure 9.15) while the other has a period equal to half the lunar day and is called *L*, the *lunar* magnetic variation. The amplitude of *L* is much smaller than that of *Sq*, which is typically 20 gammas at midlatitudes.

Balfour Stewart postulated in 1882 that the observed *Sq*—a decrease in the field near local noon in one hemisphere with a corresponding increase in the conjugate hemisphere—must be due to an electric current flowing in the upper atmosphere. The general idea in this "atmospheric dynamo" theory is that the sun causes the ionosphere to rise, and that the charged particles therefore rise with a velocity **v** through the geomagnetic field **B**. This will result in a force on the ions of magnitude **v** × **B** in a horizontal direction. This horizontal force then produces a horizontal current component as the ions respond to the force. The horizontal current is directed in such a sense that it opposes (decreases) the main field at sea level. It will, of course, add to the main field at altitudes in excess of those at which the current flows. If we consider this valid for the northern hemisphere, say, then another overhead current system is required to explain the corresponding sea level increase observed in the southern hemisphere. Thus one arrives at a picture of two huge current systems (fixed with respect to the sun) in the ionosphere. The current systems that have been deduced are shown in Figure 9.16.

These systems have been directly detected during rocket-borne measurements. The peak current flow is at an altitude of about 120 kilometers. A similar, but different, system of currents can be mapped to explain *L*, but the details are less certain because of the small size of the effect.

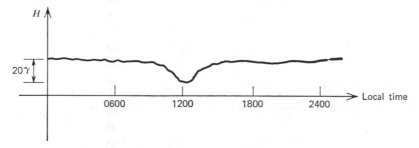

Figure 9.15   A typical midlatitude *Sq* observation.

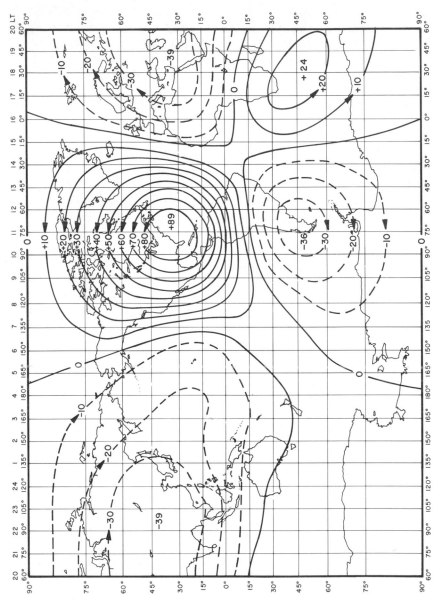

Figure 9.16   The global ionospheric current systems responsible for the *Sq* effect. A current of 10,000 amperes flows between ovals.

Problems

9.1. Find an approximate expression for the geographic coordinates of points on the geomagnetic equator. That is, an expression that relates the geographic latitude of such points to their geographic longitude is desired.

9.2. Find the modification to the geomagnetic field at an altitude of $2R_E$ (above the surface) and at a geomagnetic latitude $\lambda = 45°$, produced at the center of their orbits by a proton and an electron that travel at $v = 1 \times 10^9$ centimeters per second normal to the magnetic field.

9.3. Calculate the Alfven speed at the peak of ionization in the F1-layer of the ionosphere.

9.4. Estimate the magnitude of the electrojet over Huancayo, Peru, if it is assumed that the current flows at a height of 100 km and if the geomagnetic (surface) field is observed to increase by 4×10^{-3} gauss.

9.5. Show for a dipole field that the inclination of lines of force in the upper atmosphere near the auroral zone changes at a rate of approximately $0.1°$ per 100 km of altitude.

9.6. A once-popular theory of planetary magnetism held that the magnetic moment of a planet is proportional to its rotational angular momentum. Assume that this is correct and that all planets have uniform but different densities. Find the stand-off distance of the solar wind for Neptune. Express the result in planetary parameters such as the Neptunian day and radius.

9.7. A magnetic field is described everywhere on the surface of a sphere of radius a by a potential of the form

$$V = -B_0 \frac{\cos \theta}{r^2} a^3,$$

where r and θ are spherical coordinates in an (r, θ, φ) coordinate system. Find the horizontal and vertical field components at a position r, θ, φ, and the total field magnitude. Describe the field in words, or draw a picture.

9.8. A satellite in an unknown orbit measures magnetic field components of B (radial) $= \sqrt{3/27}B_0$ and B (horizontal) $= 1/54B_0$, where B_0 is the magnetic field at the equatorial surface. Assuming a centered, nontilted dipole, find the latitude and altitude of the satellite.

9.9. A magnetic field is described everywhere on the surface of a planet of radius R_0 by a potential of the form
$$V = B_0 r \cos \theta,$$
where r and θ are spherical coordinates in an (r, θ, ϕ) system. This potential satisfies $\nabla \times B = 0$ everywhere on the surface of the planet.

(a) What are the horizontal and vertical magnetic field components (in terms of r, θ, ϕ), and what is the total field magnitude at any point (r, θ, ϕ)?

(b) Describe the field configuration, or draw a picture.

9.10. A magnetic field is described everywhere on the surface of a sphere of radius R_0 by a potential given by

$$V = -\frac{B_0}{\sqrt{2}} r(\sin \theta \cos \Phi + \cos \theta).$$

This potential satisfies $\nabla \cdot B = 0$ and $\nabla \times B = 0$ everywhere.

(a) Find the north, east, and vertical components of the field at any point (θ, Φ) on the surface of the sphere in terms of B_0, θ, and Φ.

(b) Find the total field magnitude at any point (θ, Φ) on the surface of the sphere. Reduce all expressions in θ and Φ to their simplest possible form.

(c) From parts (a) and (b), can you describe the magnetic field configuration in words? (*Hint:* Examine the field components X, Y, and Z at $\theta = 45°$, $\Phi = 0$.)

Chapter 10

COMETS, METEORS, AND THE INTERPLANETARY MEDIUM

We have remarked before that space is not empty. Interplanetary space contains the weirdly beautiful objects called comets ("long-haired stars" or, more liberally, "the hairy ones") as well as meteors (which provide our only direct samples in bulk of extraterrestrial matter) and also dust, in addition to the solar wind. There is also, of course, electromagnetic radiation; the phenomena called the zodiacal light and the gegenschein are of interplanetary origin.

10.1 Introduction to Comets

The hairy ones were at first thought to be symbols of great disaster. Thus we hear of slogans in Europe in the year 1456 such as "God save us from the Devil, the Turk, and the Comet," since "the Comet" was obviously associated with the capture of Constantinople three years earlier. Halley's comet (the same comet) made its most recent appearance in 1910 (see Figure 10.1), and some enterprising Americans became wealthier by selling "comet pills" to their more gullible fellows; the pills were to ward off the dire effects confidently expected to occur when the earth passed through the comet's tail. This great comet reappears every 76 years; it was first recorded in 467 B.C. That there were no mass poisonings every 76 years seems to have occurred to very few.

Comet research has become more systematic since Halley's comet was last seen. A systematic attack is being made on these objects with spectroscopic, photographic, and photometric methods. Perhaps when Halley's comet reappears (orbit calculations call for this to happen in 1986) we shall be in a position to enter the next phase of research: cometary rendevous. A sampling might be made by launching a spacecraft into the nucleus of a comet.

10.2 Designations

Comets are temporarily designated at their time of discovery by the year of discovery followed by a small letter representing the order of discovery in that year. Thus the first three comets found in 1953, say, were called 1953a,

(a)

Figure 10.1*a*   Head of Halley's comet, photographed with a 60-inch telescope on 8 May 1910. Courtesy Hale Observatories, California Institute of Technology.

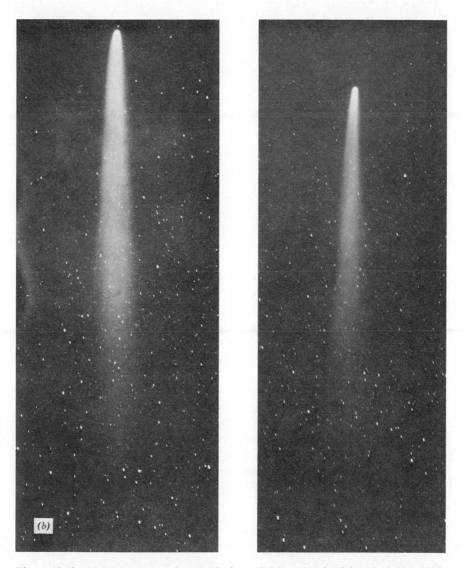

Figure 10.1b Halley's comet, photographed on 12 May 1910 (*right*) and 15 May 1910. The tail increased in angular extent from 30 to 40° in three days. Courtesy Hale Observatories, California Institute of Technology.

1953b, and 1953c. After two or three years this designation is discarded and the permanent one adopted. The permanent designation consists of the year in which the comet reaches perihelion, and the order of perihelion passage during that year is denoted by a Roman numeral that follows the year. Thus the three comets discussed above are now officially listed as 1953 III, 1953 V, and 1954 VII. A system similar to the temporary system is being adopted for satellites.

In addition to the official designation described above, a comet is usually given the name of its discoverer or orbit computer. If the comet has a sidereal period of less than 200 years (so that it is called a "short-period comet"), it has a P/prefixed to its name; for example, 1949 III is also known as comet P/Wilson-Harrington.

10.3 Cometary Statistics

About a thousand comets have been observed since the beginning of observational astronomy, and detailed records exist on about 600 of these. About 20% of this number are short-period comets. With the exception of seven, all the short-period comets share the orbital direction of the planets and have orbital inclinations $< 15°$. The orbit of a typical short-period comet has an eccentricity of ~ 0.5 and an aphelion distance of ~ 5 A.U., so that it passes through the asteroid belt and nears Jupiter.

Approximately 80% are long-period comets. The orbits of 290 of these are listed as parabolic, even though this type of orbit probably never occurs in reality. For 117 the orbits are definitely elliptical ($0.963 \leqslant e \leqslant 1.0$). The periods of this latter group range from 250 years to 3×10^7 years.

There are 65 comets of the long-period kind that have orbits characterized as hyperbolic. This hyperbolic description is the result of too few positional measurements and the action of the sun and of Jupiter. At large distances from the sun the comet moves in a slightly elliptical orbit under the attraction of both these bodies. At small distances, however, the sun alone attracts the comet and its orbit becomes hyperbolic.

10.4 Physical Characteristics

A comet far from the sun appears rather like a star surrounded by a luminous gas. As the comet approaches the sun this "head" grows a "tail," which disappears again when the comet moves away from the sun. The head itself consists of a "nucleus" of solid material and a "coma" composed of dust, neutral molecules, and atoms. The diameter of the nucleus is extremely small. It may be perhaps only 1 km and no more than about 50 km. The coma is much larger; the diameter of the coma is in the range 10^5 to 10^6 km. The tail, on the other hand, may grow to huge dimensions. It may extend for up to 10^8 km, an astronomical unit.

10.5   The Nucleus

That the nucleus is a small structure may be ascertained by its starlike appearance in telescopes. Its brightness usually averages between 1% and 6% of the total brightness of the entire head. The spectrum of a nucleus (when it can be obtained) is continuous and is crossed by many absorption lines. This is essentially a reflected solar spectrum. Such a spectrum can be produced by diffuse reflection from a solid nucleus, by scattering from small solid particles (dust), or by a combination of both. It appears at this time that the nucleus is solid, but it may have a very small cloud of dust particles surrounding it.

The mass of a nucleus is difficult to estimate accurately. One method involves the consideration of the perturbations produced by close passage of a massive body. Direct measurements of the nuclear diameter with a telescope and a photometer are difficult; most of the data are from the gravitational-perturbation method.

To illustrate the perturbation method, let us write the tidal attraction produced by the sun at a radius R from the center of the nucleus. This is

$$g = \frac{2GM_\odot}{r^3} R, \tag{10.1}$$

(see Section 8.6), where $M_\odot$ is the solar mass and r is the distance of the nucleus from the sun. For a nucleus of density ρ and radius R_c, the tidal disrupting force or tensile force F_r is given by

$$F_r = \int_0^{R_c} \rho g \, dR, \tag{10.2}$$

so that

$$F_r = \frac{GM_\odot \rho R_c^2}{r^3} . \tag{10.3}$$

Neglecting the gravitational coherence of the nucleus, we may therefore express the maximum radius of the nucleus in terms of r, the nuclear density ρ, and the maximum tensile strength of the material F_r. Let us assume that $\rho \approx 1.3$ gm cm^{-3} and that $r = 0.005$ A.U.; the maximum radius is then

$$R_{max} = 1.56 \times 10^{-2} F_r^{1/4},$$

where R_{max} is in kilometers. Now an icicle in the shape of a right circular cone of length 3 meters, such as a stalactite, requires a tensile strength of $\sim 10^5$ dynes per cm^2 in order to exist in the earth's gravitational field. If we adopt this value for F_r, we find that $R_{max} \sim 5$ km for a sungrazing comet. On the other hand, the tensile strength of dirty ice or an icy soil is $\sim 10^7$ dynes per cm^2, which yields $R_{max} \sim 50$ km.

The diameters of nuclei can sometimes be measured photometrically. They appear to fall in the range 1–50 km. Using the density given above, we may then estimate the mass; the mass of a comet turns out to be in the range 10^{17} to 10^{19} gm.

The model of the nucleus most generally accepted today is called the "icy conglomerate" model and was proposed by Whipple in 1950. In this model the nucleus is a porous structure that consists of chunks of the ices of methane, ammonia, and water. Particles of meteoritic matter—iron, nickel, and calcium—are embedded in them; the amount of solid material may exceed the amount of icy material. The average density of the combination is about 1.3 gm cm^{-3}. It may also be that some radioactive substances are present.

As the nucleus approaches the sun, it becomes heated and the surface begins to crumble, releasing dust particles which then move outward from the nucleus. In some cases a crack may develop in the surface that results in the sudden release of more gas, or the surface crust may "explode" as the gas pressure builds up inside from the solar heating. Some comets exhibit outbursts or brightenings as they near the sun. The presence of radioactive matter in the nucleus could cause a crust to form, for comets spend most of their lives near aphelion, where they receive little solar heat. Under these conditions internal heating by radioactivity would tend to heat the icy conglomerate, and the volatiles such as methane would diffuse outward and recondense on the nuclear surface. A thick, brittle crust thus forms. One comet, Wirtanen, was observed to break up far out, about 4.5 A.U. from the sun. Differential heating of the crust by even weak solar radiation might explain this.

10.6   The Coma

The coma, for the most part, does not shine by reflected sunlight. Rather, atomic and molecular emission spectra are observed. The strongest spectra are from the molecules CN, C_3, and C_2; weak atomic spectra of Na and O have also been detected. The spectrum of CN is detectable at 3 A.U. from the sun; as the comet approaches closer, the others become detectable, with the atomic emissions observed only close-in, at less than 0.1 A.U. It appears now that we are observing fluorescence radiation that is stimulated by solar radiation.

The brightness of a comet depends on the distance r from the sun to the comet and on the distance from the comet to the earth. If Δ is the earth-comet distance, we may write that

$$H = \frac{H_0 \varphi(\alpha)}{\Delta^2 r^n}, \tag{10.4}$$

where H is the brightness and H_0 is the brightness of the comet at $\Delta = r = 1$ A.U. Also, $\varphi(\alpha)$ is the appropriate phase law, which may be neglected here.

Now if the comet were to shine by reflection only, we would have $n = 2$. Observation indicates that this is not the case; values of n range between 2 and 6. Thus we infer that the comet emits light as it nears the sun, as well as reflecting the solar spectrum.

10.7 The Tail

Aside from their spectacular appearance, cometary tails have played a prominent role in the development of space science. There are two types of tails. Type I tails are composed of ions (CO^+, CO_2^+, N_2^+, CH^+, OH^+) and electrons; they sometimes are called the "gas tails." Type II tails are composed of dust particles. The two types are distinguished by the accelerations acting on their component particles; "knots" of higher-density material may be observed to move in the tail, as a function of time.

Accelerations are usually measured in units of solar gravity at the distance of the comet. This quantity is given by $(a/a_\odot)r^2$, where a is the acceleration in centimeters per second squared at the comet and $a_\odot = 0.59$ cm per second² is the acceleration due to solar gravity at 1 A.U. These accelerations are of the order of 10^2 for type I tails; type II tails exhibit accelerations of about unity.

In addition to the distinction based on acceleration, one may distinguish the two types on the basis of their activity. Type I tails exhibit a great deal of variability, while the dust tails show none. Comet Mrkos, for example, showed both types (see Figure 10.2).

Solar radiation pressure is adequate to explain the accelerations observed for the dust tails; it is totally inadequate to explain the observed values for the tails composed of ions. The very fact that stars can be seen through the tails means that photons do not appreciably interact with them. The tail would have to be opaque to even begin to account for the observations through electromagnetic pressure. Biermann therefore suggested that the corpuscular radiation from the sun is responsible. It is possible to estimate the density of the solar wind particles, which we shall assume to be ionized, since the tail particles are ionized. The idea here is that the particles liberated from the head as neutral atoms and molecules become ionized whenever the sun exhibits some activity, and the presumed process is charge exchange.

The lifetime τ of C_2 molecules is about 10^6 seconds, and the ionization potential of C_2 is about the same as the ionization potential of hydrogen. Therefore the charge exchange process

$$C_2 + H^+ \rightarrow C_2^+ + H$$

will be quite efficient. The effective cross-section q for this process is about 10^{-15} cm². Now if n is the number density of solar wind protons,

$$nqv \sim \frac{1}{\tau}.$$

Figure 10.2 Photograph of comet Mrkos clearly showing a type I
(plasma) and a type II (dust) tail. Courtesy of Hale Observatories,
California Institute of Technology.

If we assume a high but not unreasonable value of 1000 km per second for the solar wind velocity v, we obtain $n \sim 10$ protons cm^{-3}, which agrees rather well with values obtained from interplanetary probes and from the theory of the solar wind. Also, since type I tails are found at all heliocentric latitudes, we conclude that the solar wind is blowing in all directions outward from the sun.

10.8 Cometary Lifetimes

Each time a comet comes near the sun, it loses mass; the tail forms from material ejected by the head, and the tail material is lost into the interplanetary space along its orbit. Thus a comet has only a finite lifetime. Whipple estimates that as much as 0.5 % of the mass of a comet evaporates during each perihelion passage. Such a comet could not survive for more than 200 orbits, and it is believed that we lose roughly three comets per year.

If the earth should pass through the comet's orbit and this cometary debris is present, a meteor shower will result. When the orbits of meteors are computed (before they enter the earth's atmosphere) it is found that some of these meteor showers are indeed related to comets. An excellent example is the Leonid meteor shower (visible on 16 November). The Leonids are apparently the debris of a temporary comet, comet 1866 I. There are a great number of particles in such a swarm; 10^5 meteors per hour may be seen during the Leonid display.

The more the meteoritic particles are spread out over the entire cometary orbit, and the older the meteoroids in the swarm, the more uniform the distribution of particles in the stream. The planets perturb the orbits of the particles, spreading them out and reducing the frequency of intense meteor showers. Some showers are quite old; the Perseids (visible from 20 July to 20 August and associated with P/Swift-Tuttle) have a stream width of 0.5 A.U.

10.9 The Origins of Comets

Since no comets are known to execute truly hyperbolic orbits with respect to the sun, we conclude that comets must be members of the solar system. An additional piece of information that supports this idea is that no excess number of comets is observed from the constellation Hercules, toward which the solar system is moving. Oort consequently concluded that a more or less uniformly distributed "shell" of long-period comets surrounds the solar system. The short-period objects come from this cloud through perturbations of the orbits.

If there really is a shell of comets surrounding the system, the dimensions of this shell must be huge; comets probe a vast volume of space. The inner "radius" of the shell must be greater than the Pluto-sun distance, for the shell is not seen at planetary distances. The outer dimensions of the shell may be estimated by recalling that a body in an elliptical orbit about the sun spends most of its time near aphelion. Since a comet has a finite lifetime and cannot survive many perihelion passages, we require that the comet's aphelion be sufficiently great for the cometary reservoir not to be significantly depleted during the age of the solar system ($\sim 5 \times 10^9$ years). We therefore arrive at aphelia of about 150,000 A.U., or about two light-years, which is half the distance to the nearest star. Any comets that have aphelion distances significantly larger than this may be captured into orbits about other stars. It is possible that there are some 10^{11} comets in the cloud surrounding the sun. This figure is arrived at by assuming that the comets that come in close to the sun do so because of the perturbative effects of the other stars, the average densities of velocities and masses of which are known. One also uses the known annual numbers of comets passing perihelion. In addition, an average cometary mass of 10^{16} gm is assumed.

10.10   Origin of Meteors

As we have previously remarked, it appears that meteors bear a direct relationship to comets. Comets are occasionally observed to break up. The most celebrated case is that of Biela's comet. This was observed to split in two in 1846 as it neared Jupiter; the two portions, still traveling together, were seen again in 1852. The fragments had disappeared by 1872, only to be replaced by the Bielid meteor shower. Thus meteoritic swarms are believed to be small particles that once formed comets and are now traveling along the (former) cometary orbit. Sporadic meteors, on the other hand, are thought to come from the collisional breakup of asteroids or the lunar surface, and it is these nonswarm meteors that we now turn to.

10.11   Meteoritics

An extraterrestrial object is called a meteorite when a detectable mass is recoverable on the earth's surface. It is not an easy matter to determine whether an object on the surface of the planet is of extraterrestrial origin. The object may well be a normal terrestrial rock; only a careful examination of the chemical and physical composition of the specimen will decide, unless (by great good fortune) the fall to earth has been observed. About 45% of the known meteorites are observed falls; the remaining cases are "finds."

10.12   Meteor Masses

The recovered mass is always smaller than the "pre-atmospheric" mass, for a number of reasons. First, meteorites are liable to break up during entry into our atmosphere. The recovery of the resulting fragments may be incomplete. Second, ablation takes place during entry, with concomitant loss of mass. Of the ~ 1500 meteorites now recognized as such, the recovered masses range from ∼0.1 gm to several tens of metric tons; a typical meteor is smaller than a pinhead. By convention, particles with diameters of less than 100 μ are referred to as *dust* particles; they are meteors if their size exceeds 100 microns.

Massive, crater-forming meteorites are largely destroyed by explosion impact with the ground. An example is the Canyon Diablo meteorite, which produced Meteor Crater near Winslow, Arizona. The crater—truly impressive in its proportions—is nearly a mile across and about 500 feet deep. The largest piece of the Canyon Diablo so far recovered weighs only 140 pounds, but it is estimated from the crater dimensions that the mass of the meteorite was 10^4–10^6 tons, depending on its speed. For comparison purposes, and to estimate the amount of energy required to form such a crater, about 20 megatons of TNT would cause a similar crater if the explosive were buried 400 feet below the surface. Another way of estimating the mass is to note that at 15 km per second the kinetic energy of a 1-ton meteorite is equal to the explosive energy of 30 tons of TNT; giants such as Canyon Diablo almost certainly explode on impact.

Although the surface of a meteorite is heated to above the melting point of iron (>1700°C) during entry, the heat-altered zone near the surface of the recovered object is usually remarkably shallow; it does not exceed a few millimeters. No loss of the volatile inert gases arising from heating during entry has ever been detected in the unaltered portions of a meteorite.

When a meteoroid enters the atmosphere with some velocity v, a momentum exchange occurs on the front side of the incoming object between this body, which decelerates, and the air molecules, which are given a forward velocity. If ρ is the atmospheric density, C_D is the drag coefficient (= 1.00 for a sphere), and s is the effective cross-sectional area of the meteoroid, we have for the momentum exchange that

$$m \frac{dv}{dt} = -C_D s \rho v^2. \tag{10.5}$$

Usually the cross-section s is approximated by $\rho_m^{-2/3} m^{2/3}$, where ρ_m is the density of the meteoroid. Thus we may write the *drag equation*,

$$\frac{dv}{dt} = -C_D \rho_m^{-2/3} m^{-1/3} \rho v^2. \tag{10.6}$$

The meteoroid loses mass as well as velocity as it penetrates the atmosphere. This mass loss, called *ablation*, can occur through vaporization, melting, or fragmentation. From conservation of energy, we equate the energy given to the meteor by the colliding molecules to the energy required for ablation and find the mass loss of the meteor:

$$\xi \frac{dm}{dt} = -\tfrac{1}{2}\Lambda \rho_m^{-2/3} m^{2/3} \rho v^3, \tag{10.7}$$

where ξ is the energy required to ablate 1 gm of the meteor and Λ is the efficiency of energy exchange. Dividing (10.7) by (10.6), we have the *mass equation*,

$$\frac{1}{m}\frac{dm}{dt} = \frac{\Lambda}{2C_D\xi}v\frac{dv}{dt}.$$ (10.8)

The ablated meteoroid atoms may ionize and excite the air particles so that light is emitted. If τ is the percentage of their kinetic energy converted into light, we have the *luminosity equation*

$$I = -\tfrac{1}{2}\tau v^2\frac{dm}{dt},$$ (10.9)

where I is the total luminous energy (ergs per second) emitted by the meteor; meteoric brilliance is a function of the particle mass. The surface-area to volume ratio of a particle is inversely proportional to the size of the particle; accreted interplanetary dust may radiate heat so effectively that it may slow down and settle into the stratosphere with little or no mass loss.

10.13   Meteor Influx

Various authors have estimated the fall frequency and mass per year of meteorites colliding with the earth. The estimates are greatly encumbered by sociological factors such as population density and working habits. The presently accepted values are one meteorite fall per square kilometer per 10^6 years and a mass influx of about 1000 tons per day. It is not known what the fall frequency was in the past; if it has been constant throughout the age of the earth, our planet has accreted some 3.6×10^{13} tons during the past 4.6×10^9 years. The amount accreted from interplanetary dust is at most a few times this figure.

10.14   Meteor Trajectories

Unfortunately only two meteorites have been recovered that were photographed during their fall to earth. Thus we can only infer, more or less directly, meteoric origins in only these two cases. (Radar studies of the luminous trails of meteorites are also adding to our knowledge of these objects, and there are now about 10 such cases accepted as reliable.)

On the night of 3–4 January 1970, many residents of a large area in the midwestern United States reported having seen a big fireball. The luminous track was photographed (see Figure 10.3) by two of the 16 cameras that constitute the Prairie Network, a system of automatic camera stations set up in 1964 and, as Figure 10.4 shows, covering seven states. The 10-kg stony meteorite was recovered that same night (only about 1 km from the impact point predicted from the photographs), near the hamlet of Lost City, Oklahoma (see Figure 10.5). The chemical and radioactivity analyses of the Lost City meteorite have not yet been completed.

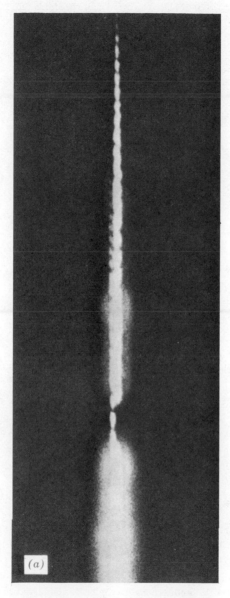

(a)

Figure 10.3a The Lost City meteorite
falling through the atmosphere. Courtesy
Smithsonian Astrophysical Observatory.

Figure 10.3*b* The Lost City meteorite plunging to earth, 3 January 1970. Photograph from the Prairie Network, Smithsonian Astrophysical Observatory. Star trails are visible in the background.

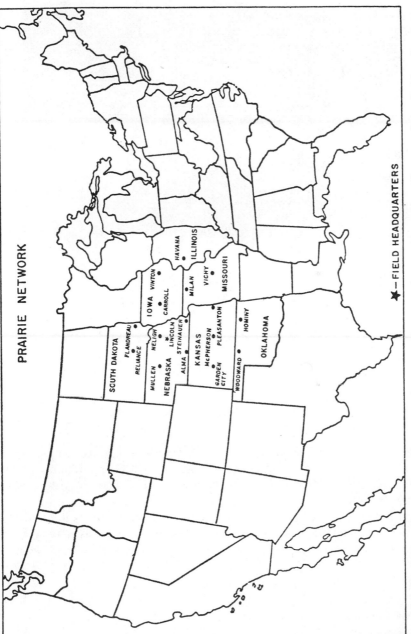

Figure 10.4 Map showing the locations of Prairie Network automatic camera stations. Courtesy of the Smithsonian Astrophysical Observatory.

Figure 10.5 Impact site of Lost City, near the town of the same name in Oklahoma. The hole in the snow is less than 1 km from the predicted impact point. Courtesy Smithsonian Astrophysical Observatory.

The other photographed and recovered meteorite was also stony. It fell near Pribram, Czechoslovakia, on 7 April 1959. Computations of the orbital parameters of Pribram have been made. The data from the photographs form the bases for the calculations, together with relationships such as the luminosity and mass equations and the laws of planetary motion. The orbital parameters that result from the calculations necessarily apply only to the final partial revolution of the meteor about the sun. Pribram's parameters are:

Semimajor axis = 2.42 A.U.
Perihelion distance = 0.790 A.U.
Aphelion distance = 4.06 A.U.
Eccentricity = 0.590.
Sine (inclination) = 0.181.
Entry velocity = 16.9 km per second.

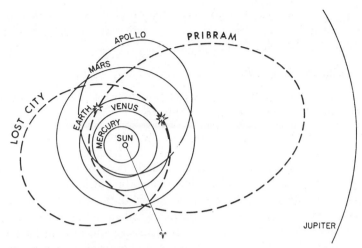

Figure 10.6   Calculated trajectories of the two meteors photographed in their fall to earth at Pribam, Czechoslovakia, and Lost City, Oklahoma.

Figure 10.6 illustrates the orbits computed for the two meteorites. Both have an aphelion point near the asteroid belt between Mars and Jupiter.

It appears that the speeds at the earth's orbit range from 11 to 73 km per second for those photographed and studied by radar. Now 11 km per second is the escape velocity from the earth and 42 km per second is the velocity required to escape from the solar system at the orbit of the earth. Since the earth's orbital speed is 30 km per second and $42 + 30 = 72$, we may conclude that all meteors are permanent members of the solar system if their maximum speed is less than 72 km per second.

The gravitational field of the earth, however, accelerates the meteors somewhat. It can be shown that the total energy of a particle of mass m moving in a hyperbolic orbit of semitransverse axis a in the field of a mass M is $+GMm/2a$. A consequence of this is that the hyperbolic equivalent of the vis-viva equation (2.41) is

$$v^2 = GM\left(\frac{2}{r} + \frac{1}{a}\right). \tag{10.10}$$

Now meteors approaching the earth are in hyperbolic orbits with respect to the earth, because they have speeds in excess of 11 km per second, the planetary escape velocity. Hence, with respect again to the *earth*,

$$v_{\max}^2 = (72)^2 + \frac{2GM_E}{R_E} \text{ (kilometers per second)}^2,$$

so that $v_{max} \approx 73$ km per second. Since the observed speeds are within the two limits (11 to 73 km per second), we conclude that meteors are indeed permanently within the solar system.

10.15   Chemical Composition

Chemical composition constitutes the principal basis of classification of meteorites into four major groups, as follows:

1. *Iron* meteorites, which are alloys of nickel and iron, with Ni ranging from 5 to 50%.
2. *Stony irons.* These are either a metal lattice with numerous silicate inclusions, or a silicate lattice with sizable metal inclusions.
3. *Chondrites* (stony meteorites) are so called because of the occurrence of small spherical or near-spherical particles called chondrules in a fine-grained ground mass. The major constituents are silicates, metal particles, and iron sulfide. The average composition of chondrites is remarkably close to the composition of the nonvolatile elements of the sun. The chemical and isotropic analysis of chondrites has played a dominant role in delineating the so-called "cosmic abundances" of the elements.
4. *Achondrites.* These are also stony meteorites, but without chondrules. The chemical composition strongly suggests that these are products of differentiation processes involving perhaps melting or partial melting. In this respect it is significant that achondrites do not contain metallic particles, whereas chondrites do.

Although it is believed that perhaps 90% of the falls are stony meteorites, less than half the confirmed meteorites are stony. This reflects the difficulty associated with distinguishing them from earth rocks. A major criterion is the presence of chondrules, but of course one must first "suspect" a rock to be a meteorite before an examination will be conducted.

There are, of course, many subgroups within each of these four major groups. The irons, for example, may be further subdivided on the basis of nickel content or on the presence or absence of an octahedral crystal structure (known as the Widmanstätten pattern) that rather resembles hieroglyphics. Some investigators believe that a rare type of chondrites, known as the *carbonaceous chondrites*, which contain substantial amounts of carbon and carbonaceous matter, represent the most primitive solid matter of the solar system known to us.

10.16   Cosmic Ray Bombardment

While in space meteorites are exposed to bombardment by the cosmic radiation, which consists primarily (see Chapter 13) of protons. The integral energy spectrum of the radiation depends on the energy E as $E^{-1.6}$; the flux is 0.25 proton per cm²-second-steradian at $\sim$1 BeV.

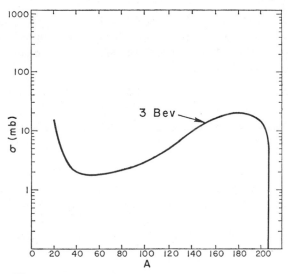

Figure 10.7 Spallation cross section in millibarns (1 mb = 1×10^{-27} cm²) for protons of cosmic ray energies incident on bismuth. Approximately the same shape of curve is observed for other elements.

The primary cosmic rays penetrate into the meteorite from all sides. Beneath the surface the intensity of the primary rays decreases exponentially, with a half-thickness (in iron) of ~25 cm. The interaction of the protons with the nuclei of the meteorite leads to nuclear reactions of a type called *spallation reactions.*

Spallation reactions can be described in terms of two stages. During the first stage a small number of protons and neutrons are ejected in the forward direction, and these secondary particles can themselves cause further interactions, so that a "nucleonic cascade" develops. During the second stage, however, the residual nucleus is de-excited by "evaporation" of a larger number of particles such as p, n, H^3, and He^4. The mass yield curve of proton-induced spallation reactions of nuclei in the mass region of iron ($A = 56$) typically shows a maximum between atomic masses 40 and 50, a minimum around neon ($A = 20$), and a pronounced rise toward very light masses, as though elements like helium were "chipped off" from the parent nucleus. Figure 10.7 illustrates a typical case.

Both radioactive and stable end products may be the result of spallation reactions. To illustrate, the reaction

$$Fe^{56} + p \rightarrow Cl^{36} + 2He^3 + 7p + 8n \qquad (10.11)$$

or

$$Fe^{56} + p \rightarrow Sc^{45} + He^4 + 4p + 4n \qquad (10.12)$$

may occur, where the stable nuclides are italicized.

10.17   Meteoritic Ages

The formation of radioisotopes by cosmic ray bombardment leads to a method for determining the duration of the bombardment, when the following assumptions are made:

1. Initially the meteorite was in a large object, well shielded against cosmic radiation. This requires an assumed depth of at least a few meters beneath the solid surface of the parent object.

2. At τ years ago the meteorite was broken off by an impact on the surface and placed in an orbit around the sun.

3. The average cosmic ray intensity remained constant during the exposure time.

4. The size and shape of the meteorite did not change during the exposure time.

5. The exposure time τ was long compared to the half-life of the radioactive nuclide in question.

The fifth assumption assures that the radioactive product chosen (say Cl^{36}, where $T_{1/2} = 300,000$ years) is in secular equilibrium with the radiation at the time of the meteorite's fall to earth. That is, we assume that the time in space has been sufficiently long that the rate of decay is equal to the production rate of Cl^{36} by the radiation.

Typical intensities are very small. The radioactivity ranges from 1 to 20 disintegrations per minute for a 1-kg sample. Such small activities require the use of special counters having backgrounds as low as 0.05 count per minute.

Radioactive chlorine decays into argon; a stable product such as Ar^{36} has integrated the radiation over the entire time τ. Typical values range from 1×10^{-8} to 100×10^{-8} cm^3 (at standard temperature and pressure) for a 1-kg sample. This can be detected by an ultrasensitive mass spectrometer.

Now if $[Ar^{36}]$ is the number of argon-36 atoms per gram and $R\,Cl^{36}$ is the rate of decay Cl^{36} in atoms per minute-gram,

$$\tau = \frac{[Ar^{36}]}{R\,Cl^{36}} \left(\frac{P\,Ar^{36}}{P\,Cl^{36}} + 1 \right)^{-1}, \tag{10.13}$$

where $P\,Cl^{36}$ is the production rate of Cl^{36} by cosmic rays in atoms per gram-minute and $P\,Ar^{36}$ is the production rate of Ar^{36}, in the same units. Our fifth assumption requires that $R\,Cl^{36} = P\,Cl^{36}$.

The ratio $P\,Cl^{36}/P\,Ar^{36}$ must be measured in the nuclear laboratory. High-energy protons bombarding iron targets provide the necessary data that allow us to correct for the fact that some argon is made directly through spallation.

There is some discrepancy between the various ages that have been measured (i.e., the Ar^{36}-Cl^{36} "age" and the K^{41}-K^{40} "age"), but some salient

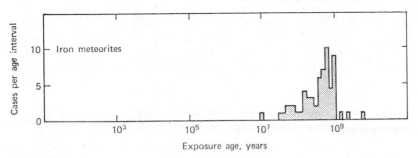

Figure 10.8 Cosmic ray exposure ages of iron meteorites. These data, based on the cosmogenic isotopes of potassium, refer to the amount of time the meteroid was in interplanetary space.

features have emerged:

1. Iron meteorites have ages from a few tens of millions of years to about 2×10^9 years (see Figure 10.8). The break-off of irons from parent objects seems to have been more or less continuous, but there occurred at least two major events, one about 6×10^8 years ago and the other about 9×10^8 years ago. These are explained as major fragmentations of parent objects that had metallic portions such as cores.

2. The ages of stony meteorites are systematically shorter than the irons; the ages range from 0.1 to 60×10^6 years. As shown in Figure 10.9, there

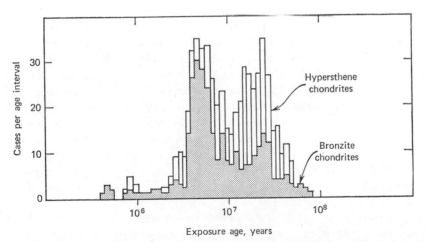

Figure 10.9 Exposure ages of various types of chondrites. These ages are based on the levels of cosmogenic Ne^{21} in the meteorite. A short exposure age (between 0 and 10 million years) implies an orbit that has a high probability of collision with the earth; orbit calculations indicate that the Apollo asteroids are the most likely source of the short-age chondrites.

occur distinct age clusters, especially when subgroups are considered by themselves. For instance, most bronzite-chondrites seem to have departed from their parent object during a single event 4–6×10^6 years ago. It is significant that ages of less than 1 million years rarely occur.

10.18   Origins

The source strength of meteorites must be greater than the integral flux on earth. Debris produced in the solar system is swept by not only the earth but by other planets as well. Indeed, ejection from the solar system (orbit eccentricity > 1) is possible. Also, close encounters with the sun may result in destruction by vaporization.

One would like to be able to work backward from the observed orbital elements (at the time of fall) to determine the origin of the meteors. However, computer calculations show that encounters with other planets and the asteroids can drastically change the orbits; Pribram's apparent asteroidal origin might also result if the meteorite had come from the lunar surface. Most investigators presently feel that the irons originated in the asteroids; collisions between these objects injected the meteorites into solar orbit. It is not clear whether the major source is the asteroids in the asteroid belt, the "Mars asteroids" (asteroids whose orbits cross that of Mars), or the "Apollo asteroids" (the asteroids that cross the orbit of the earth).

It was believed for many years that the stony meteorites may come from the lunar surface; the composition of the chondrites, in this view, was representative of the composition of the surface of the moon. This origin for the chondrites would have explained the observed short ages of the stony meteorites, for any debris broken off from the moon, say 500 million years ago, has long since been swept up by the terrestrial planets.

Direct *in situ* measurements, however, indicate that the lunar surface composition is not similar to the chondritic composition. The first data were obtained with the unmanned spacecraft Surveyors 5, 6, and 7. Surveyor 5 carried an alpha-scattering experiment to the lunar surface in September 1967. This experiment, based on the principles of Rutherford scattering, was designed to conduct a crude elemental analysis of the lunar surface. Six-million-electron-volt alpha particles from a radioactive source consisting of Cm^{242} irradiated a 10-cm diameter patch of the moon's surface. Some of the alpha particles were scattered by $\sim 180°$, and the energy of the back-scattered radiation was measured by two solid-state detectors. An analysis of the relative abundance of elements present in the lunar surface may be made from a measurement of those energies present in the back-scattered particles. (Hydrogen cannot be detected by such an experiment, but the samples of maria surface returned later by the parties of astronauts show no evidence of any hydration; see Table 8.2.)

The Surveyor 6 and Surveyor 7 results at different lunar sites were about the same as reported by Surveyor 5. The elemental composition was found to be as listed in Table 10.1. This composition is quite similar to that of the terrestrial basalts, the material that on earth appears as volcanic flow to the surface from the mantle.

The Apollo 11 and Apollo 12 flights to the moon, in July and November 1969, respectively, were even more definitive than the Surveyor missions, because the astronauts returned samples of the lunar surface for detailed

TABLE 10.1. SURVEYOR MEASUREMENTS OF LUNAR SURFACE COMPOSITION

Element	Amount
Carbon	$<3\%$
Oxygen	$59.0 \pm 5\%$
Sodium	$<2\%$
Magnesium	$3.0 \pm 3\%$
Aluminum	$6.5 \pm 2\%$
Silicon	$18.5 \pm 3\%$
All heavier elements	$13.0 \pm 3\%$

analysis on earth (see Figure 10.10). These analyses confirmed the basaltic character of the surface material. In addition, the Apollo lunar samples contain glassy spheres throughout them, which strongly suggests that heating took place at some time in the history of the moon.

This fact, together with the basalts, indicates that there was a molten stage in lunar history. The most reasonable and likely source of heating was volcanic action in the lunar past. While some of the heating was no doubt supplied by meteoritic impacts, it appears that a significant amount of meteor heating would have resulted in minerals different from those found in the Apollo 11 samples.

Since radioactive dating measurements show that many of the Apollo 12 samples may be 10^9 years younger than the Apollo 11 material, the volcanic action may have continued for a billion years or more early in lunar history. The moon is certainly composed of substances that are not significantly younger than the earth; Apollo 11 ages range up to four billion years. The oldest age, 4.66×10^9 years, was detected in an Apollo 12 rock. In any event, the lunar surface at the Apollo 11 landing site in the Sea of Tranquility is probably representative of the elemental abundances that existed early in the history of the solar system.

Some information on the history of the sun has been obtained from the Apollo samples. Measurements of the tracks left in lunar surface matter by

Figure 10.10*a*   Apollo 12 astronaut deploying ALSEP experiments on the lunar surface. Notice the spacecraft on the horizon. Courtesy National Aeronautics and Space Administration.

cosmic rays (galactic and solar) indicate that the solar particle output has been relatively constant for the past 10 million years.

10.19   Particle Loss Mechanisms

That a large mass spectrum of solid particles exists within the solar system is demonstrated by meteorites, the zodiacal light (Section 10.20), and the gegenschein (Section 10.21), along with direct spacecraft measurements of interplanetary dust. These particles may be produced by cometary break-up or by asteroid grinding, but which of these two sources predominates is unknown.

Figure 10.10b   Apollo 12 astronaut removing parts from Surveyor 3 spacecraft on the surface of the moon. The arm that scooped the first samples of lunar surface still extends from Surveyor 3. The Apollo 12 spacecraft is visible high on the rim of Surveyor Crater. Courtesy National Aeronautics and Space Administration.

The early direct experiments were plagued by instrumental problems, so that it was at one time thought, on the basis of these experiments, that a "dust cloud" surrounds the earth. Careful laboratory investigations on the dependence of the output of an acoustical sensor on particle momentum (at high speeds) resulted in a recalibration that considerably lowered the near-earth flux value; they now agree with the influxes derived by Whipple in 1963 from studies of visual meteors.

At one astronomical unit in the ecliptic plane, the more recent data, obtained from punctures of the unmanned Pegasus spacecraft and from pitting of the windows of the manned Gemini satellites, have found that the flux of particles with mass greater than m grams and density ρm may be represented by Whipple's expression

$$\log N = -1.34 \log m + 2.68 \log (0.44/\rho m) - 9.5 \text{ [particles/meter}^2\text{-day]};$$

the flux nearly, but not quite, goes inversely as the mass. The total rate is only $\leqslant 3$ m^{-2} day^{-1}, a value about equal to that reported by Mariner 4 out

(c)

Figure 10.10c   Part of the samples of lunar surface material taken from the moon by the Apollo 11 astronauts. Courtesy Smithsonian Astrophysical Observatory.

to 1.5 A.U. This rate is also consistent with particle densities inferred from studies of the brightness of the solar corona and of the zodiacal light, if those particles are assigned a 6% geometrical albedo and a density distribution that depends upon distance r from the sun as $r^{-1.5}$, for distances of less than one A.U.

It is interesting to estimate the fate of these particles. As we have seen, the planets accrete quite a bit; "showers of shot" falling from the skies have been reported on the earth and layers of "spherules" (magnetic spheres) have been found in glaciers and the sea floor. Extraterrestrial particles are also mixed with planetary dust in the stratosphere.

More recently sounding rockets have carried mass spectrometers aloft and layers of meteoritic ions have been detected in the E-layer (around 100 km). The layers are quite thin ($\sim$1 km) and dense ($\sim$10^5 cm^{-3}). They may cause the "sporadic E" phenomenon observed by ionosondes (see Section 3.16).

Another loss mechanism is the *Poynting-Robertson* effect. A particle in solar orbit will experience radiation and solar wind pressures that tend to blow it outward. However, because of the particle's motion, there is a small component of the radiation pressure tangential to the orbit (perpendicular to the radial direction) because of the aberration of light. The particle reradiates this energy, but because of the Doppler effect more energy comes at the particle than it isotropically reradiates. The net effect is that the particle loses angular momentum by this "braking mechanism," and it spirals into the sun. This assumes that the particle is sufficiently small that radiation pressures of the photons and solar wind overcome gravity and blow the particle away. One result of the Poynting-Robertson effect is that elliptical orbits are circularized. It is possible to calculate, for a circular orbit, the time required for a particle originally at a astronomical units to spiral into the sun. The result is $\tau = 7 \times 10^6 \, \rho s a^2$ years, where ρ is the particle density and s is its radius. A particle originating at $a = 1$ A.U. that has $\rho = 1 \, gm \, cm^{-3}$ and $s = 1$ cm will survive for only 7 million years; replenishment must therefore take place.

10.20 The Zodiacal Light

The zodiacal light is a feeble glow (see Fig. 10.11) seen along the ecliptic plane; it is not fixed with respect to the stars. At an elongation of $\sim 40°$ the brightness is only about 4×10^{-13} the brightness of the mean solar disc; it is down to $\sim 1 \times 10^{-13}$ at an elongation of $70°$. Figure 10.12 depicts the dependence on elongation. The light appears to be concentrated—if we may use the word for so weak a source—in a belt $\sim 20°$ wide on either side of the ecliptic; that is, in the zodiac. From satellites the light may be detected at ecliptic latitudes that range up to $90°$ (the ecliptic pole). Figure 10.13 summarizes data obtained during 1965 with photometers aboard the spacecraft OSO-B2.

The phenomenon is now understood as the result of scattering of sunlight by interplanetary dust particles. The spectrum is consistent with sunlight scattered from dust particles averaging about $1 \, \mu$ in diameter. These particles presumably represent the small-size portion of the meteoritic spectrum, which tells us that the number of meteors increases very rapidly with decreasing size. An analysis on the basis of Mie scattering is difficult, but number densities of the order of 10^{-14}–10^{-15} particles cm^{-3} are indicated.

Gustav Mie studied the theoretical behavior of particles with sizes a near the wavelength of light λ in the early part of the twentieth century. The calculations of the scattered and absorbed light as a function of (a/λ) are quite difficult; the scattering, absorption and diffraction depend critically, for example, on whether the scattering center is a dielectric or a metal. It is generally found that for a given kind of particle the variation of scattering with λ follows a curve of the general form shown in Figure 10.14; that is, there is a flat portion where $a \gg \lambda$, a maximum for a wavelength near a, and a

Figure 10.11   The zodiacal light with star trails photographed by Van Biesbroeck in 1928. Courtesy Yerkes Observatory, University of Chicago.

fall (proportional to λ^{-4}) when $\lambda \gg a$. This last is known as Rayleigh scattering; atomic and molecular scattering is of this latter type.

The size of the particles in a slight atmospheric haze is of the same order as the wavelength of visible light. It is consequently found that penetration by size of the particles in a slight atmospheric haze is of the same order as the wavelength of visible light. It is consequently found that penetration by infrared is much better than penetration of the haze by optical wavelengths; in a fairly clear atmosphere (visibility, 5–10 miles) the quality of infrared photographs is superior to those taken with visible light (see Figure 10.15). This is not true, however, in a fog. In a typical fog the particle size is large compared to infrared wavelengths (as well as to optical wavelengths); infrared penetration is little better than visible penetration.

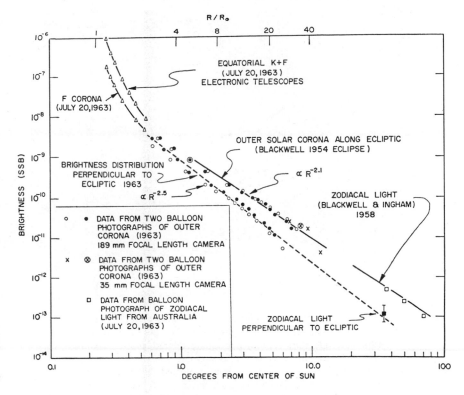

Figure 10.12   Results of brightness distribution measured by Ney and his co-workers during the eclipse of 20 July 1963 with balloon-borne detectors at 110,000 feet. Brightness is given in units of the average surface brightness of the sun and is estimated to be accurate to ±10%. The electronic telescopes measured the inner corona three days earlier at a wavelength of 4750 Å. From F. C. Gillett, W. A. Stein, and E. P. Ney, *The Astrophysical Journal*, University of Chicago Press, 1967.

To illustrate further, small amounts of water in molecular form produce negligible absorption until small aggregates are formed in haze, fog, or clouds, when the same mass of material becomes nearly opaque. The same phenomenon is responsible for the great efficiency for absorption by the fine dust found in dust storms; the absorption coefficient becomes extremely high when $a \sim \lambda$. If Rayleigh scattering of sunlight by small particles within the earth's atmosphere is the reason for the blueness of our sky, then Mie scattering of atmospheric light by the larger particles found in clouds is the reason for the whiteness of the clouds.

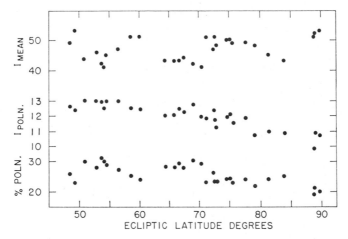

Figure 10.13   Plots of mean intensity of zodiacal light ($(I_\parallel + I_\perp)/2$ and amplitude of the polarized component $(I_\parallel - I_\perp)/2$, in units of tenth-magnitude stars (visual) per square degree, plotted against ecliptic latitude at a constant elongation of 90°. Percentage polarization $(I_\parallel - I_\perp)/(I_\parallel + I_\perp)$ is also plotted. These data were obtained with five photometers, each having a 10° view-field and looking in various directions from the OSO-B2 satellite. From J. G. Sparrow and E. P. Ney, *The Astrophysical Journal*, University of Chicago Press, 1968.

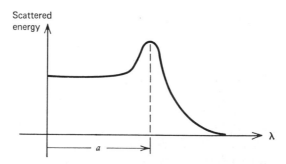

Figure   10.14   Typical   relation   between   energy scattered and wavelength for particles of radius a.

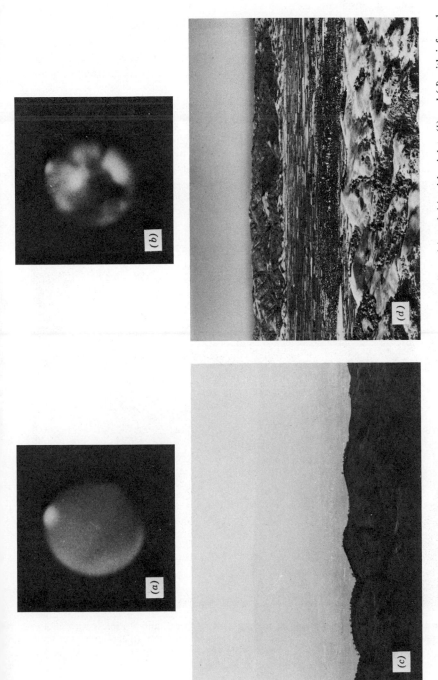

Figure 10.15 Mars (11 September) and San José photographed from Mt. Hamilton (*a*) and (*c*) with violet, (*b*) and (*d*) with infrared light. The obliteration in (*c*) is due to the earth's atmosphere, and the comparison is suggestive of the presence of an atmosphere on Mars. San José is distant 13½ miles. Greenwich mean time of Mars observations: (*a*) 18ʰ50ᵐ, (*b*) 18ʰ30ᵐ. Lick Observatory photographs.

10.21   The Gegenschein

We discussed the Lagrangian points in Section 2.20. These are the points, in a two-body system that is gravitationally interacting and also in rotation, where no net forces are found. Dust particles, once placed at one of these points, will remain there if stability is associated with the particular Lagrangian point.

The *gegenschein* ("counterglow") can only be seen on very clear, dark nights, and then with great difficulty (see Figure 10.16). It is an oval patch of light from a diffuse source exactly in the antisolar direction; it is centered 180° away from the sun and is perhaps 20° wide along the ecliptic. The glow may be explained in terms of Lagrangian points, for it is consistent with

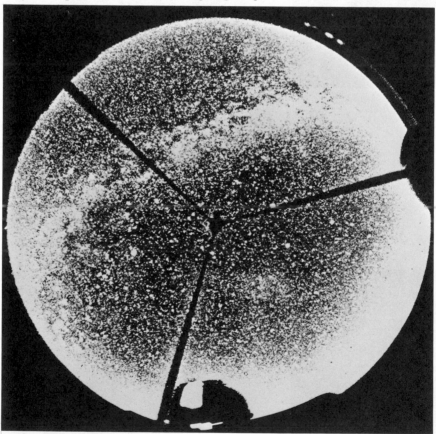

Figure 10.16   The Gegenschein, photographed by Osterbrock and Sharpless with a 140° camera in 1950. Exposure time 15 minutes. The Gegenschein is the small faint oval of light at approximately the 5 o'clock position in the photo.

sunlight scattered from interplanetary dust that has collected at one of the synodic Lagrangian points in the earth-sun system.

It is very difficult to understand the gegenschein quantitatively on this basis, however. A synodic, colinear solution should be unstable; small perturbations will remove the dust from this Lagrangian point. To add to the difficulty, although the light is feeble to the naked eye, sizable dust concentrations are required to produce it by reflection, should the dust be located much further than the 0.001 A.U. ($10^2 R_E$) expected for a Lagrangian point.

Hence many investigators currently believe that the brightness of the gegenschein is due to the "back-scatter" peak found in Mie's scattering theory; the probability for 180° scattering of optical electromagnetic radiation by 0.1–1-μ-diameter dust particles is higher than at other scattering angles outside the forward lobe. Little or no additional concentration of interplanetary dust in the antisolar direction is required by such a mechanism.

10.22 The Solar Wind

It was shown by Parker in 1958 that Biermann's idea of a continuously blowing solar wind is the result of a hydrodynamic expansion of the solar corona. The corona is hot; temperatures of $\sim 2 \times 10^6$°K have been deduced, from the radio spectrum of the quiet-time emissions of the corona. The spectrum is thermal in character (i.e., it is well fit by the Rayleigh-Jeans law). This two-million-degree temperature is much hotter than the 5770°K temperature of the photosphere that lies below it and hotter than the chromosphere that lies between the photosphere and the corona.

Figures 10.17 and 10.18 show the results of various calculations on chromospheric conditions. It appears that the quiet chromosphere has an electron temperature rising from about 4500°K at its base to about 7000°K about 1500 km up. At that altitude the temperature suddenly rises to about 1 × 10^6°K. The small, slender, pointed structures called *spicules* stretch up into the corona; the upper chromosphere is mostly composed of these small vertical spikelike columns or prominences.

It is thought that the primary reason for the high coronal temperature (relative to the chromosphere) is that the chromosphere is much more efficient at radiating energy than is the corona. Thus in this model both regions are heated by acoustic waves coming up from below in the photosphere, but the chromosphere gets rid of the heat more readily than does the corona. Aside from radio radiation, the heated corona produces particles that escape into interplanetary space.

In fact, the corona is so hot that solar gravity cannot contain it; the corona simply explodes continuously. Alternatively the coronal gas pressure nKT is greater than the weight of the overlying atmosphere, leading to a continuing

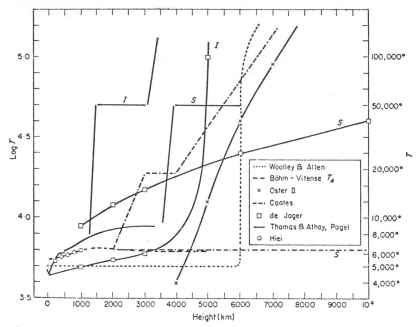

Figure 10.17   Distribution of electron temperature with chromospheric height after various authors. S = spicules, I = region between spicules. From B. E. J. Pagel, *Annual Review of Astronomy and Astrophysics*, 1964.

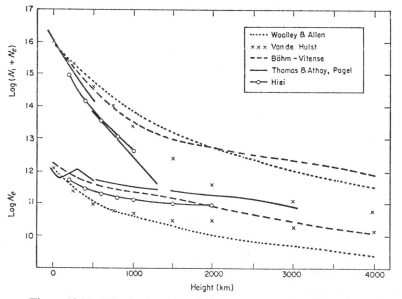

Figure 10.18   Distribution of total density $(N_1 + N_e)$ and electron density N_e with chromospheric height after Woolley and Allen (1950), Van de Hulst (1953), Bohm-Vitense (1955), Thomas and Athay (1961) above 500 km, Pagel (1961) below 500 km, and Hiei (1963); N_e as shown for Thomas and Athay above 1500 km is a weighted mean between spicules and interspicule regions. From B. E. J. Pagel, *Annual Review of Astronomy and Astrophysics*, 1964.

expansion, even though the average thermal velocity of particles in the lower corona is less than the weight of the overlying atmosphere.

The whole corona is blown off and replaced in a period of about a day. The resulting loss of mass, equivalent to 10^{-10} $M_\odot$ per year, has only a negligible effect on solar evolution, but stellar winds may be a more important factor for other stars.

10.23   Supersonic Flow

That the solar wind is supersonic may be seen by examining the gas flow in a rocket engine, the basis of which is called the deLaval nozzle. Consider a compressible gas flowing down a tube of decreasing cross-sectional area (Figure 10.19). The mass flow (a constant for a steady state) is

$$A\rho V = \text{constant,} \tag{10.14}$$

where A is the cross-sectional area of the tube, ρ is the mass density, and V is the flow velocity. The Bernoulli equation of hydrodynamics says that

$$dP = -\rho V\,dV; \tag{10.15}$$

a change of pressure is balanced by a time rate of change of momentum. We may rewrite (10.15) as

$$\frac{dP}{\rho} = \frac{dP}{d\rho}\frac{d\rho}{\rho} = -V\,dV. \tag{10.16}$$

Let us assume that we are dealing with an adiabatic process, although this assumption is not critical. This tells us that

$$\frac{dP}{d\rho} = V_s^2, \tag{10.17}$$

where V_s is the speed of sound. Equation 10.17 is one of the simplest equations of state available. Inserting (10.17) into (10.16), we find that

$$\frac{dP}{\rho} = -\frac{V}{V_s^2}\,dV. \tag{10.18}$$

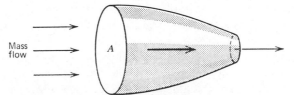

Figure 10.19   A gas flowing through a tube with a constantly decreasing cross-sectional area.

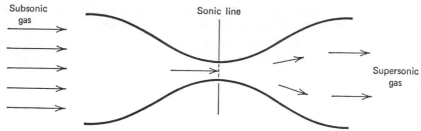

Figure 10.20   A deLaval nozzle. It is also a Venturi tube; which it is depends on the inlet-to-outlet pressure ratio.

Now let us differentiate the logarithm of (10.14); we find that

$$\frac{dA}{A} + \frac{d\rho}{\rho} + \frac{dV}{V} = 0, \tag{10.19}$$

into which may be substituted (10.18). We obtain

$$\frac{dA}{A} = \left(\frac{V^2}{V_s^2} - 1\right)\frac{dV}{V}. \tag{10.20}$$

Thus, as long as the tube is converging and the fluid velocity increasing (dA/A negative and dV/V positive), V must be less than V_s. For $V = V_s$, $dA/A = 0$, or the tube must stop converging. Furthermore, if V is to exceed V_s, (dA/A) must *increase*. Such a configuration (see Figure 10.20) is called a deLaval nozzle. If the flow is so slow that the speed of sound is not reached in the narrowest part of the tube (the throat), the tube acts as a Venturi tube: the gas slows down again as it passes the throat. Whether or not the converging-diverging tube acts as a Venturi tube or deLaval nozzle depends on the pressure ratio of the upstream to downstream sections. If the downstream end opens into a vacuum, the flow will always be supersonic.

10.24   Supersonic Wind

To apply some of these ideas of gas dynamics to the solar corona, let us assume that the coronal gas is ideal and that we can treat it as a hydrodynamic fluid. (This means that collisions are sufficiently frequent that local thermodynamic equilibrium prevails. Also, we ignore viscosity and magnetism; only inertial and gravitational forces will be considered.)

Equation 10.20 shows that to get supersonic flow, a converging-diverging solution is required. For a spherically symmetric expanding corona with a radial (or zero) magnetic field, the streamlines of the gas flow diverge monotonically as r^2, so that supersonic flow would not seem at first to be possible. However, the solar gravitational force acts in such a way as to choke or constrict the flow, as does the converging section of a deLaval nozzle, which in turn permits the development of sonic flow. Once this is achieved, the corona can expand supersonically. The point is that if we can arrange a situation such that the density ρ falls off faster than $1/r^2$, the wind velocity will increase and eventually become supersonic. Gravity does this arranging for us.

To demonstrate this, we rewrite (10.15) to include gravitational forces:

$$dP = -\rho V \, dV - \rho g \, dr = -\rho V \, dV - \rho \frac{GM_\odot}{r^2} \, dr. \tag{10.21}$$

Proceeding as before, we substitute (10.21) into (10.15) and obtain

$$\frac{dA}{A} = \left[\frac{V^2}{V_s^2} - 1 \right] \frac{dV}{V} + \left[\frac{GM_\odot}{V_s^2 r} \right] \frac{dr}{r}. \tag{10.22}$$

For expansion with spherical symmetry, $A \propto r^2$, or

$$\frac{dA}{A} = \frac{2}{r} \, dr. \tag{10.23}$$

When we substitute (10.23) into (10.22) we see that

$$\left[2 - \frac{GM_\odot}{V_s^2 r} \right] = \left[\frac{V^2}{V_s^2} - 1 \right] \frac{dV}{V}. \tag{10.24}$$

When $GM_\odot/(V_s^2 r) > 2$, we have $V < V_s$; when $(GM_\odot/V_s^2 r)$ is less than 2, $V > V_s$. If the coronal expansion velocity increases monotonically, the wind achieves sonic velocity at $r = r_c$, that radius at which $(GM_\odot/V_s^2 r_c) = 2$, or at

$$r_c = \frac{GM_\odot}{2V_s^2}. \tag{10.25}$$

This becomes

$$\xi_c = \frac{r_c}{R_\odot} = \frac{g_0 R_0}{2V_s^2} \tag{10.26}$$

in solar radii. In (10.26) g_0 is the acceleration due to solar gravity at the sun's surface. Since the velocity of escape V_e is

$$V_e = \left(\frac{2GM_\odot}{r} \right)^{1/2} = (2gr)^{1/2}, \tag{10.27}$$

(10.26) may be rewritten as

$$\xi_c = \left(\frac{V_e}{2V_s} \right)^{1/2} \tag{10.28}$$

for the distance in solar radii at which the velocity of the wind becomes sonic. If we assume a coronal temperature of $1.0 \times 10^6 °K$, $V_s = 1.7 \times 10^5$ meters per second, so that $r_c = 3.5 R_\odot$. If $T = 2.0 \times 10^6 °K$, $r_c = 1.7 R_\odot$.

It is interesting that the sun is at just the right temperature for supersonic wind speeds to occur; supersonic flow occurs over a limited range of coronal temperatures.

If it is too cool the atmosphere will be essentially static. If it is too hot the corona will expand at subsonic speeds and a "solar breeze" will result. This latter result follows from considering a static (nonexpanding) atmosphere, the density of which, $\rho(r)$, is given by

$$\rho(r) = \rho_0 e^{-\varphi m/kT}, \tag{10.29}$$

where φ is the gravitational potential, $-GM_\odot/r$, and m is the average mass of the atmospheric constituents. If this atmosphere is heated so that it expands and starts to escape into space, the mass flow through any spherical surface of radius r is

$$\rho(r)V_r 4\pi r^2 = \text{constant.} \tag{10.30}$$

If the temperature of the atmosphere is very high, (10.29) shows that ρ will tend to remain independent of radial distance, so that, from (10.30), V_r will *decrease* monotonically as r increases. However, for a cooler atmosphere, $\rho(r)$ will decrease sharply with increasing radial distance, so that V_r must increase with increasing radial distance in order to keep the mass flow constant. This situation leads to a supersonic wind. The solar breeze (evaporative) solutions may have some applicability either for very hot coronal regions or for stars other than the sun, and may be of importance in understanding mass ejection by some stellar types.

10.25 Interplanetary Magnetic Fields

The outflowing solar wind carries with it those magnetic field lines that pass through the corona. This is because a magnetic field embedded in a highly conducting plasma tends to behave as if it were "frozen in" the material. To show that this is the case, let us start with Maxwell's equations:

$$\nabla \cdot \mathbf{B} = 0, \tag{10.31}$$

$$\nabla \cdot \mathbf{E} = 0, \tag{10.32}$$

$$\nabla \times \mathbf{E} = -\frac{\partial \mathbf{B}}{\partial t} \tag{10.33}$$

$$\nabla \times \mathbf{H} = \mathbf{j} + \frac{\partial \mathbf{D}}{\partial t}. \tag{10.34}$$

Because of the high scalar conductivity σ of a rarefied plasma such as the solar wind, we may neglect the displacement current $\partial \mathbf{D}/\partial t$. That is, there is very little to impede the motion of $(+)$ charges and $(-)$ charges. We can also write that

$$\mathbf{j} = \sigma(\mathbf{E} + \mathbf{V} \times \mathbf{B}), \tag{10.35}$$

where $\mathbf{B} = \mu\mathbf{H}$.

Equation 10.35 is a sort of Ohm's law for a plasma that has a streaming velocity **V**. If we solve (10.35) for **E** and insert this value into (10.33), we find that

$$\frac{\partial \mathbf{B}}{\partial t} = -\frac{1}{\sigma} \operatorname{curl} \mathbf{j} + \operatorname{curl} (\mathbf{V} \times \mathbf{B}). \qquad (10.36)$$

This may be rewritten with the aid of (10.34) and (10.31):

$$\frac{\partial \mathbf{H}}{\partial t} = \operatorname{curl} (\mathbf{V} \times \mathbf{H}) - \eta \, \nabla^2 \mathbf{H}, \qquad (10.37)$$

where

$$\eta \equiv \frac{1}{\mu\sigma}, \qquad (10.38)$$

the "magnetic diffusivity."

It is instructive to consider (10.37). There are two special cases. First, consider a case like the solar interior (or, for another example, the terrestrial core), where $V \approx 0$. In this case (10.37) becomes

$$\frac{\partial \mathbf{H}}{\partial t} = -\eta \, \nabla^2 \mathbf{H}, \qquad (10.39)$$

which is a *diffusion equation*, with a diffusion coefficient η. We may speak of the field lines H *diffusing* through the medium. Alternatively, one may think of the charged particles "diffusing" across the lines of magnetic flux.

This diffusion of field lines leads to a *decay* of the magnetic field; oppositely directed fields at different points "leak" together and neutralize each other. It is of interest to see how long such a neutralization takes to occur.

If we perform a dimensional analysis, we see that η has dimensions (L^2/T), say square meters per second, where L is a length comparable with the dimensions of the plasma and T is the characteristic relaxation time we are seeking. Thus we may set

$$T = \mu\sigma L^2. \qquad (10.40)$$

In the laboratory we might have a copper conductor 1 meter in radius (see Figure 10.21). The conductivity of copper is 5.8×10^{-7} mho per meter. This yields $T \sim 10$ seconds; the field lines slip fairly readily even through this large a conductor when we move it with respect to a magnetic field. Even a large piece of copper is not "magnetized" by moving through a magnetic field.

On the cosmic scale, however, matters are quite different, because L is enormous. If we take $L \approx R_\odot/3$ for the general solar field, and if $\sigma \approx 10^{-7}$, then $T \approx 10^{10}$ years! It takes 10 billion years for a general solar field to decay. This time is perhaps twice the age of the sun.

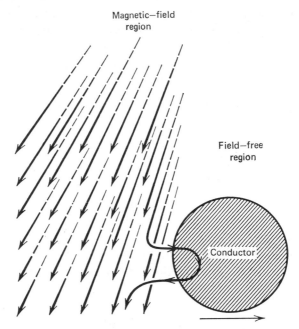

Figure 10.21 Field lines are "pulled along" in the laboratory for a short distance at most, even by large moving conductors.

There is another special case of (10.37), when $V \neq 0$ but σ is very large. The solar wind fulfills these conditions. In this situation

$$\frac{\partial \mathbf{H}}{\partial t} = \mathrm{curl}\,(\mathbf{V} \times \mathbf{H}), \tag{10.41}$$

where we recognize the right-hand side of (10.41) to be the curl of an induced electric field strength $\mathbf{E}$. Thus when a plasma moves in a magnetic field, an electric field is induced. But the conductivity $\approx \infty$; thus we must have $\mathbf{E} = 0$ (or else currents will flow that will cause the E-field to be neutralized). An observer moving with the solar wind sees no electric fields. Consequently, if $\mathbf{E} = 0$, there must be no motion of $\mathbf{H}$ relative to the ionized matter. Hence the magnetic field $\mathbf{H}$ is said to be *frozen in* the plasma.

10.26 Spiral Structure

Gas issuing in a narrow stream with constant velocity from an isolated region of the sun near the solar equator will trace out an Archimedes spiral in space. That is, although each volume element of gas moves radially outward from the sun, the rotation of the sun causes the locus of a line of moving particles to lie along a curved path. This is analogous with a stream of water coming from a moving garden hose nozzle (see Figure 10.22).

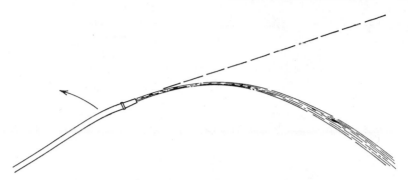

Figure 10.22   The spiral path exhibited by water issuing from a moving hose.

Now for a constant radial solar wind velocity starting at a heliocentric distance b, we may write two parametric equations in time that describe the spiral structure in the equatorial plane:

$$r = Vt + b \tag{10.42}$$

and

$$\varphi = \varphi_0 + \Omega t, \tag{10.43}$$

where V is the solar wind velocity, Ω is the angular velocity of the sun, and φ is the heliocentric longitude measured from a reference longitude φ_0. Eliminating t, we obtain

$$r = \frac{V}{\Omega(\varphi - \varphi_0)} + b, \tag{10.44}$$

which describes the spiral structure shown in Figure 10.23. The constant b is the distance at which V reaches its maximum; it should be somewhat larger than r_c, so that it may be 5–10 $R_\odot$.

We may therefore expect that the interplanetary field lines will be stretched out along the Archimedes spiral (10.44), even though the plasma itself moves radially. Now if B_0 is the field strength at heliocentric distance b, we find the field strength in the solar equatorial plane to have components,

$$B_\parallel = B_0 \left(\frac{b}{r}\right)^2 \tag{10.45}$$

and

$$B_\perp = B_0 \frac{\Omega}{V} \left(\frac{b}{r}\right)^2 (r - b). \tag{10.46}$$

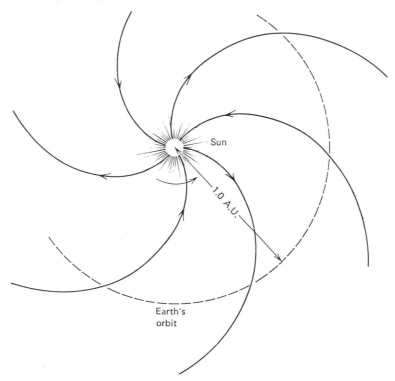

Figure 10.23 Schematic representation of the Archimedes spiral structure of the interplanetary magnetic field in the ecliptic plane.

If r is sufficiently large compared to b,

$$B_r = B_0 \left(\frac{b}{r}\right)^2 \left(1 + \frac{\Omega^2 r^2}{V^2}\right)^{1/2}, \tag{10.47}$$

$$B_{\parallel} = B_0 \left(\frac{b}{r}\right)^2, \tag{10.48}$$

$$B_{\perp} = B_0 \frac{b^2 \Omega}{rV} . \tag{10.49}$$

In these relations B_r is the total magnetic field strength at a distance r from the sun, $B_{\parallel}$ is the component of B_r parallel to the radius vector, and $B_{\perp}$ is the perpendicular component. We see that for small r, B_r varies as $1/r^2$, while it varies as $1/r$ for large values of r. The field does not follow a $1/r^2$ law throughout the solar system. Furthermore, zero flux is to be expected perpendicular to the solar equatorial plane; no magnetic field component perpendicular to the ecliptic should be found.

10.27   Interplanetary Configuration

Experiments conducted on deep space probes have confirmed the general spiral structure of the interplanetary field. Cosmic ray data obtained previously at the earth's surface also supported these ideas; solar charged particles emanating from near the western limb of the sun are detected more frequently than particles originating elsewhere on the solar disc. The particles behave as though they were guided in space by a twisted magnetic field.

A typical solar wind velocity, as we saw in Chapter 9, is 400 km per second. If we let $b = R_\odot$ and let B_0 be the photospheric magnetic field (~ 1 gauss), we may evaluate (10.47) to find the magnetic field strength at $r = 1$ A.U., the distance of the earth's orbit. We find $B_r \sim 8$ gammas, in good agreement with the measured values, which typically range between 5 and 10 gammas near the earth, but in front of the bow shock. Figure 10.24 illustrates the range of magnitudes; the median field measured near 1 A.U. in 1963–1964 by Imp-1 was 5.5 gammas.

Furthermore, the speed of rotation of the sun determines the "garden hose angle" θ, where

$$\tan \theta = \frac{\Omega r}{V}. \tag{10.50}$$

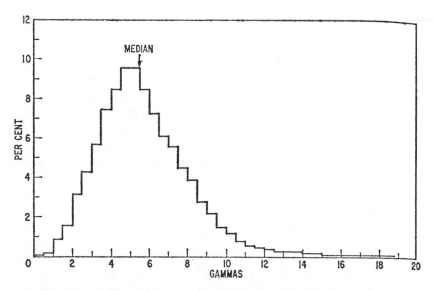

Figure 10.24   Histogram of the magnitude of the total field $\mathbf{B}_T$ measured in interplanetary space by the IMP-1 spacecraft during 1963–64. The median field strength was 5.50 ± 0.25 gammas at 1 A.U. N. F. Ness, *Annual Review of Astronomy and Astrophysics*, 1968.

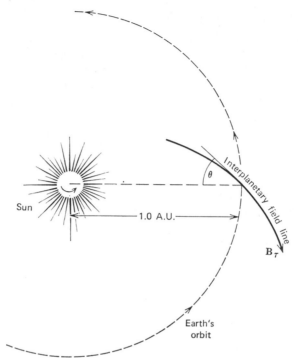

Figure 10.25   The garden-hose angle θ. It is typically $\sim$45° at the earth's orbit.

This angle is depicted in Figure 10.25. At $r = 1$ A.U. (and $1/\Omega = 24.7$ days, the solar equatorial rotation period), we find $\theta \approx 45°$; the average interplanetary field vector makes about a 45° angle with the earth-sun line. This is also in agreement with direct measurement. It is fortunate that $V \approx \Omega r$ near the earth, for this makes detection of the effect easier; a 45° angle is more obvious than 0° or 90°.

Direct measurements of interplanetary magnetic fields have been made by magnetometers carried by various spacecraft, such as the early Mariners, the Interplanetary Monitoring Platforms, and the Pioneers. These have revealed that the fields were structured into sectors of unequal size (see Figure 10.26). The average fields within each sector were directed either toward or away from the sun within 25° of the theoretical spiral angle. Within each sector the ratio of the number of fields along the spiral angle in one direction to that in the opposite direction is typically 100:1. As the sun rotates, these sectors sweep past the earth, reappearing approximately every 27 days. Thus it appears that the interplanetary field is connected to the sun.

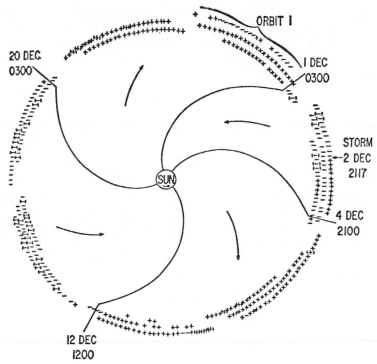

Figure 10.26   The structure, in the ecliptic plane, of the interplanetary magnetic field measured by instruments aboard IMP-1 near solar minimum 1963. Successive three-hour points are shown. A plus sign means that the field points away from the sun; a minus sign, that it points toward the sun along the Archimedes spiral. The sectors appear to rotate with the sun. These observations are consistent with the idea that the sources of the interplanetary field are unipolar magnetic regions on the solar surface. The sector pattern may change as the sun becomes more active. J. M. Wilcox and N. F. Ness, *Journal of Geophysical Research*, 1965.

The interplanetary field is predominantly in the ecliptic plane. Data on the sector boundary thickness are difficult to obtain with only one spacecraft, since both the boundaries and the vehicle are in motion and only the time that a boundary crossing occurs may be measured. These durations are typically of the order of 10 minutes.

10.28   Termination of the Interplanetary Field

The solar wind expands spherically. Therefore the mass density of the wind decreases as $1/r^2$. At some heliocentric distance r_h the solar wind pressure is balanced by the pressure of the interstellar medium. We expect that the wind will there undergo a shock transition to subsonic flow; the subsonic plasma beyond the shock forms a boundary shell. The interplanetary field terminates at the shock; tangled, disordered field lines are to be expected in the subsonic plasma. It is possible that some of them might even connect with field lines reaching out from other stars. It is not easy to evaluate r_h, but if we balance the interplanetary and interstellar pressures as outlined above, we find that

$$\frac{\rho_{so}V^2}{r_h^{\,2}} = P_i \tag{10.51}$$

where ρ_{so} is the solar wind mass density at 1 A.U. and r_h is in astronomical units. The quantity P_i is the interstellar pressure. For example, were we to assume that $P_i = B_i^2/2\mu_0$ (where B_i is the interstellar magnetic field, so that the interstellar pressure was entirely magnetic), we would find $r_h = 60$ A.U. for an interstellar magnetic field strength B_i of 1 gamma.

We should not expect the interplanetary field to be symmetric about the sun. The solar system is moving with respect to the interstellar medium; an asymmetric configuration rather reminiscent of the earth's field configuration may result. Thus the "heliosphere" should extend to greater distances in the antapex direction than it does in the solar apex, just as the earth's tail is longer than the stand-off distance in the solar direction. Figure 10.27, due to Kovar and Dessler, indicates one possible general configuration of the heliosphere.

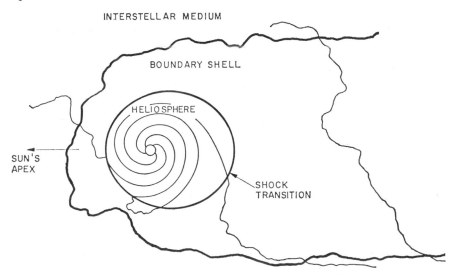

Figure 10.27   The magnetic heliosphere of the sun. The solar apex, the direction toward which the solar system is moving at $\sim$20 km per second, is in the constellation Hercules; the tail-like structure shown above therefore points toward the antapex at $\alpha = 6^h$, $\delta = -30°$. R. P. Kovar and A. J. Dessler, *Astrophysical Letters*, 1967.

It is to be hoped that future space experimenters will shed more light on the size of r_h. For example, studies of Lyman alpha radiation ($\lambda = 1216$ Å) will be useful; a "map" of the intensity of this hydrogen atom radiation around the solar system can help to determine r_h. Lyman alpha in interplanetary space is caused by sunlight scattered by hydrogen atoms. If one looks at the *Doppler-shifted* Lyman alpha, one can deduce where the maximum concentrations of energetic hydrogen atoms are to be found. Energetic hydrogen atoms are presumably concentrated at the shock surrounding the solar system, where the energetic wind protons undergo charge exchange with cold interstellar hydrogen atoms. It may be that the greatest flux of interplanetary Lα emanates from the solar apex.

Problems

10.1. Laboratory experiments reveal that the isotopes Ar^{38} (stable) and Ar^{39} (unstable; $T_{1/2} = 260$ years) are produced in equal amounts when protons of cosmic ray energies bombard iron targets. If a 1-kg sample of an iron meteorite is found to have 1×10^{14} atoms of Ar^{38} and 20 decays of Ar^{39} per minute, how long (in years) was the sample in space? Is this age typical of iron meteorites?

10.2. Show that a spherical particle of density ρ in interplanetary space will experience a net repulsive force from the sun if the radius r of the particle fulfills the inequality

$$r < \frac{L_\odot}{\frac{4}{3}GcM_\odot}\left(\frac{1}{\rho}\right),$$

where $L_\odot$ is the electromagnetic solar power putput.

10.3. Assuming that the brightness of the zodiacal light is due to scattering by free electrons (because of the state of polarization), we find that the surface brightness B is given by

$$B = \frac{I}{2\sin\epsilon}\left(\frac{e^2}{mc^2}\right)\int_\epsilon^\pi \rho\sin\theta\, d\theta,$$

where I is the solar flux at 1 A.U., ϵ is the angle of elongation, θ is the angle through which the sunlight is scattered so that it reaches the earth, and ρ is the number density of the scattering particles. Find ρ at about 1 A.U.

10.4. Assume an isothermal flow of an ideal gas rather than an adiabatic one, in deriving (10.16). Does the speed of sound change drastically?

10.5. Find the maximum temperature T_m at which supersonic expansion can occur. The speed of sound is given by $V_s = (2\gamma kT/m)^{1/2}$.

10.6. Calculate the relaxation time for fossil fields in the core of the earth, assuming it to be all iron.

10.7. Determine the interplanetary field strength and direction to be expected near Jupiter.

10.8. On a particular day the *Kp* values were as follows: 5o, 3+, 2−, 4+, 4o, 3+, 5+, and 5−. Find the garden hose angle at the earth for that day.

10.9. (a) If α is the angle between the prolonged radius vector r from the sun and the axis of a type I comet tail, show that for a radial solar wind

$$\cot \alpha = \frac{\omega + V \cos \theta}{V \sin \theta},$$

where the solar wind speed is ω, V is the cometary speed, and θ is the angle between **r** and **V**.
(b) What is cot α at perihelion?

10.10. Can particles ejected from the nucleus of a 10^{18}-gm comet at ~ 1 km per second escape from the comet if the nuclear radius is ~ 1 km?

10.11. Suppose that 10^{12} photons cm^{-2} second^{-1} are detected on the ground from a meteor at 100 km of altitude. What is the maximum rate of mass loss for the meteoroid if 1% of the kinetic energy of the ablated mass is converted to light energy?

10.12. Verify (10.13).

10.13. Assume that Mars has an intrinsic magnetic dipole field. If the solar wind velocity is 400 km per second and the average solar wind density at Mars is 3 protons per cm^3, what is the minimum surface magnetic field magnitude required to shield the surface from the solar wind? (In MKS units, $\mu = 4\pi \times 10^{-7}$ henry meter.)

Chapter 11

INTRODUCTION TO STELLAR STRUCTURE

There are a great many stars in the universe; our own galaxy alone contains some 10^{11} of them, and countless millions of other galaxies have been observed. Each star represents a challenge to space science; the attempt to understand how they shine, how they generate energy and transport it to the visible surface, and what happens to the intervening material must all be ascertained from radiation rather than direct measurement. One of the major problems of theoretical astrophysics has been to elucidate the nature of stellar structure. We can only attempt an introduction of this complex and elegant theory here.

11.1 The Sun

Since our sun is the star closest to us, it seems reasonable to use observations of it as a basis for our calculations, if we can show that the sun is a typical star. In fact, it is a very typical star; measurements of its brightness may be made, and these reveal it to be quite representative.

These measurements amount to determining the solar constant at the earth and also to finding the value of the astronomical unit in absolute units. From these measurements, and utilizing the fact that the brightness decreases with distance r as r^{-2}, we find that the solar brightness is equal to that of a star with an *absolute magnitude* of $+4.72$. The absolute magnitude of a star is the apparent brightness (integrated over all electromagnetic wavelengths, so that we refer here to the so-called *bolometric* absolute magnitude) of a star at a distance of 10 parsecs ($= 33.3$ light-years); it is a (logarithmic) measure of the power output of the star.

The solar absolute magnitude is less than two magnitudes fainter than the average magnitude of all the other stars. It may therefore be concluded that our sun is very normal and typical, a conclusion that is further supported by measurements of its mass and diameter. (Astronomers say that the sun is nearly in the middle of the *Main Sequence*; see Chapter 12.) It therefore seems safe to use the sun as a model of the other stable, normal stars. We postpone a discussion of the peculiar stars until later.

The sun is *hot*; the photospheric average temperature determinations reveal that $T_e = 5570°\text{K}$. This figure is deduced from the angular diameter and, as remarked above, from the values of the solar constant and the astronomical unit.

The sun also rotates its own axis. The period of rotation may be measured by measuring the time required for sunspots to reappear on the solar disc at their original position. Spots are just what their name suggests; relatively black regions on the solar disc that appear to be fixed in the sun and therefore rotate with it (see Figure 11.1).

The results of the period measurements provide some insight as to the nature of the closest star. At a solar latitude of $0°$ (the solar equator) the period is $24^\text{d}16^\text{h}$. At $\pm 30°$ the period is $27^\text{d}04^\text{h}$, and at $\pm 75°$ we find 33^d. Near the solar poles ($\sim \pm 90°$) the period is about 35^d, over 10 days longer than required for the equatorial material to complete a rotation. The mean period is 25.4 days, leading to a mean synodic period of rotation of 27 days.

The fact that the period varies with latitude means that we are viewing a nonrigid body. The sun is a self-luminous globe of hot gas.

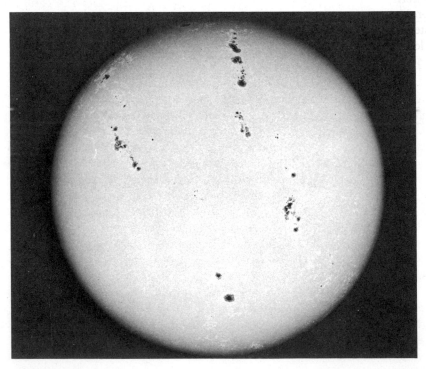

Figure 11.1   The sun, photographed at sunspot maximum, 21 December 1957. Courtesy Hale Observatories, California Institute of Technology.

11.2   Sunspots and the Solar Cycle

Among the many observational facts that a successful theoretical model of the sun must explain is the cyclical variation in the number of spots as measured by the Wolf number R, where

$$R = k(10g + f). \tag{11.1}$$

In (11.1) f is the number of individual spots detectable on the solar disc by the observer. There are also complex spot groups; many unresolved spots (perhaps $\sim$10, depending on the observing equipment) are present in each group. The number of groups is denoted by g in (11.1). The factor k is inserted so that various observatories and telescopes may be compared on a common basis; the comparison factor usually nears unity.

The Wolf number varies from $\sim$0 at *solar minimum* up to $\sim$200 at *solar maximum*. After passing through maximum, R slowly declines back toward zero, but may not quite reach it. It seems that a minimum following an unusually large maximum is also relatively high.

The time between maxima is about $11\frac{1}{4}$ years; the solar cycle is quite periodic. The "rise time," however (solar minimum to maximum), is shorter than the "decay time" from maximum to minimum. On the average the decay time is about 6.7 years, but the rise time is only 4.6 years. Thus a plot of R versus time looks more like a "sawtooth" than a sine wave (see Figure 11.2). The last minimum at this writing occurred in 1964, and the last maximum may have occurred in late 1969.

11.3   Spot Characteristics

A spot is dark compared to surrounding stellar material. We conclude, therefore, that is is cooler than the surrounding gas. Observations using thermocouples that integrate over wavelength reveal that the light intensity from the center of the spot is only $\sim$0.4 that of the surrounding material. Thus the effective temperature of the center of the spot will be $\sim(0.4)^{1/4} \times 5770°K$, or about $4600°K$.

Some spot groups (see Figure 11.3) are huge. Group lengths of more than 1×10^5 km ($\sim\frac{1}{7} R_\odot$) have been reported. Individual sunspot diameters may be as large as 10% of this. It is difficult to detect spots that have diameters of less than 10^3 km.

Hale discovered in 1908 that spots have intense magnetic fields associated with them. The Zeeman effect tells us that an observer looking along the magnetic field sees a splitting of a spectral line into two components. The amount of the splitting, $\delta\lambda$, is determined by the field strength. The frequency difference is a constant times the cyclotron frequency; it is given by $geB/2\pi mc$, where g is the Landé g-factor and B is the intensity of the field component.

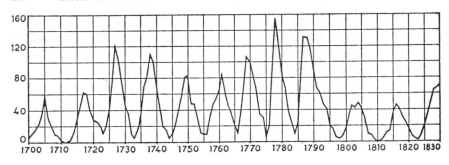

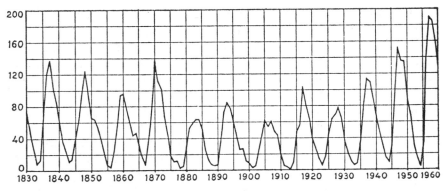

Figure 11.2 Variation of sunspot number with time for the years 1700–1960. M. Waldmeier, *The Sunspot: Activity in the Years 1610–1960*, Schulthess and Co. A. G., 1961.

Thus for the wavelength λ in centimeters and the field B in gauss we find that

$$\delta\lambda = 9.4 \times 10^{-5} \, g\lambda^2 B. \qquad (11.2)$$

A field of 200 gauss will cause a red spectral line to be separated into two circularly polarized components about 0.1 Å apart in wavelength. Modern spectrographs can resolve lines that are only 0.01 Å apart in wavelength (see Section 11.5).

About 84% of the groups are bipolar, but most of the shortlived groups (i.e., those lasting only a single day) are unipolar. An important feature is that the arrangement of magnetic polarities in essentially bipolar groups in the northern hemisphere is opposite to that in the southern hemisphere. The entire system of polarities reverses itself each cycle.

We also should note that individual spots occur in *pairs* on either side of the solar equator. Magnetic fields presumably link the oppositely polarized members of each pair.

We have seen that the number of spots on the sun varies in time with the 11-year cycle. It is interesting that the *latitude* of spots also depends on the

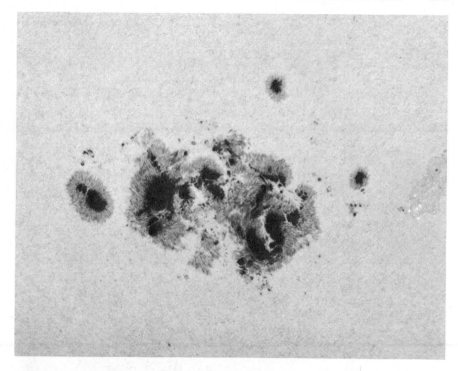

Figure 11.3 Large sunspot group, photographed on 17 May 1951. Courtesy Hale Observatories, California Institute of Technology.

solar cycle. At the beginning of a new cycle (when $N \sim 0$) spots appear at relatively high latitudes; the average latitude at that time is $\pm 30°$. At solar maximum the average latitude decreases to $\pm 15°$, and the average latitude continues to diminish until it reaches $\pm 8°$ at the end of the cycle. Spots on the succeeding "new" cycle may then be observationally distinguished from spots left over from the old cycle, because the new spots appear at high latitudes. This steady progression in spot latitude results in what is known as the "butterfly diagram"; a typical butterfly diagram is shown in Figure 11.4. It is remarkable that spot groups are rarely if ever observed at latitudes in excess of $\pm 40°$ or less than $\pm 5°$.

11.4 Solar Flares

Small portions of the visible disc occasionally sharply brighten; these brightenings are called *solar flares*. One is shown in Figure 11.5.

Flares are classified into various "importance" categories on the basis of their area and brilliance. This dual-importance scheme was adopted by the

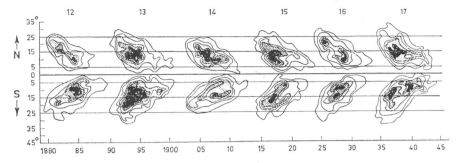

Figure 11.4 "Butterfly diagram" of sunspot latitude versus time. The "isolines" represent lines of equal spot frequency. Spots in a new solar cycle (cycle numbers are at the top of the figure) appear at relatively high solar latitudes. U. Becker, *Z. Astrophys.*, 1955.

International Astronomical Union as of 1 January 1966, and the criteria are summarized in Table 11.1. Each importance figure (as given in the table) is followed by "f," "n," or "b"; the letters stand for "faint", "normal," and "brilliant," respectively. Thus we might have an importance 2 flare described as a 2f, 2n, or 2b. The brightness factor is assigned by the observer for the time of maximum brilliance.

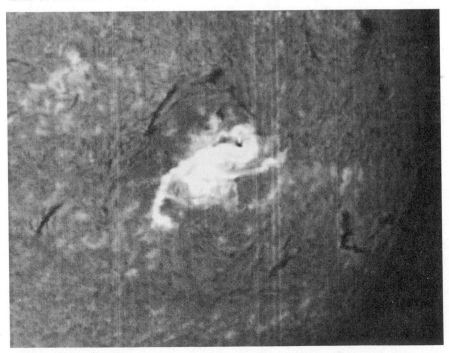

Figure 11.5 Solar flare photographed in the red light of the Hα line, 16 July 1959. Courtesy Hale Observatories, California Institute of Technology.

TABLE 11.1

AREA OF SOLAR FLARES

Importance	Area (fraction of solar disc)
S	$0–200 \times 10^{-6}$
1	$200–500 \times 10^{-6}$
2	$500–1200 \times 10^{-6}$
3	$1200–3600 \times 10^{-6}$
4	$>3600 \times 10^{-6}$

A "typical flare" (if there is such a thing) rises to its maximum brilliance in a few minutes and then slowly fades away over a period of 1–3 hours. The optical decay curve seems to be approximately exponential in time.

Flares generate a considerable amount of optical energy; for example, if an importance 3n flare were to produce 10^{27} ergs per second for $\sim 10^3$ seconds, some 10^{30} ergs would be liberated during this (rather large) solar flare.

The occurrence-frequency spectrum of flares is an inverse function of their importance. There are perhaps 10 times as many importance 1 flares as there are importance 3 events.

Flares usually occur near sunspot groups. Therefore we expect and observe that the number of flares per unit time depends on the 11-year solar cycle.

We should note here that the sun is not the only star observed to exhibit flares. There is a class of stars sometimes called UV Ceti stars that are also seen to flare up; they are called "flare stars." Flare stars are relatively cool ($\sim 2700°$K surface temperature), red objects. A UV Ceti star is occasionally observed to brighten by a factor of about 100 in several seconds and then decline in luminosity to 'normal" in a period of an hour or so. Several flares may be seen in a typical week. Now the additional light must emanate from a small region of the stellar surface (probably only about 1%), because the blue-white color of the "flare" is equivalent to a temperature of approximately 15,000°K; a large fraction of the surface area would radiate far more energy than is observed.

The star that is often closest to the sun is a flare star. Proxima Centauri is a member of a stellar class called "red dwarfs," and it is a component of the three-star system called Alpha Centauri, some 4.3 light-years distant from the sun. The orbit of Proxima is such that it is carried somewhat closer to the sun that either of its companions.

Proxima, like most red dwarfs, has about a tenth the mass of the sun, its radius is only about 10% of the solar radius, and it is underluminous; it requires a telescope for observation, since its apparent visual magnitude is only +10.7. It is interesting and possibly significant that all known flare stars are red dwarfs in close gravitational association with other stars; it may be that the flares in the dwarfs are induced by gravitational tidal perturbations on it by its companions.

It is possible that flares are characteristic of many other stars also, but flares would be extremely difficult to detect in stars hotter than the UV Ceti stars. That is, the increased brilliance would not be as noticeable.

Interest in solar flares became heightened when it was realized that, in addition to their optical output, flares also produce much energy in the radio and x-ray regions of the electromagnetic spectrum, as well as in corpuscular emissions.

We shall discuss this corpuscular emission, sometimes called *solar cosmic rays*, in Chapter 13. It is interesting that some 3b flares have been observed where $\sim 10^{32}$ ergs have apparently been released in the form of particles, an energy comparable to or in excess of that found in electromagnetic radiation and on time scales short compared to the electromagnetic flare. Perhaps the most remarkable characteristic of solar flares is that they are not destroyed by the emission of 10^{32} ergs.

A flare may be thought of as one (i.e., optical) manifestation of an explosion in the solar *chromosphere*, that region of the solar atmosphere that extends from the visible surface (the photosphere) outward to about 20,000 km, where the solar corona starts. The chromosphere is so named because of its red color, caused mainly by the glow of hydrogen.

The chromosphere's lowest stratum, sometimes called the "reversing layer," is the source of most of the dark lines of the solar spectrum. Photospheric continuum radiation is absorbed selectively by the gases in the reversing layer.

Figure 11.6 shows the radio spectra detected in solar flares. Various types of radio signals exist; the classification into types II, III, IV, and V is made on the basis of the radio frequency spectrum. Type I radio noise bursts do not appear to be associated with solar flares. That the flare-associated emissions are nonthermal follows from the observation that T_b [see (4.12)] may reach $10^{15} °$K during a noise burst!

It is now believed that radiation types II and III are plasma oscillations that occur at the plasma frequency ω_p (see Section 3.19). Type III, the first (historically) to be recognized as a plasma oscillation, is interpreted as due to a disturbance traveling outward through the solar atmosphere. Since ω_p depends on the solar altitude, the frequency "drifts." The rate of drift then gives the velocity of the disturbance (an ion cloud, say); observations indicate that $v \sim c/4$.

Type II bursts contain the most power. Since their rate of frequency drift is much slower than found for type III, the outward velocity of a disturbance appears to be "only" ~ 1000 km per second.

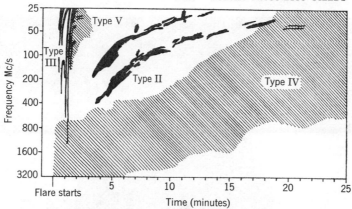

Figure 11.6   The spectra of solar radio bursts, showing how they vary with time (dynamic spectra). From J. P. Wild, S. F. Smerd, and A. A. Weiss, *Annual Review of Astronomy and Astrophysics*, 1963.

Both types IV and V are polarized, but type V is much more so. The two types are thought to be the result of synchrotron radiation. Type IV, however, probably represents a mixture of plasma oscillations and synchrotron radiation. Types IV and V are strikingly well correlated with the emission of high-energy particles known as "solar cosmic rays." We shall defer a discussion of the corpuscular radiation until Chapter 13.

11.5   Solar Magnetic Fields

During an eclipse of the sun the dazzling "K-corona" becomes visible, out to perhaps $3R_\odot$. (The K-corona is what we have previously referred to as the solar corona; astronomers call the zodiacal light the "F-corona.") Photometers, however, have detected the K-corona during eclipses out to much greater elongations than $3R_\odot$. Figure 11.7*b–d* illustrates some of these data, as obtained by T. J. Pepin. These photometers were mounted in a jet flying with the sun above most of the clouds in order to increase the duration of totality.

Plumes are visible in the K-corona around the solar poles. The whole effect strongly suggests the existence of a general solar magnetic field. It is an extremely difficult matter to measure this (weak) surface field, however. Controversy raged during the first half of the twentieth century concerning its very existence, to say nothing of its magnitude. It has now become established that the general field does exist; the field strength near the solar poles is about 1 gauss.

A field this small will be extremely difficult to detect; a glance at (11.2) shows that the Zeeman splitting to be expected for $B = 1$ gauss is hopelessly small. Modern heliomagnetographs, however, can detect fields down to 0.3

Figure 11.7*a* The solar corona as photographed from the earth. (Courtesy Hale Observatories, California Institute of Technology.)

gauss. The key elements of such a magnetograph are a high-dispersion grating, a polarization analyzer, and a photoelectric detector.

To illustrate with an actual example, if the dispersion of the grating is 11 mm per Å, it can resolve lines only 0.009 Å apart in wavelength. Now the intrinsic width of the Fraunhofer lines in the solar spectrum is about 0.1 Å, 10 times as great as the resolution. Therefore the grating can be used to give a very good reproduction of the line *profile* (i.e., shape). The two Zeeman components in the longitudinal Zeeman effect are circularly polarized. When a polarizer is inserted in front of the grating, the line profile changes slightly. A detector with a slit that looks at the light in one of the wings of the spectral line will thus see a change in the light level in the wing; this "signal" may be then amplified, recorded, and interpreted in terms of the splitting (see Figure 11.8). It is possible to mechanize the instrument so that it scans the solar disc, thus mapping the magnetic fields on the sun.

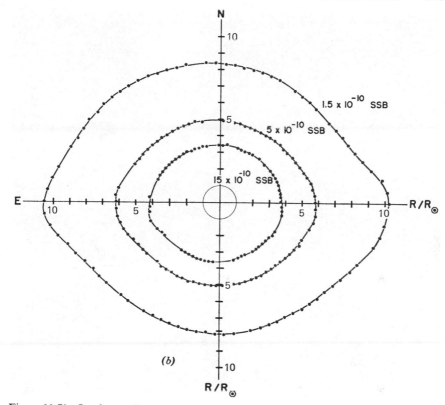

Figure 11.7*b*   Isophotes observed (corrected to the top of the atmosphere) during the 12 November 1966 eclipse. The unit SSB refers to solar surface brightness. The observations were taken from an aircraft at an altitude of 40,000 feet. Courtesy University of Minnesota, School of Physics and Astronomy.

The general field predominates near the poles; opposite polarities are found at the two poles. Near the solar equator, however, a tangled web of field lines is seen; they appear to principally emanate from the "centers of activity" that also harbor sunspots.

The general surface field apparently reverses polarity every 11 years. The reader who recalls Section 10.25, however, will remember that a field of solar dimensions should take billions of years to decay, to say nothing of reversing itself. The problem, if not the solution to it, is obvious.

One clue to resolving the dilemma comes from the curious manner in which the field appears to reverse itself. For example, in 1957–1958 (a recent reversal) the south pole reversed itself first; the north pole didn't switch polarity for about another 18 months. We suspect, therefore, that the field

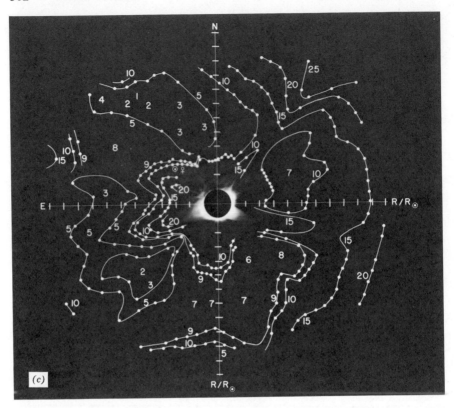

Figure 11.7c Lines of constant percentage polarization. Superimposed is a photograph of the inner corona.

reversal is only apparent. The surface field seems to change, but the general field does not. It is now believed that the principal agency responsible for the apparent reversal is the differential rotation of the sun.

The differential rotation causes the field lines, which are frozen into the plasma, to be carried around at different rates. The result of this is that the field lines "wind up" (much like a rubber band) into "magnetic ropes" of flux. When field lines are wound up like this, they are more closely packed, so that the field is in effect amplified in the ropes.

Now (the magnetic pressure $P_B = B^2/2\mu$) the rope steadily becomes more "buoyant" as B increases. Finally, at some critical value of B (perhaps $\sim 10^3$ gauss) P_B becomes sufficiently great that the rope actually "pops up" above the solar surface. The two places where the field lines enter and leave the surface and presumably the two magnetically linked sunspots across the equator from each other. This is the basic idea underlying the

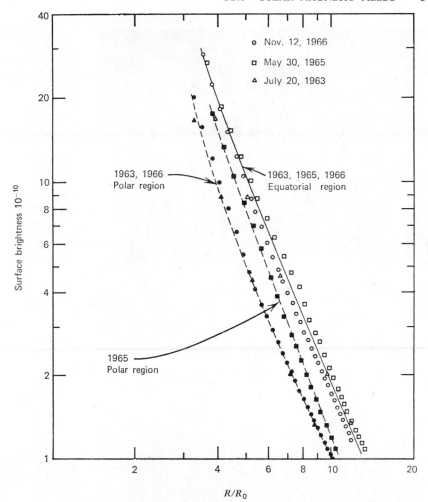

Figure 11.7d   Eclipse observations of the K-corona made by T. J. Pepin from aircraft. A temporal variation of the coronal surface brightness apparently exists, mainly near the solar poles.

current ideas of the formation of spots on the sun. That the spots are relatively cool also follows from this idea; magnetic fields inhibit the convective process, so that spot gas is cooler. There is independent evidence (see Section 11.14) that the solar surface material is in convective equilibrium, so that it is heated by the upwelling of hot gases from below.

As time passes, the number of spots increases so that the tangled field lines near the middle of the disc become increasingly complex. Also, the field lines that rise out of the spot come into "contact" with the general field lines that

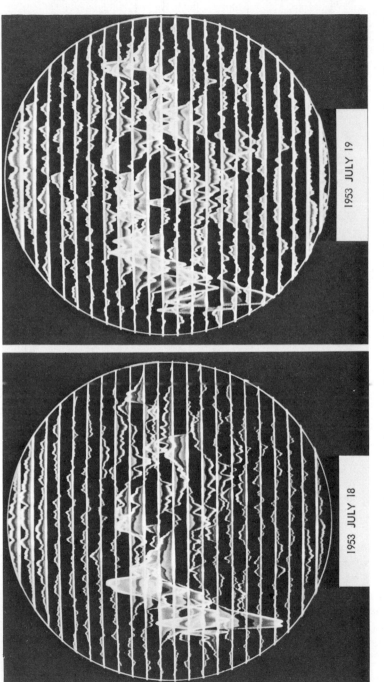

Figure 11.8 Solar magnetograms. These magnetic maps of the sun's disc show the location, field intensity, and polarity of weak magnetic fields in the photosphere of the sun, apart from sunspots. The records are made automatically by a scanning system that employs a polarizing analyzer, a powerful spectrograph, and a sensitive photoelectric detector for measuring the longitudinal component of the magnetic field by means of the Zeeman effect. A deflection of one trace interval corresponds to a field of about 1 gauss. The small deflections of opposite magnetic polarity near the north and south poles are indicative of the sun's "general magnetic field." The extended fields near the equator arise from characteristic bipolar magnetic regions that sometimes produce spots. North is at top; east, at right. (Courtesy Hale Observatories, California Institute of Technology.)

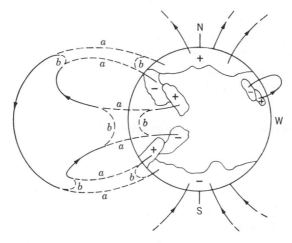

Figure 11.9 Apparent reversal of the solar magnetic field. Parts labeled *a* are replaced by parts labeled *b* and a partial neutralization occurs; a continuation of the process produces the reversal. From H. W. Babcock, *The Astrophysical Journal*, University of Chicago Press, 1961.

arch above the solar surface. If magnetic merging occurs, severing and reconnection of the field lines take place, as shown in Figure 11.9. The general field thus *appears* to diminish and eventually even reverses. This explains why the reversal occurs at sunspot maximum and also why it occurs in piecemeal fashion.

There are many effects of local, intense solar surface magnetic fields. One such apparently is the stability of the solar *filaments* (dark, linelike structures on the visible disc); they apparently remain more or less stationary for time scales of months. They may be like solar prominences viewed from "above," so that we are looking down on the prominences. Some prominences arch above the solar surface for distances as great as $R_\odot/2$; it is believed that the magnetic pressure of the localized fields suspends these great masses of heavily ionized material above the solar surface.

The sun is not the only star to show a magnetic field. Fields have been observed that range all the way up to the 34,000 gauss reported for the peculiar, spectral class A star known as HD 215,441 (see Section 12.3). (The HD refers to the Henry Draper catalog, which gives the spectral classification, visual and photographic magnitudes, and coordinates of hundreds of thousands of stars.) These stellar fields, like that of the sun, vary and even reverse. The "magnetic stars" so far identified have sharp and unusually intense spectral lines of the metals and rare earths. It is unknown whether

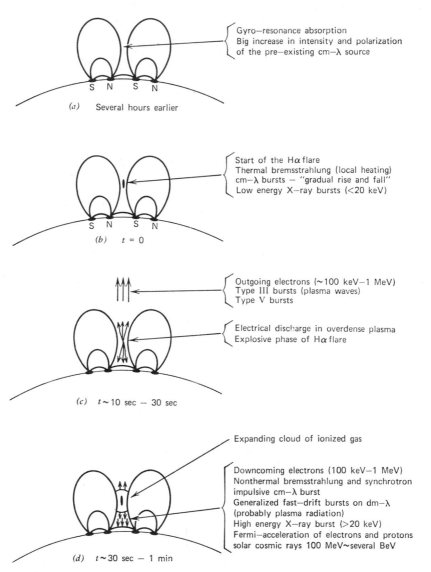

Gyro–resonance absorption
Big increase in intensity and polarization
of the pre–existing cm–λ source

(a)   Several hours earlier

Start of the Hα flare
Thermal bremsstrahlung (local heating)
cm–λ bursts – "gradual rise and fall"
Low energy X–ray bursts (<20 keV)

(b)   t = 0

Outgoing electrons (~100 keV–1 MeV)
Type III bursts (plasma waves)
Type V bursts

Electrical discharge in overdense plasma
Explosive phase of Hα flare

(c)   t ~ 10 sec – 30 sec

Expanding cloud of ionized gas

Downcoming electrons (100 keV–1 MeV)
Nonthermal bremsstrahlung and synchrotron
impulsive cm–λ burst
Generalized fast–drift bursts on dm–λ
(probably plasma radiation)
High energy X–ray burst (>20 keV)
Fermi–acceleration of electrons and protons
solar cosmic rays 100 MeV~several BeV

(d)   t ~ 30 sec – 1 min

Figure 11.10   Schematic view of Sweet's mechanism of solar flares. Two regions of conducting fluids containing oppositely directed magnetic fields are compressed from the sides so that the fields approach each other more closely. The increase in field gradient

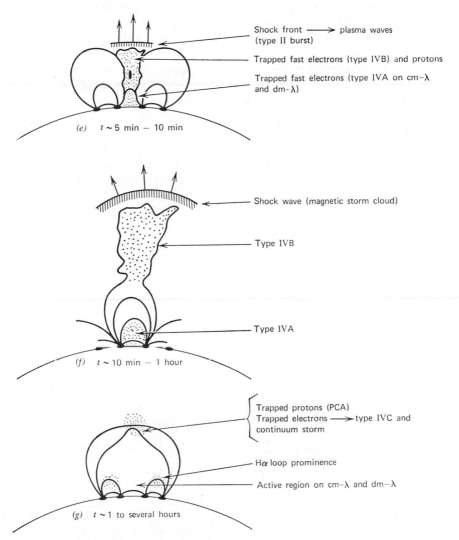

Shock front ⟶ plasma waves
(type II burst)

Trapped fast electrons (type IVB) and protons

Trapped fast electrons (type IVA on cm-λ
and dm-λ)

(e) $t \sim 5$ min $-$ 10 min

Shock wave (magnetic storm cloud)

Type IVB

Type IVA

(f) $t \sim 10$ min $-$ 1 hour

Trapped protons (PCA)
Trapped electrons ⟶ type IVC and
continuum storm

Hα loop prominence

Active region on cm-λ and dm-λ

(g) $t \sim 1$ to several hours

increases the Joule dissipation. The rate of energy release is no doubt connected with standing magneto-hydrodynamic waves, which may convert magnetic energy to plasma energy.

stars of other spectral classes also possess magnetic fields, although circularly polarized light has been detected from one peculiar white dwarf star (Section 12.10) at this writing. This polarization seems consistent with plasma emission of light when the plasma is in a magnetic field of nearly one hundred million gauss, the surface field strength expected, should a normal star collapse to white dwarf dimensions.

Solar magnetic fields may also provide the explanation for solar flares, although there is no generally accepted theory of flares. It is noticed that flares occur in regions of strong magnetic field. Furthermore, the first bright points of a flare normally lie close to the "neutral line" that separates a region of north magnetic polarity from a region of south magnetic polarity. Figure 11.10 illustrates Sweet's mechanism of solar flares.

In Sweet's model of flares, the flare energy is supplied by the thin sheet of current (perhaps only ~100 meters thick) that is present at the boundary between oppositely directed field lines. In the flare region (within the solar atmosphere) the field lines from, say, the north polarity region, which otherwise extend out into interplanetary space (permitting the escape of the solar wind), reconnect with the south polarity lines. Plasma is ejected outward from the region and bright filaments are formed by plasma from the region that travels downward along the connected field lines.

11.6   Stellar Energy Sources

We know from observations of the sun that it is hot and, indeed, self-luminous. Therefore a source of energy must be present within the star to compensate for (at least) this energy radiated as heat and light. We seek now to determine the nature of the energy source.

One might suppose that gravitational collapse is adequate to provide the required energy. That is, we suppose that the sun originally consisted of a large, extended mass of tenuous gas that contracted under the influence of self-gravity; the matter "collapse" from infinity to $R_\odot$ would make a great deal of potential energy available, presumably in the form of radiation.

The virial theorem of classical mechanics provides us with the basis for answering the question as to whether gravitational collapse provides an adequate amount of energy. The theorem states that one-half the total gravitational energy is available for radiation. To provide this, consider the "spherical moment of inertia" I of an assemblage of gas particles, where

$$I \equiv \sum_{j=1}^{n} \tfrac{1}{2} m_j \mathbf{r}_j \cdot \mathbf{r}_j. \tag{11.3}$$

The particles each have a mass m and are at a position $\mathbf{r}$. Therefore

$$\ddot{I} = \sum_{j=1}^{n} m_j \dot{\mathbf{r}}_j \cdot \dot{\mathbf{r}}_j + \sum_{j=1}^{n} m_j \ddot{\mathbf{r}}_j \cdot \dot{\mathbf{r}}_j. \tag{11.4}$$

Now we may reason physically that if we wait long enough $\ddot{I} \to 0$; the time average of $\ddot{I}$ approaches zero. If the particles are assumed always to move in a closed space, the magnitude of $\mathbf{r}_j$ can never exceed a certain maximum value. In addition, the velocities $\dot{\mathbf{r}}_j$ are likewise bounded; the kinetic energy of a collection of particles cannot become infinitely great, so that $\dot{I}$ is bounded and, as time increases, $\bar{\ddot{I}} \to 0$. Therefore

$$\overline{\tfrac{1}{2} \sum_{j=1}^{n} m_j \dot{\mathbf{r}}_j \dot{\mathbf{r}}_j} = \overline{-\tfrac{1}{2} \sum_{j=1}^{n} m_j \mathbf{r}_j \cdot \ddot{\mathbf{r}}_j}. \tag{11.5}$$

The left-hand side of (11.5) is the time average of the kinetic energy K of the system. If F_j is the force acting on the jth particle,

$$\bar{K} = \overline{-\tfrac{1}{2} \sum_{j=1}^{n} \mathbf{F}_j \cdot \mathbf{r}_j}. \tag{11.6}$$

The right-hand side of (11.6) is called the *virial* Ω of the system; the time average of the kinetic energy equals the virial of the system.

Consider an inverse square force such as gravity. If V is the potential energy of a gravitationally interacting system,

$$\bar{\Omega} = -\tfrac{1}{2}\bar{V}. \tag{11.7}$$

Therefore the kinetic energy of a system of particles interacting through gravity is equal to half the potential energy of the system. The total energy of the system, E, is the sum of the kinetic and potential energies, so that

$$E = \frac{V}{2}. \tag{11.8}$$

If V is negative (an attractive potential), the total energy is also negative.

If the solar particles started their "collapse" from infinity, E was originally zero; it is now negative, by an amount $-V/2$. Thus $V/2$ is lost in the collapse, presumably through radiation, and it remains to calculate V for this star.

The potential energy is defined as the work done on the system to bring the matter from infinity into our (assumed) spherical distribution of matter. If we already have brought a mass $M(r)$ in from infinity, the work required to bring an additional amount of matter $dM(r)$ as a spherical shell of thickness dr is

$$-GM(r)\, dM(r) \int_r^\infty \frac{dr}{r^2} = -\frac{GM(r)\, dM(r)}{r},$$

so that

$$V = -G \int_0^R \frac{M(r)\, dM(r)}{r}. \tag{11.9}$$

The integral in (11.9) may be evaluated; it turns out that $V \approx GM^2/R$, where M and R are the mass and radius of the star, respectively. Thus, if $M = M_\odot$ and $R = R_\odot$, we find that $V_\odot \approx 4 \times 10^{48}$ ergs.

Now the present rate of energy generation by the sun may be measured; it is known as *luminosity* of the sun, $L_\odot$, where $L_\odot = 4 \times 10^{33}$ ergs per second. Let us assume that the solar luminosity has been constant over the lifetime τ of the sun, and ask what τ turns out to be.

We see that if we assume gravitational collapse to be valid,

$$\tau = \frac{V_\odot}{2} \cdot \frac{1}{L_\odot}. \tag{11.10}$$

Thus $\tau = 5 \times 10^{14}$ seconds $= 15$ million years for the solar lifetime under this assumption.

This lifetime is, of course, much too short. Studies of radioisotopes in the earth and in meteorites tell us that the solar system has been around for some 4.6×10^9 years. Also, fossil studies on the earth reveal that the sun must have been producing approximately the same amount of energy per unit time for the past several hundreds of millions of years.

We conclude, therefore, that *gravitational collapse is totally inadequate as the solar energy source*, and we must search for other mechanisms. As we shall see, *thermonuclear fusion* does provide an adequate energy source, for (see Sections 11.7 and 11.8) the central pressure and temperature of the sun must exceed 10^8 atm and $2 \times 10^{6\circ}$K, respectively. At these high densities and energies nuclear reactions occur. The fusion of 4 hydrogen atoms to form 1 helium atom liberates 4×10^{-5} erg per helium atom formed. Thus 6.4×10^{18} ergs of energy are liberated for each gram of helium that is produced.

This is more than enough to supply $L_\odot$ for 10^{10} years. The excess energy is presumably carried off by the neutrinos that are formed in the nuclear reactions. Since neutrinos do not interact appreciably with matter, their energy is not available to the gas for heating. It should of course be pointed out that while thermonuclear fusion would provide an adequate energy source, it is not yet proven that this is actually the energy source for stars.

11.7 Central Pressure of the Sun

We saw in the last section that if we could demonstrate that the pressure of the material near the center of the sun was $\sim 10^8$ atm and if the gas were at a high temperature, nuclear reactions would set in and provide an adequate source of energy.

If $M(r)$ is the mass of gas enclosed within a shell at a distance r from the stellar center,

$$M(r) = \int_0^r \rho 4\pi r^2 \, dr, \tag{11.11}$$

where ρ is the density of the gas and is itself a function of r. We may there-fore write a differential equation for the variation of mass with radius; it is

$$\frac{dM(r)}{dr} = 4\pi r^2 \rho. \tag{11.12}$$

Now the star is assumed to be stable: it is neither expanding nor contracting, so that the pressure P just balances the weight of overlying material. Our familiar condition of hydrostatic equilibrium tells us that we may write, if the local acceleration of gravity is $GM(r)/r^2$,

$$\frac{dP}{dr} = \frac{-GM(r)}{r^2} \rho. \tag{11.13}$$

If we now combine (11.12) and (11.13), we obtain

$$\frac{dP}{dr} = -\frac{GM(r)}{4\pi r^4} \frac{dM(r)}{dr}. \tag{11.14}$$

Let us digress for a moment and consider the function $P + [GM^2(r)/8\pi r^4]$; it will be helpful to prove that this function decreases outward from the stellar center. To demonstrate this, we take the derivative of the function with respect to r:

$$\frac{d}{dr}\left[P + \frac{GM^2(r)}{8\pi r^4}\right] = \frac{dP}{dr} + \frac{GM(r)}{4\pi r^4}\frac{dM(r)}{dr} - \frac{GM^2(r)}{2\pi r^5}. \tag{11.15}$$

But according to (11.14), the first two terms on the right-hand side of (11.15) are equal and opposite, so that

$$\frac{d}{dr}\left[P + \frac{GM^2(r)}{8\pi r^4}\right] = -\frac{GM^2(r)}{2\pi r^5}, \tag{11.16}$$

which is a negative quantity; the derivative is negative and therefore the function does indeed decrease from the center of the star.

Hence if P_c is the pressure at the center of a star of mass M and radius R,

$$P_c > P + \frac{GM^2(r)}{8\pi r^4} > \frac{GM^2}{8\pi R^4}, \tag{11.17}$$

so that

$$P_c > \frac{GM^2}{8\pi R^4}. \tag{11.18}$$

Let us insert numbers into (11.18) and evaluate it for our sun, so that $M = M_\odot$ and $R = R_\odot$. The result is that $P_c > 4.4 \times 10^{14}$ dynes per cm^2; the pressure at the center of the sun must exceed 4.5×10^8 atm.

11.8 Stellar Composition and the Central Temperature of the Sun

It is also possible to estimate the minimum temperature at the center of the sun. Let us assume that we are dealing with an ideal gas composed of ionized hydrogen atoms, so that we are dealing with a gas (plasma) made up of protons and electrons. An ideal gas is one in which there are no forces between particles; they are completely free to move about.

The equation of state of an ideal gas is commonly written as $PV = nRT$, where V is the volume of n moles of the gas and R is the universal gas constant. The gas temperature is denoted by T. The equation of state may be rewritten as

$$P = \frac{k}{\mu M_p} \rho T, \tag{11.19}$$

where k is Boltzmann's constant (1.38×10^{-16} ergs per degree) and M_p is the mass of the proton, 1.67×10^{-24} gm. The quantity μ is the mean molecular weight of the material in the star. For neutral hydrogen atoms, $\mu = 1$; for completely ionized hydrogen, however, $\mu = \frac{1}{2}$ because the number of particles has doubled and the electron mass is negligible compared to that of the proton.

We can write a simple expression for the mean molecular weight of the material in a star, if we take it to be *completely ionized*. Consider 1 gm of such a mixture. Let X be the fraction of this gram that is hydrogen, Y the fraction that is helium, and Z the fraction that is everything else. It is clear then that $X + Y + Z = 1$.

To get at the mean molecular weight of such a mixture, take the total mass (in atomic weight units) and divide by the total number of particles. If M_p is the mass in grams of 1 hydrogen atom, there are X/M_p hydrogen atoms in X grams of hydrogen and there are $2X/M_p$ particles arising from the hydrogen, because of our complete ionization assumption. We have Y grams of helium; this tells us that we have $Y/4M_p$ helium atoms and that the number of particles arising from the complete ionization of the helium is $3Y/4M_p$, since a completely ionized helium atom produces three particles. Assume that the heavy material (Z grams of it are present) has an atomic weight A; we have Z/AM_p heavy atoms, each of which contributes approximately $A/2$ particles upon complete ionization, giving rise to $Z/2M_p$ individual particles. (This last is not even approximately true, of course, for atoms lighter than carbon. However there aren't many such, compared to hydrogen and helium.) Thus in 1 gm of completely ionized substance there are

$$\frac{2X + \frac{3}{4}Y + \frac{1}{2}Z}{M_p} \text{ particles.}$$

We must divide this into the total mass expressed in atomic mass units. The total number of atomic mass units in 1 gm of matter is $1/M_p$. Assuming only 1 gm to be present, all that we have to do is divide $(1/M_p)$ by our result for the number of particles in a completely ionized gram and simplify the result.

Table 11.2 shows the results of measurements of the solar composition. Photospheric abundances are given logarithmically, relative to the abundance of hydrogen, and are determined from the measurements of the "equivalent widths" of photospheric absorption lines. The equivalent width of a line is the width of the continuum adjacent to the spectral line in question which just equals the intensity subtracted from the solar spectrum by the absorption line.

In general the mean molecular weight will vary throughout the star, even if the matter comprising the stellar atmosphere is well mixed. This is because the temperature varies throughout the star and we therefore expect the degree of ionization (Section 11.13) to vary accordingly with distance from the stellar center. In addition to this effect, nuclear reactions take place more rapidly near the center of the star than elsewhere; the resulting transmutation of elements thus affects the value of the mean molecular weight near the center.

If $\bar{T}$ is the average temperature of the gas, we may write

$$M\bar{T} = \int_0^R T \, dM(r), \qquad (11.20)$$

so that

$$M\bar{T} = \frac{\mu M_p}{k} \int_0^R \frac{P}{\rho} \, dM(r), \qquad (11.21)$$

from the equation of state. If $d\tau$ is an element of volume within the gas, (11.21) may be rewritten as

$$M\bar{T} = \frac{\mu M_p}{k} \int_0^R P \, d\tau. \qquad (11.22)$$

If we recall the definition of the potential energy V of a gas and utilize the fact that the sun is very nearly spherical, we may integrate (11.22) and find that

$$M\bar{T} = -\frac{1}{3} \frac{\mu M_p V}{r}. \qquad (11.23)$$

We now wish to find a minimum value for V so that we may then find $\bar{T}$. This may be accomplished by proving that the function $(\pi P_c R^3 + GM^2/2R)$ must exceed $-V$, which is defined by (11.9). The reader is referred to the book *Stellar Structure* by Chandrasekhar for details of the proof. The minimum value of V found in this way, if inserted into (11.23), gives $\bar{T}$; T_c, the central temperature, must exceed $\bar{T}$.

TABLE 11.2. PHOTOSPHERIC ABUNDANCES OF
VARIOUS ELEMENTS GIVEN RELATIVE TO
HYDROGEN ON THE BASIS OF LOG $(N_H) = 12.00$

Atomic No.	Element	Log (N_{El}/N_H)
3	Li	1.54
4	Be	2.34
6	C	8.62
7	N	7.88
8	O	8.86
11	Na	6.30
12	Mg	7.36
13	Al	6.20
14	Si	7.45
15	P	5.34
16	S	7.30
19	K	4.70
20	Ca	6.04
21	Sc	2.80
22	Ti	4.58
23	V	4.12
24	Cr	5.07
25	Mn	4.80
26	Fe	6.70
27	Co	4.40
28	Ni	5.44
29	Cu	3.50
30	Zn	3.52
31	Ga	2.51
32	Ge	2.49
37	Rb	2.48
38	Sr	2.60
39	Y	3.20
40	Zr	2.65
41	Nb	2.30
42	Mo	2.30
44	Ru	1.82
45	Rh	1.37
46	Pd	1.27
47	Ag	1.04
48	Cd	1.66
49	In	1.28
50	Sn	2.05
51	Sb	0.42
56	Ba	2.50
70	Yb	2.28
82	Pb	1.63

Adapted from L. Goldberg, E. A. Müller, and L. H.
Aller, *Astrophysical Journal Supplement*, University
of Chicago Press 1960.

We may estimate the minimum central solar temperature differently, through a calculation rather simpler than that indicated above. Let us again use the virial theorem; the kinetic energy may be taken as $\sim GM^2/2R$ (as we saw previously, this amounts to 2×10^{48} ergs). Therefore we set

$$\frac{GM^2}{2R} = \tfrac{3}{2}Nk\bar{T}, \tag{11.24}$$

where N is the total number of gas atoms, assumed to be monatomic. If we now assume our star to be entirely composed of hydrogen, $N = M_\odot \times N_A/\mu$, where N_A is Avogadro's number (6.023×10^{23}). Thus $N = 2.4 \times 10^{57}$ hydrogen atoms and (11.24) tells us that $\bar{T} \simeq 6 \times 10^6\,°\mathrm{K}$. The more rigorous calculation tells us that $\bar{T}_\odot > 2 \times 10^6\,°\mathrm{K}$. The central solar temperature must be greater than $2{,}000{,}000\,°\mathrm{K}$; as we shall see, most detailed calculations place the central temperature around $16{,}000{,}000\,°\mathrm{K}$.

11.9   Radiation Pressure

We have so far considered a star to be composed of an ideal gas and have entirely neglected effects due to radiation and to degeneracy. The only pressure we have considered is the gas pressure, P_g; we must now inquire into the effects of radiation pressure, P_r.

When photons are normally incident on a unit area, they impart momentum to that area, so that "radiation pressure" is exerted on the surface. In this case $P_r = h\nu/c$, where ν is the frequency of the radiation. Also, $h\nu/c$ is equal to the energy density u of the electromagnetic radiation; $P_r = u$.

When, however, the photons are isotropically distributed over directions, we must average over the angle of incidence, θ. The component of momentum normal to the surface is $P_r \cos \theta$, and the effective area also varies as $\cos \theta$. Therefore

$$P_r = \frac{h\nu}{c} \cos^2 \theta = \frac{h\nu}{3c} = \frac{u}{3}. \tag{11.25}$$

From atomic physics, we know that the energy density of the radiation emitted by a black body is related to the temperature T of the body. The relation is

$$u = aT^4. \tag{11.26}$$

so that the radiation pressure is given by

$$P_r = \frac{a}{3} T^4, \tag{11.27}$$

where a is a constant $(7.7 \times 10^{-15}$ in cgs units). We see that radiation pressure depends on temperature to the fourth power, while gas pressure depends on

T to the first power. Therefore P_r is unimportant at low temperatures; it becomes important only for very hot stars. To find the temperature T_1 at which $P_r = P_g$, we set

$$\frac{a}{3} T_1^{4} = \frac{N_A k}{\mu} \rho T_1,$$ (11.28)

so that

$$T_1 = \frac{3 N_A k^{\frac{1}{4}}}{a} \frac{\rho^{\frac{1}{3}}}{\mu}.$$ (11.29)

Equation 11.29 may be evaluated numerically; the result is

$$T_1 = 3.2 \times 10^7 \frac{\rho^{\frac{1}{3}}}{u}.$$ (11.30)

We see that radiation pressure is unimportant in our sun, since any reasonable model of the sun results in $T_1 > T_{\odot}$.

11.10   Degeneracy Pressure

There are other occasions when the pressure as given by the ideal gas law cannot be expected to be correct. If the density is extremely high, for example, quantum mechanical effects must be considered. According to quantum mechanics, electrons obey Fermi-Dirac statistics, which means that they are subject to the Pauli exclusion principle. This principle says that electrons cannot have all their quantum numbers the same. For free electrons in a momentum range p to $p + dp$ (in a given volume) quantum mechanics says that there are only a finite number of momentum states available. (This number is given by $[8\pi/3]p^2 \, dp$.) When all the available momentum states are filled up, we say that the electron gas is *degenerate* and any additional electrons must have different (higher) momenta. This means that a dense star will have higher pressure than we might ordinarily expect from considering the energy kT; the additional pressure is called the degeneracy pressure, P_D, and is due to the dense electron gas.

The uncertainty principle tells us that $\Delta p \, \Delta x \gtrsim h/2\pi$, where p and x are conjugate momenta and coordinates of the electron, respectively. Hence we may write (in the nonrelativistic case) that $mv \sim \hbar/\Delta x$, where $\hbar \equiv h/2\pi$. The left-hand side of this is $mv = m\sqrt{kT/m} = \sqrt{mkT}$. If n_e is the electron density of a degenerate gas, we may estimate that the electrons are so tightly packed together that they are only $\sim \Delta x$ apart. We may write $\Delta x \sim 1/n_e^{\frac{1}{3}}$, so that we will have a degenerate gas if $n_e^{\frac{1}{3}} \gtrsim (mkT/\hbar^2)^{\frac{1}{2}}$, or $n_e \gtrsim (mk/\hbar^2)^{\frac{3}{2}} T^{\frac{3}{2}}$. A rigorous calculation based on considerations of the number of momentum states available in a completely degenerate gas of free electrons would have yielded the result that $n_e \geqslant (20m/h^2)^{3/2}(\pi/3)(kT)^{3/2}$. Now

$$n_e = \frac{\rho N_A}{\mu_e}, \tag{11.31}$$

where, as before ρ is the gas density, N_A is Avogadro's number, and μ_e is the number of atomic mass units per electron ($\mu_e = 1$ for hydrogen). Therefore we may rewrite our condition as

$$\rho \geqslant \frac{\mu_e(20)^{3/2}(\pi/3)}{N_A h^3}(kT)^{3/2}, \tag{11.32}$$

so that degeneracy is important only if

$$\rho \geqslant 2.4 \times 10^{-8}T^{3/2} \tag{11.33}$$

in a star composed of hydrogen. If we take for T the temperature believed to exist in the solar core, $16 \times 10^{6\circ}$K, we find $\rho \geqslant 1.5 \times 10^3$ gm cm^{-3} in order for degeneracy to be important. Densities this high are not met in our sun or any other typical main sequence star except, perhaps, their cores.

There is, however, a class of stars that have very high densities. The *white dwarfs* have densities ranging up to 10^6 gm cm^{-3}, three orders of magnitude higher than the condition expressed in (11.33). The white dwarfs may therefore be expected to be nearly completely degenerate; the ideal gas laws will not apply to such stars. The ideal gas equation of state (or some more suitable relation such as van der Waals') may still apply to the ions, but we must add P_D to the gas pressure.

11.11 Energy Transport

We now ask how energy is transported from the central energy source of a star to the stellar surface. There are in general three modes to consider: radiation, convection, and conduction. Conduction, however, is not important in the atmospheres of normal stars. That is, conduction is an inefficient heat transfer mechanism in gases. (We should, however, remark that conduction is important in the white dwarfs; it plays a dominant role in the very slow cooling of these dense objects.)

11.12 Radiative Equilibrium

Radiation is the primary process to be considered in most stellar envelopes. Convection will only set in when the temperature gradient is high, as we shall see.

Let us consider a slab of stellar material in radiative equilibrium. The slab is assumed to be dr thick; the hotter (inner) edge is at a temperature $(T + dT)$, and the cooler surface of the slab is at a temperature T. For simplicity's sake, let us assume the "flat star" approximation, so that we ignore the fact that the star is spherical (see Figure 11.11). The "walls" (surfaces) of the slab are of unit cross-sectional area.

The walls radiate heat; the hotter surface radiates $\sigma(T + dT)^4$, and the outer surface radiates σT^4 back to it. The net flow of heat, H, is therefore given by

$$H = 4\sigma T^3 \, dT. \tag{11.34}$$

Since $\sigma = ac/4$ (from radiation theory), we may rewrite (11.34) as

$$H = acT^3 \left(\frac{dT}{dr}\right) dr, \tag{11.35}$$

where H is now expressed in terms of the temperature gradient, dT/dr.

When photons pass through matter, some are absorbed. The absorption law is

$$I = I_0 e^{-K\rho x}, \tag{11.36}$$

where I_0 is the flux incident upon the absorber, I is the flux that survives through a slab of thickness x, ρ is the density of the absorbing material, and K is the absorption coefficient of the medium for the wavelength being considered. We see that $(K\rho)^{-1}$ is the $1/e$ length for the flux, sometimes called the "optical depth" of the medium.

Let us set

$$dr = \frac{1}{K\rho} \; ; \tag{11.37}$$

the thickness of our slab is taken equal to the optical depth. This gives

$$H = \frac{ac}{K\rho} T^3 \frac{dT}{dr} . \tag{11.38}$$

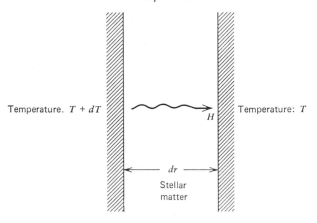

Figure 11.11   Illustration of the energy H radiated across a flat slab of stellar material.

If we now assume the star to be spherical rather than flat, we may relate the luminosity of a shell of gas at distance r from the center, L_r, to the energy passing through the shell. The luminosity of the entire star may be defined as

$$L = \int_0^R \epsilon\rho 4\pi r^2\, dr. \tag{11.39}$$

In the present case this becomes

$$L_r = 4\pi r^2 H, \tag{11.40}$$

yielding

$$L_r = \frac{4\pi r^2 ac T^3}{K\rho}\left(\frac{dT}{dr}\right). \tag{11.41}$$

A more rigorous calculation yields

$$L_r = \frac{16\pi r^2 ac T^3}{3K_\nu\rho}\left(\frac{dT}{dr}\right) \tag{11.42}$$

for the luminosity of a shell of gas that is in radiative equilibrium and at a temperature gradient dT/dr; the constant is 5.33 rather than the 4.00 calculated above.

11.13 Opacity and LTE

The quantity K_ν in (11.42) is called the opacity by astrophysicists. It is in general extremely difficult to calculate, for it depends critically on wavelength and composition.

A few general remarks may be made, however, concerning K_ν. Stellar temperatures, as we have seen, are in the range 10^6 to $10^{7\circ}$K. The maximum in the Planck distribution falls in the soft x-ray region at these temperatures, so that we need only concentrate—for a given composition—on the opacity at wavelengths from perhaps 2 Å to around 50 Å. Thus the value of K_ν will be determined principally by the photoelectric cross-section of the elements comprising the medium, since that absorption process dominates at short wavelengths.

Nevertheless, while it is true that the principal wavelength region is restricted, we must still integrate the opacity over frequency in order to use it in (11.42). We may define a *harmonic mean* of K; it is given by $\bar{K}$, where

$$\frac{1}{\bar{K}} = \int_0^\infty \frac{1}{K_\nu(1 - e^{-h\nu/kT})}\frac{d\Phi}{dT}\,d\nu \bigg/ \int_0^\infty \frac{d\Phi}{dT}\,d\nu. \tag{11.43}$$

The function $\bar{K}$ is called the *Rosseland mean absorption coefficient*, where K_ν is the absorption coefficient at a frequency ν for a material at temperature T. The function $\Phi = \Phi(\nu, T)$ is just the Planck distribution,

$$\Phi(\nu, T) = \frac{2h\nu^3}{c^2}\frac{1}{e^{h\nu/kT} - 1}. \tag{11.44}$$

The expression (11.43) for the Rosseland mean absorption coefficient may be numerically integrated over frequencies. A useful approximation in this integration is to set $K_\nu =$ a "constant" $\times \nu^{-3}$ in the x-ray region; the absorption between x-ray edges goes inversely as the cube of the frequency. The value of the "constant," however, will only be truly constant at frequencies between the discontinuities in K_ν caused by the energy levels of the atoms. Different values of the "constant" are required between different sets of discontinuities.

It is difficult to overstate the importance of the opacity to our understanding of stars. A high value of K leads to a high value of the temperature gradient; an opaque stellar atmosphere leads to a high temperature deep within the star. As we shall see in Section 11.18, the nuclear reactions that supply energy to stars are extremely temperature-sensitive; a small change in the core temperature will produce a tremendous change in the number of ergs per second that are generated, and this will greatly affect the stellar luminosity.

It is still necessary to determine how the opacity coefficient K depends on the composition of the star. The heavy elements are the most important, since the absorption coefficient is proportional to Z^2 in the photoelectric process. Clearly the absorption coefficient is also dependent on the state of ionization of the various atoms that comprise typical stellar mixtures.

We expect that the state of ionization will depend on several factors. First, it is clear that the greater the total number of atoms, the greater the number of ionized atoms; each neutral atom has a chance of becoming ionized. Next, it depends on the free electron density, since this factor tends to decrease the number of ionized atoms. Then, too, the number of ionized atoms must depend on the Boltzmann statistical factor, which tells us how the various atomic energy levels are populated as functions of temperature.

Under most conditions in stellar atmospheres, only two stages of ionization prevail at any one time, say the rth and $(r + 1)$st. We may calculate the state of ionization of the gas with the aid of the *Saha equation*, which is derived from statistical theory and may be written as

$$\log \frac{N_{r+1}}{N_r} P_e = - \frac{5040}{T} I_r + \tfrac{5}{2} \log T - 0.48 + \log \frac{2U_{r+1}(T)}{U_r(T)} . \quad (11.45)$$

In this equation N_{r+1} and N_r refer to the number of atoms per cubic centimeter in the $(r + 1)$st and rth stages, respectively. The quantity I_r is the energy in electron volts necessary to ionize the atoms from the rth stage of ionization to the $(r + 1)$st stage. The functions denoted by $U(T)$ in (11.45) are called the *partition functions* of the two ionization stages. They are defined through $U_r(T) = \sum_j g_{r,j} \exp(-I_{r,j}/kT)$, where $g_{r,j} = 2J + 1$ is the statistical weight of the level. The statistical weight of an atomic level that has a quantum number J may be determined observationally by finding the number of Zeeman states into which the level is split by a magnetic field.

Partition functions are not too easy to compute, since the indicated summation has to be truncated somewhere. Books on astrophysics, such as the one by Aller, contain tables of the values of $U(T)$ for various temperatures and elements of astrophysical interest.

In any event, the spectrum of an ionized atom is quite different from that of an un-ionized atom, and so, as the body of gas is subjected to higher and higher temperatures and more and more atoms become ionized, the stellar spectrum is modified by the emergence or strengthening of certain lines that are referred to by astronomers as "enhanced lines." As we have seen, the temperatures and pressures at which a given percentage of different elements is ionized are very different; the relative intensity of the enhanced and ordinary lines in the spectra provides a clue to these conditions.

In Saha's equation the electron pressure P_e is $N_e k T$ (where N_e is the electron density). The electron pressure may be obtained from an analysis of the Stark effect. The value of the electron density is in itself complicated; it depends on the formation of negative ions and hence on the composition and quantities known as *electron attachment coefficients*. The coefficients and their temperature dependence may be measured in the laboratory.

Only the elements hydrogen and helium need be given in calculating stellar opacities, for they comprise about 98 % by mass of the solar composition; as we have seen, the sun is a typical star. If X and Y are the abundances of hydrogen and helium, respectively, the Rosseland mean absorption coefficient per gram of stellar material may be written for "bound-free transitions" (i.e., photoelectric ionization) as an expression that is proportional to the electron density N_e and inversely proportional to the temperature T as $T^{-3.5}$, at least for photon energies greater than the energy level of the atom. Since $N_e = \rho[X + Y/2 + \frac{1}{2}(1 - X - Y)]/M_p = \frac{1}{2}\rho(1 + X)/M_p$, a good approximation for the "constant" we referred to in our discussion of K_v is $(65/M_p)(g/t) \rho/T^{3.5} (1 + X)(1 - X - Y)$. Numerically we have that

$$\nu^3 \times K_v = 3.9 \times 10^{25} \rho T^{-3.5}(1 + X)(1 - X - Y)\frac{g}{t}. \tag{11.46}$$

In (11.46) g is a quantum mechanical correction to the opacity formula called the *Gaunt factor*, and t is known as the *guillotine factor*. In most cases $g \approx 1$; the opacity, however, is quite sensitive to t, which depends on the ionization state of the atoms. The greater the number of ionized atoms present, the smaller the number that are capable of absorbing radiation at the previous (un-ionized) rate. The guillotine factor t contains the principal quantum number of the atom; $\nu^3 K_v$ is discontinuous at the various energy levels.

A rough formula for the Gaunt and guillotine factors is $g/t = A[(1 + X)\rho]^n$, where A and n are adjustable constants. Thus, since the density ρ is a function of P and T, we see that the opacity is a function of P, T, X, and Y; the concentrations of hydrogen and helium are crucially important to solutions of the equations of stellar structure.

The form of (11.43) shows that if the opacity is due to two or more substances, the absorption coefficients cannot simply be added. Strömgren introduced an empirical rule that seems to work. The opacity due to the sum of the two absorbers equals the larger of the pair plus $\frac{3}{2}$ times the smaller of the two; it is such an opacity value that is described by the approximation given in (11.46).

It is interesting that values of the opacity coefficient K are such that stellar matter is an extremely efficient absorber of radiation. Matter at any point is effectively screened from all other matter even a small fraction of a centimeter away; a photon emanating from the center of the sun has $\sim 10^{11}$ mean free paths to go before it can reach the solar surface. We may therefore think of a "photon gas" "diffusing" through the solar matter.

An enclosure whose contents have come to a perfectly steady state at a uniform temperature T is said to be in *thermodynamic equilibrium*; T will remain constant for all time in such an enclosure. The principles and ideas of classical thermodynamics may be applied to such enclosures.

Since the solar opacity is great, and since the mean free paths for collisions of atoms and ions with other particles are even smaller than the photon mean free paths, we conclude that the material within the sun is well shielded from its surroundings. Also, we shall see in the following section that the average solar temperature gradient is quite small; approximately the same temperature exists on opposite sides of a parcel of stellar matter.

These considerations imply that solar material is adiabatically enclosed material; it is (nearly) at a constant temperature and is therefore locally in thermodynamic equilibrium. We say that the condition of Local Thermodynamic Equilibrium, LTE, prevails. This is not true in and above the photosphere, where a very non-equilibrium situation exists. Because of LTE, however, it appears that we may locally apply the ideas of classical thermodynamics to a star, even though the star as a whole is not in equilibrium.

11.14 Convective Equilibrium

We indicated in Section 11.12 that convection of the stellar material will set in when a large temperature gradient is formed. At low temperature gradients radiation provides the heat transfer mechanism. In the sun the central temperature is $\sim 20 \times 10^{6}°\text{K}$. The solar radius $R_{\odot}$ is 7×10^{10} cm, so that the average solar temperature gradient is only $3 \times 10^{-4}°\text{K}$ per cm; radiation is expected to predominate in most of the sun.

There are, however, regions even in the sun where convection is important; the surface is such a region. Our sun is not particularly hot, as stars go; the vast majority of its photospheric radiation is in the optical region of the spectrum. Therefore, even though hydrogen atoms are easily the most numerous species in the photosphere, they are not the chief source of opacity near the solar surface, since transitions from the ground state of hydrogen (the Lyman series) lie in the ultraviolet.

Negative hydrogen ions are the chief contributions to the surface opacity of the sun. There are plenty of free electrons in the photosphere, and some of these attach themselves to the neutral atoms, forming these ions. The extra electrons are loosely bound, however; even photons with energies as low as a few electron volts (the optical) may readily detach them. The detachment takes place, of course, at the expense of the photon; photon absorption occurs, and so a stellar atmosphere containing negative hydrogen ions is opaque to optical radiation.

Neutral hydrogen atoms *are* important sources of opacity in the surfaces of stars where the temperatures are higher than in the sun. If the star radiates appreciably in the far ultraviolet, absorption will occur. Indeed, most

Figure 11.12   Photograph revealing "supergranulation" of the solar surface, taken from the ground. Courtesy Hale Observatories, California Institute of Technology.

important increases occur in the opacity long before we get to that temperature. If the star is sufficiently hot that appreciable numbers of the hydrogen atoms are in excited states rather than in the ground state, the atmosphere will be opaque to optical photons; a star with twice the solar temperature may be 20 times as opaque in the optical as is the sun.

Since the negative ion H⁻ has a high coefficient of opacity K, we expect from (11.42) that the temperature gradient will be high in regions that are opaque, at a given temperature T. Thus the regions near the solar surface should be in convective equilibrium. This prediction has been confirmed by balloon-borne telescopes. Even a small (~12-inch) telescope that is carried aloft to ~80,000 feet can resolve many features not detectable from the ground, for the telescope will be above the earth's turbulent atmosphere (see Section 3.8), so that the "seeing" is good.

These telescopes, flown during Project Stratoscope I, reveal the solar surface to be finely "granulated." The granules rise and fall, just as though we were looking at the tops of convective cells. To a good approximation, the visible surface of the sun may accurately be described as "boiling." *Supergranulation* of the solar surface may be observed from the ground, as in Figure 11.12. The dark edges of the cells are believed to be at the bases of chromospheric spicules; it may be, therefore, that these relatively dark regions mark the places where hot material moves vertically to and from the corona.

The first law of thermodynamics, when applied to a gram of gas in the sun, may be written as

$$dQ = dU + dW = dU + P\,dV, \tag{11.47}$$

where Q is heat, dU represents the internal energy of the gram of gas, and $P\,dV$ is the work done on the gas. For an ideal gas

$$U = N(\tfrac{3}{2}kT), \tag{11.48a}$$

or

$$U = \frac{N_A}{\mu} \times \tfrac{3}{2}kT, \tag{11.48b}$$

so that

$$dU = \frac{3}{2}\frac{N_A k}{\mu}\,dT. \tag{11.48c}$$

Our equation of state is $P = (N_A k/\mu)\rho T$; if V is the volume of 1 gm, then

$$P = \frac{N_A kT}{V}. \tag{11.49}$$

The first law (11.47) becomes

$$dQ = \frac{3}{2}\frac{N_A k}{\mu}\,dT + \frac{N_A kT}{\mu}\frac{dV}{V}. \tag{11.50}$$

Let us assume that a bubble of gas that tries to expand as it rises upward has no heat put into it by the surroundings; this is known as the adiabatic assumption and we have met it before, when we studied the earth's atmosphere. In other words, the rising bubble will transport heat (energy) from the lower to the upper reaches of the star through convection, by physically moving hot gas from below to the cooler layers above.

If no heat is supplied, $dQ = 0$ and (11.50) becomes

$$\frac{dV}{V} = -\frac{3}{2}\frac{dT}{T}. \tag{11.51}$$

Equation 11.51 may be integrated. The result is

$$T^{2/3}V = \text{constant.} \tag{11.52}$$

(We may rearrange terms with the aid of the equation of state to see the more familiar

$$PV^{5/3} = \text{constant,} \tag{11.53}$$

the well-known adiabatic law for an ideal gas.) In general, if γ is the ratio of specific heats for the gas,

$$PV^{\gamma} = \text{constant.} \tag{11.54}$$

Now a bubble expanding adiabatically from an interior layer where the pressure and density are P_1 and ρ_1, respectively, to a higher layer where P_2 and ρ_2 exist, may have a bubble pressure and density at the upper layer of $P_2{}^*$ and $\rho_2{}^*$. We see that

$$P_2{}^* = P_2. \tag{11.55}$$

In adiabatic convection there is a simple relation between the pressure and density of stellar material; it is $P = \text{constant} \times \rho^{\gamma}$, so that

$$\rho_2{}^* = \rho_1\left(\frac{P_2}{P_1}\right)^{1/\gamma}. \tag{11.56}$$

If it happens that $\rho_2{}^* > \rho_2$, the bubble is "heavy" and sinks back down; we say that the gas is *stable* against convection. However, if the bubble is "light," the gas is "unstable" and the bubble continues to rise; convection is an important energy transport mechanism when the bubble is light.

11.15   Convective Stability Criterion

A star is stable against convection if the adiabatic temperature gradient (lapse rate) exceeds the temperature gradient of the star. Radiation is the only important heat transfer mechanism for such locations within a star. Convection can be maintained inside a star only if the rising and therefore expanding material in a convective current cools off because of its expansion so slowly that its interior temperature is always higher than the surrounding temperature of the neighboring gases. Since this condition is realized only where the temperature gradient is fairly steep, most of the stellar material is not in convective equilibrium, and the energy transport is due primarily to radiation. Close to the center of hot stars, however, the temperature gradient becomes sufficiently steep that convective equilibrium does apply. Thus in general we may expect the deep interior of a star to have two zones: (a) a central core in convective equilibrium, and (b) a radiative envelope (consisting of most of the star) where radiative equilibrium exists.

It is interesting that an unstable condition tends to correct itself by making the upper layers hotter, thus reducing the magnitude of $|dT/dr|$. Thus we now wish to find the adiabatic temperature gradient $|dT/dr|_{\mathrm{ad}}$; it is the maximum gradient for radiative equilibrium. We rewrite our expression for the pressure gradient of a star as

$$\frac{dP}{dr} = -g\rho. \tag{11.57}$$

We may substitute the ideal gas equation of state into (11.57) for ρ and find that

$$-\frac{dP}{dr} = \frac{gM_p\mu}{kT}, \tag{11.58}$$

or, multiplying both sides of (11.58) by dT/dP, that

$$-\frac{dT}{dr} = \frac{gM_p\mu}{kT}\left(\frac{d\log T}{d\log P}\right). \tag{11.59}$$

Now $P = \mathrm{constant} \times \rho^\gamma$; we may insert this into the equation of state $P = N_A k/\mu(\rho T)$ and thus eliminate ρ. The consequent relation between P and T may be logarithmically differentiated, as indicated in (11.59). Equation 11.59 will thus result in the adiabatic temperature gradient, $|dT/dr|_{\mathrm{ad}}$, where

$$\left|\frac{dT}{dr}\right|_{\mathrm{ad}} = \left(1 - \frac{1}{\gamma}\right)\frac{T}{P}\frac{dP}{dr}. \tag{11.60}$$

The stability criterion thus becomes

$$\left|\frac{dT}{dr}\right| < \left(1 - \frac{1}{\gamma}\right)\frac{T}{P}\frac{dP}{dr}. \tag{11.61}$$

If the temperature gradient does not exceed the right-hand side of inequality (11.61), convection will not occur; the star will be stable and radiation alone will be important.

11.16 Summary of Stellar Structure

In the previous sections we have seen that we have four differential equations that describe the structure of a star. These may be listed as follows:

Pressure gradient:

$$\frac{dP}{dr} = -\frac{GM_r}{r^2}\rho; \tag{11.13}$$

Mass conservation:

$$\frac{dM_r}{dr} = 4\pi\rho r^2; \tag{11.12}$$

Energy loss:

$$\frac{dL_r}{dr} = 4\pi\rho r^2\epsilon; \tag{11.39}$$

Temperature gradient:

$$\frac{dT}{dr} = \begin{cases} \dfrac{3K_v L_r\rho}{16\pi acr^3 T^3} & \text{radiative equilibrium (11.42)} \\[2ex] \left(1-\dfrac{1}{\gamma}\right)\dfrac{T}{P}\dfrac{dP}{dr} & \text{convective equilibrium (11.61)} \end{cases}$$

Equation 11.39 for the loss of energy by a star applies only in a static configuration. That is, we assume that no mass ejection (such as a stellar wind or a stellar explosion) occurs, and that the star is not pulsating.

We thus have four equations and six variables (P, r, ρ, M_r, L_r, T), which is not too satisfactory a situation. There is, however, a fifth equation, the equation of state. This might be, for example, the equation of state of an ideal gas. We have seen cases, however, in which the ideal gas laws cannot be expected to apply and we must seek another equation of state.

There are two parameters in the four equations that remain to be described. These are the rate of energy generation, ϵ, and the opacity, K. Both ϵ and K depend on the composition of the star, as well as on the density and temperature of the stellar material. Hence ϵ and K are prescribed by two *constitutive relations*,

$$\epsilon = \epsilon(\rho, T, \text{composition}),$$
$$K = K(\rho, T, \text{composition}),$$

that are most difficult to write in closed form. Some physical approximations must be employed in the constitutive relations for ϵ and K.

The procedure, then, in calculating stellar models is to eliminate one of the six variables (e.g., ρ) with the equation of state. One may then pick one of the remaining five—the radius, say—an as independent variable, and express the pressure, mass, luminosity, and temperatures as functions of that variable.

Since the quantities mass and luminosity are in principle observable from the earth, we may also set some boundary conditions. The appropriate boundary conditions are $T(R) = T_e \approx 0$, $P(R) = 0$, $M(R) = M_*$, and $L(R) = L_*$, where M_* and L_* are the stellar mass and luminosity, respectively. We also have that $M\ (r = 0) = 0$ and $L(r = 0) = 0$ to guide us in constructing a model of a star. The equation of state will provide the means for calculating the composition at various depths.

Thus the general procedure for calculating a model of a star is as follows. We use our four equations to determine P, T, M, and L for an assumed value of the mean molecular weight μ, since it enters through the equation of state. The solution of the equations may start at the stellar surface, where, as we have seen, the physical conditions are best known, and may be used as boundary conditions. The four equations are then used to calculate how the values of the mass, luminosity, pressure, and temperature change over a short distance inward, thus yielding the values of these quantities at the new depth. Next the equations are used to calculate the changes over the next short distance inward. Thus, step by step, P, T, M, and L are found at successively deeper layers within the star, until the center is reached.

Our equations tell us the amounts by which P, T, $L(r)$, and $M(r)$ change as we go from, say, the nth shell, which is at a distance r_n from the center, to the shell $n - 1$ (the next innermost); we find a set of numerical values for the $n - 1$ shell by using a set of assumed quantities for the nth shell. In this way, from our initial set of values for the nth shell, we can find values for the four physical parameters everywhere in the star, always first evaluating K and ϵ at each point (which we can do from the new values of T and P that we obtain). In evaluating ϵ we should take all methods of nuclear energy generation (such as the CNO and P-P chains) into account, even though either or both may contribute very little.

Since neither the physics required for calculations of K and ϵ nor the chemical composition is known accurately, the initial solution of the four equations may not lead to zero values of the mass and luminosity at the center. One then makes trial adjustments to the chemical composition (that is the values of X, Y, and Z) as a function of depth until $M(r = 0)$ and $L(r = 0)$ do equal zero at the center of the star. Clearly this can be a tedious task; it is best accomplished with electronic computers.

When a model is found that does give the correct central boundary conditions, we have the runs of P, T, M, and L throughout the star. This model also tells us that we have the "correct" chemical composition of the star, with the reservation that our knowledge of ϵ and K may be sufficiently inaccurate to cause errors in the knowledge of the composition.

When calculations are made like this on the structure of main sequence stars, hydrogen and helium are usually found to predominate within the star; 70% hydrogen and 28% helium (by mass) are found in most models of the sun, for example. This has some implications for our understanding of stellar evolution (see Section 12.11) and has been our only means so far for determining the composition of stellar interiors. Also, present models of the sun yield a central temperature of about 16,000,000°K and a central density of perhaps 160 gm cm^{-3}.

11.17   Nuclear Energy Sources

Thermonuclear fusion provides an energy source adequate to account for the observed luminosity of the sun (and therefore presumably for other stars as well). It is perhaps well, however, to stress at this point again that it is by no means clear as yet that the processes of thermonuclear fusion have anything to do with stars; the whole idea that thermonuclear fusion is the stellar energy source must still be regarded as an attractive hypothesis. Let us now explore some of the consequences of the hypothesis.

Since hydrogen is the most abundant nucleus in the sun (some 90% by number of the nuclei present are hydrogen nuclei), we look to "hydrogen burning"; that is, to the transmutation of hydrogen into helium. The amount of energy released is just $c^2(4M_p - 4M_{He})$, or about 6.4×10^{18} ergs per gm. If we assume, for the moment, that all of the energy produced by hydrogen burning is available to supply the observed luminosity of the sun, we may estimate how long this supply of energy will last.

The luminosity is 4×10^{33} ergs per second. Thus $4 \times 10^{33}/6.4 \times 10^{18} = 6 \times 10^{14}$ gm of hydrogen are burned per second. But the solar mass is 2×10^{33} gm; if we take it to all be hydrogen, then 3×10^{18} seconds ($= M_\odot/6 \times 10^{14}$) will be required to burn all of the hydrogen, or about *100 billion years*. Of course, not all of the sun will be involved; only the core "burns," so that the lifetime of the sun is probably about 10 billion years. The sun is about 4.6 billion years old. It therefore has a long time to go before it exhausts its supply of hydrogen.

11.18   The Carbon Cycle

The first series of nuclear reactions to be proposed that burn hydrogen to form helium were given by H. Bethe in 1939. This cycle, which requires hydrogen and carbon for its operation, is as follows:

$$C^{12} + H^1 \rightarrow N^{13} + \gamma, \tag{11.62}$$

$$N^{13} \rightarrow C^{13} + e^+ + \nu_e, \tag{11.63}$$

$$C^{13} + H^1 \rightarrow N^{14} + \gamma, \tag{11.64}$$

$$N^{14} + H^1 \rightarrow O^{15} + \gamma, \tag{11.65}$$

$$O^{15} \rightarrow N^{15} + e^+ + \nu_e, \tag{11.66}$$

$$N^{15} + H^1 \rightarrow C^{12} + He^4, \tag{11.67}$$

where γ denotes a photon released in the reaction. The kinetic energy of the charged particles and the energies of the photons formed in this cycle of six reactions are available to heat the gas.

The net effect of this *carbon cycle*, as it is called [some people refer to reactions (11.62) through (11.67) as the *CNO cycle*], is to combine 4 protons and form 1 helium nucleus. The carbon consumed in (11.62) is replaced in (11.67); the carbon is not used up, and C^{12} may be regarded as a catalyst for the reactions.

The carbon cycle generates 6.4×10^{18} ergs per gm of helium that is formed. A little over half this energy, however, is taken up by the neutrinos formed in reactions (11.63) and (11.66). Neutrinos have a very small cross section for interaction with matter (typically, $\sigma \sim 10^{-43}$ cm^2), so that the neutrinos formed in the core of a star leave the star without interacting. The neutrino energy therefore is not available for heating the stellar material.

The rate of energy generation, per unit mass, ϵ_{CN}, by the carbon cycle depends on several factors. Since carbon is required to catalyze the reactions, it will depend on the concentration of that element as well as on the concentration of hydrogen, X. Another important factor is the cross-section σ. σ is in general energy-dependent, so that we expect ϵ_{CN} to be dependent on temperature.

The cross-sections may be measured in the nuclear laboratory as a function of energy; the result is that ϵ varies as T^{20} at $T \approx 16,000,000°$K, and it varies as T^{18} at $20,000,000°$K. More precisely, the rate ϵ_{CN} is given by $\epsilon_{CN} = B\rho X Z_{CN} T^m$, where B is almost constant (it varies only slightly with T). Z_{CN} is the carbon-nitrogen fraction of the heavier material concentration Z; $Z_{CN} \sim Z/3$ for sunlike stars, and m is an exponent that also depends slightly on T. It is found that m goes from 20 to 13 as T changes from $12 \times 10^6°$K to $50 \times 10^6°$K. For deep portions of stellar interiors, where $T \sim 12$–$15 \times 10^6°$K, $m = 20$ and $B = 1.6 \times 10^{-142}$, so that $\epsilon_{CN} = 1.6 \times 10^{-142}(\rho XZ/3)T^{20}$ in such places.

The rate of energy generation by the CNO cycle goes up two orders of magnitude if the temperature increases by 30%, and it therefore appears that the CNO cycle will be extremely important (dominant) for hot stars. Indeed, it is doubtful that sufficient carbon will be present in a star for the CNO cycle to operate until the stellar temperature becomes quite high; C^{12} may be formed by helium fusion at high temperatures (see Section 11.21).

11.19   The Proton-Proton Cycle

Fowler and Lauritsen proposed another nuclear cycle for burning hydrogen to form helium. This cycle, called the proton-proton cycle, does not require carbon for its operation. The reactions are as follows:

$$H^1 + H^1 \rightarrow H^2 + e^+ + \nu_e, \tag{11.68}$$

$$H^1 + H^2 \rightarrow He^3 + \gamma, \tag{11.69}$$

$$2He^3 \rightarrow He^4 + 2H^1. \tag{11.70}$$

Once again 4 protons are combined and 1 helium nucleus is formed as a result of the reactions. It is interesting that reaction (11.68) has yet to be observed in the laboratory. This is not unexpected, because nuclear theory predicts an extremely small cross-section (3×10^{-45} cm^2) at energies of the order of a few kilo-electron volts. Indeed, the calculated cross-sections are only 2×10^{-45} cm^2 ($\sim 10^{-21}$ barn) at 1 MeV, and σ is expected to be only 3×10^{-15} barn even at $E \sim 100$ MeV. It is possible that neutrino astronomy (Section 11.20) will provide some data on (11.68).

Reaction (11.70) is not the only He^3-consuming reaction available. There is another chain of reactions possible at somewhat higher temperatures. The following competes with (11.70):

$$He^3 + He^4 \rightarrow Be^7 + \gamma, \tag{11.71}$$

$$Be^7 + H^1 \rightarrow B^{8*}, \tag{11.72}$$

$$B^{8*} \rightarrow Be^8 + e^+ + \nu_e, \tag{11.73}$$

$$Be^8 \rightarrow 2He^4, \tag{11.74}$$

so that He^4 is formed again, but an additional neutrino is also liberated, in reaction (11.73).

Now the rate of energy generation by the proton-proton cycle also depends on temperature. The interesting point, however, is that the P-P cycle operates at lower temperatures than the CNO cycle requires.

If we express the rate of energy generation ϵ_{PP} by this cycle in terms of the stellar density ρ and hydrogen concentration X, we find that

$$\epsilon_{PP} = A\rho X^2 T^n,$$

where $A = 1.05 \times 10^{-29}$ at $T = 12$–15×10^6°K and $n = 4$ in this same temperature range. At a temperature of about 2×10^7°K the P-P cycle produces about as much energy as the CNO cycle does, and the P-P cycle probably produces about 100 times as much energy at the temperature of the solar center as does the CNO cycle. Therefore most astrophysicists today believe that the P-P cycle is the energy generation process predominant in the sun and cooler stars, while the CNO cycle is responsible for the energy output of hotter hydrogen-burning stars.

Thus we have at least three thermonuclear processes that may be responsible for stellar energy generation through hydrogen burning. The three differ in the amount of energy available to the star.

The burning of 4 protons to form 1 helium nucleus liberates 26.72 MeV of energy, the value of the released binding energy. The CNO cycle, however, results in neutrinos that carry away 1.69 MeV, leaving only 25.03 MeV in the alpha particle, positrons, and gamma rays. These radiation types have only short ranges, so that their energy is available for heating the star.

Reaction (11.68) generates 2.38 MeV; (11.69) liberates 10.98 MeV, while (11.70) produces 12.85 MeV. Thus this cycle makes

$$(2.38 + 10.98 + 12.85) \text{ MeV} = 26.21 \text{ MeV}$$

available to the star (0.51 MeV is lost to neutrinos). The other cycle [(11.71) through (11.74)], however, leaves only 19.1 MeV in the star; 7.6 MeV is carried off by neutrinos and is therefore not available for radiation by the star. Thus the highest energy is retained in the star by the reactions in (11.68), (11.69), and (11.70); only 2% of the total energy release is lost to neutrinos in these reactions.

11.20   Neutrino Astronomy

We have seen in the previous sections that both the CNO and the P-P cycles generate neutrinos. If we could develop techniques for detecting solar neutrinos, we could establish experimentally which reactions proceed in the solar core; quantitative measurement would provide direct information on the conditions (pressure and temperature) in those inaccessible regions. Since neutrinos do not appreciably interact with matter such as stellar atmospheres, the detection of solar neutrinos would be equivalent to "seeing into the interior of a star." Thus neutrino astronomy is potentially quite valuable, especially if neutrino fluxes can be detected from other stars as well. For example, calculations of the supernova process indicate that stars may strongly increase their neutrino emission months before they explode.

Neutrino astronomy is not easy, because of the low interaction cross-section characteristic of neutrinos. If one accepts the proton-proton cycle as being predominant in our sun, the flux of neutrinos created in reaction (11.68) may be calculated. Some 6×10^{10} neutrinos cm^{-2} second^{-1} are to be expected at the earth; the neutrinos have maximum energies of 0.510 MeV, and "low-energy" neutrinos like these are extremely difficult to detect, in spite of the huge flux. The neutrinos resulting from the decay of boron 8 (11.73), on the other hand, have energies that extend up to 14.1 MeV; interaction cross-sections of $\sim 1 \times 10^{-42}$ cm^{-2}, nearly five orders of magnitude greater than found at lower energies, are expected. These higher-energy neutrinos should be easier to detect, even though the predicted flux is smaller (4×10^6 cm^{-2} second^{-1} at the earth from the sun).

Easier, but not simple. One method for detection is to utilize the capture by Cl^{37}: $\nu_e + Cl^{37} \rightarrow Ar^{37} + e^-$. (This reaction has a threshold at neutrino energies of 0.81 MeV.) The isotope Ar^{37} is radioactive; an electron with 2.7 KeV of energy is emitted in the decay of Ar^{37}, and the detection process then involves the detection of this isotope of argon in a large chlorine detector.

It is possible to do this in practice by assembling 390,000 liters of tetrachloroethylene and collecting any argon atoms (gas) formed by bubbling helium through the C$_2$Cl$_4$. The Ar^{37} is then frozen out of the helium onto charcoal and is deposited with carrier argon in a low-background counter. But only $\sim$2–7 counts per day are expected from the solar neutrinos incident on this large a quantity of C$_2$Cl$_4$, and a severe background problem exists from the decay of fast-cosmic-ray neutrons, which also may produce 14-MeV neutrinos. The first results from this experiment are negative; no counts above the background rate of 0.8 $\pm$ 0.3 per day were detected. These data have set an upper limit on the flux of B^8 solar neutrinos of 2×10^6 cm^{-2} second^{-1} and have already compelled a theoretical re-examination of the nuclear parameters used in the calculations.

11.21 Helium Burning

One may inquire as to what happens to a star when all its available hydrogen is used up and helium has been formed. In a sense the star "stops"; no more energy is produced, whether by the CNO or the P-P process. Gravitational collapse then sets in, because no more heat is provided to supply the necessary counterbalancing pressure.

However, as the core collapses, its temperature rises because of the gravitational energy that is made available. Eventually a point is reached where a new series of nuclear reactions is ignited, which prevents further contraction. This series involves the "burning" of helium in the stellar core. A process known as the *triple-alpha* process sets in. The reactions are

$$He^4 + He^4 \rightarrow Be^8, \tag{11.75}$$

$$Be^8 + He^4 \rightarrow C^{12*} \rightarrow C^{12} + \gamma. \tag{11.76}$$

We previously (in connection with our analysis of the proton-proton chain) said that Be^8 decays into two alpha particles, but are saying here that it is stable. The reason for the difference lies in the temperature regimes encountered in the two cases.

Any nucleus is stable against decay into two others if the masses of the two candidate nuclei exceed the mass of the potential parent. At the high temperatures encountered in helium burning the alpha particles have high speed, and the theory of relativity indicates that their mass will be greater than found in stellar regions where the proton-proton chain is operative. The relativistic mass increase is just great enough to prevent the decay of Be^8.

The end result of the helium burning is that 3 helium nuclei fuse to form a single C^{12} nucleus. The *triple-alpha process* (so named because of the fusion of 3 helium nuclei) becomes important when $T \approx 10^8 °K$ and $\rho \approx 10^5$ gm cm^{-3}.

It is clear that helium burning can produce isotopes heavier than carbon-12. For example, the reaction $C^{12}(\alpha, \gamma)O^{16}$ will proceed if the temperature and density are sufficiently high; oxygen-16 will be formed. Once O^{16} is made in the star, we may form heavier nuclides: $O^{16}(\alpha, \gamma)Ne^{20}$ may "go" (i.e., become a significant energy source). It is significant in the stellar core at around 200,000,000°K. Helium burning thus produces a core (if no mixing with the outer layers occurs) consisting of C^{12}, O^{16}, and Ne^{20} at temperatures between 100,000,000 and 200,000,000°K. Heavier nuclei may be formed from these core constituents at temperatures of about $10 \times 10^8 °K$.

We see that nucleosynthesis proceeds in stars in the following manner: initial gravitational collapse heats the gas until hydrogen burning starts; the collapse stops when sufficient pressure is built up from this energy source. When this nuclear fuel (hydrogen) is exhausted, collapse sets in once again until the "ashes" (helium) of the previous reaction start burning. And so on up the periodic table up to iron, the heaviest nucleus that can be produced in this manner. The isotope Fe^{56} is at the minimum in the packing fraction curve; it is the most stable nucleus known.

11.22   Heavy Element Synthesis and Supernovae

It is interesting to inquire as to the origin of elements heavier than iron-56. In the preceding section we saw that successive nuclear reactions will lead to the formation of iron; we expect an iron-rich core to form eventually in the star.

But at this point the star must "stop." There is no more nuclear fuel left. Therefore the core gravitationally collapses once again. This time a large collapse may occur, since conventional nuclear reactions cannot occur. However, when the temperature rises to several *billions* of degrees, an iron nucleus is dissociated into 23 alpha particles and 4 free neutrons; the iron core suddenly converts to a helium core, with a large density of free neutrons present (perhaps 10^{24} neutrons cm^{-3} in a star where $M = M_\odot$).

JUNE 9, 1950   FEB. 7, 1951

NOVA IN MESSIER 101

Figure 11.13   Photographs of M101, a spiral galaxy in Ursa Major. Photographed with supernova, 7 February 1951. The galaxy is also shown in a picture taken eight months earlier; no event was then detectable. Courtesy Hale Observatories, California Institute of Technology.

In a very short time (short compared to the beta-decay lifetimes of the heavy elements) nuclides heavier than Fe^{56} may be built up by successive neutron captures. Relatively slow capture of free neutrons, the so-called s-process, can occur up to the isotope Bi^{209}. In the s-process the capture occurs so slowly that there is sufficient time between successive neutron captures for the nuclei to settle down to stable isotopes by beta-ray emission. Thus isotopes are formed that are stable; a significant neutron excess does not exist.

When Bi^{209} captures a neutron, it quickly emits an alpha particle; s-process build-up of the heavy elements terminates here. However, trans-bismuth nuclei may be cooked up through the r (for rapid) process.

In the r-process, a nucleus may capture some of the free neutrons successively, in a very short time. It is almost as though the nucleus absorbed them all "at once" and it is clear that the r-process will only take place if an extraordinarily intense neutron beam is present.

There are theoretical expectations that the build-up by the r-process may cease at $A \sim 254$, where spontaneous fission becomes important. Nuclei this

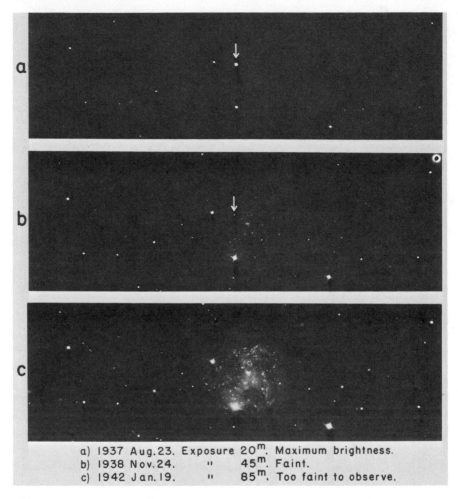

a) 1937 Aug.23. Exposure 20^m. Maximum brightness.
b) 1938 Nov.24.   "   45^m. Faint.
c) 1942 Jan.19.   "   85^m. Too faint to observe.

Figure 11.14 Photographs of supernova in IC 4182 taken with 100-inch telescope: (*a*) 20-minute exposure in 1937; (*b*), 45-minute exposure taken in 1938; (*c*) 85-minute exposure in 1942. Courtesy Hale Observatories, California Institute of Technology.

heavy are far removed from any of the nuclear "magic numbers," those values of Z and N where the nuclear levels are filled and the nuclei are spherical. These heavy nuclei are so distorted, so nonspherical, that they spontaneously break apart. It is interesting that the isotope californium-254 was first observed in the debris from the test explosion of a thermonuclear device in 1952.

Possibly the whole heavy build-up process requires only 1 minute or so; the reactions proceed at an explosive rate. It is so explosive that some astrophysicists believe that this heavy element "cook-up" represents a *supernova*

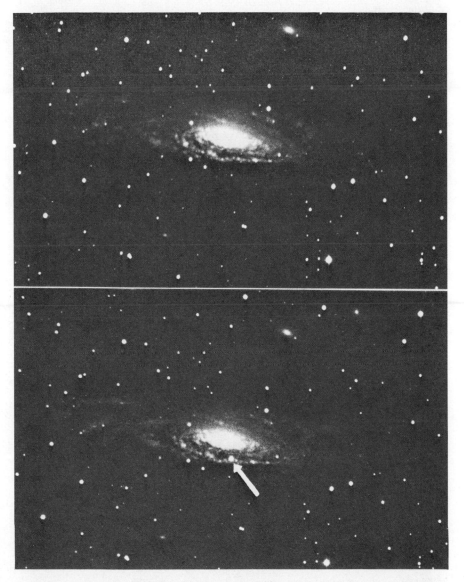

Figure 11.15 Two photographs of NGC 7331, taken with the 36-inch reflector. The top picture was taken before the supernova; the bottom was obtained near the maximum brilliance of the 1959 supernova. Lick Observatory photograph.

explosion. The products of the explosion are dispersed into interstellar space, so that the heavy elements on the earth—in this view—represent the debris of some ancient stellar explosion.

Supernovae are sudden, tremendous flare-ups in the optical output of a star. The single star may become as bright as a galaxy of stars; the stellar luminosity may increase by a factor of 10^8–10^9. Several such explosions are shown in Figures 11.13–11.15.

When the curve representing the luminosity as a function of time is integrated for Type I supernovae, it is found that $\sim 10^{50}$ ergs are liberated in the event. Type I supernovae (see Figure 11.17) exhibit exponential light-decay curves; the light curves of Type II events are much less regular (Figure 11.18).

The decay of light from a Type I event is such that the supernova is down to $\frac{1}{2}$ the light-output in 55 nights, on the average. This has reminded many of the 55-day half-life of Cf^{254}, even though supernova "half-lives" are observed to range from 30 to 100 days. In this general model, therefore, the light curve of a Type I supernova is caused by the spontaneous fission of Cf^{254}, at least for the first year or so after the peak brilliance is achieved.

Figure 11.16   Galaxy NGC 4725 in Coma Berenices. Left-hand photo was taken 10 May 1940; right-hand photo, showing supernova, was taken 2 January 1941. Courtesy Hale Observatories, California Institute of Technology.

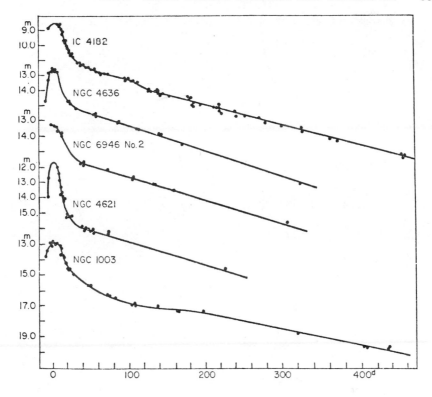

Figure 11.17 Photographic light curves of type I supernovae. There is a great similarity among them, and the "time constant" on the exponential portion of the decay averages 0.0137 magnitude daily for a "half-life" of 55 days. R. Minkowski, *Annual Review of Astronomy and Astrophysics*, 1964.

Other possibilities than the californium hypothesis have been discussed; for example, it has been found that the light curve of a Type I supernova may be explained through the hydrodynamics of a mass of expanding, decaying Ni^{56}. In turn Si^{28}, believed by nuclear astrophysicists to be produced in stellar cores by thermonuclear processes, leads to the formation of Ni^{56} through Si^{28} burning. The abundant element iron may be formed as a result of the decay of nickel. Thus both possibilities tend to link astrophysics with nuclear physics.

It is also possible to explain the light curves without invoking radioactivity at all. These models interpret supernovae as the expansion and subsequent transparency of an exploding star, the optical light curve may depend critically on presupernova structure and upon the gas density in the vicinity of the exploding star.

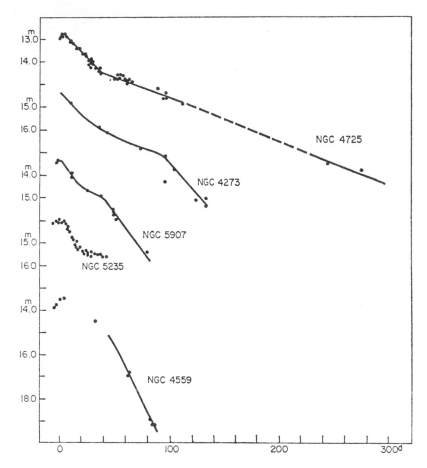

Figure 11.18 Photographic light curves of type II supernovae. Type II events are more irregular than type I explosions and may indicate that type II supernovae do not represent a uniform group of objects. From R. Minkowski, *Annual Review of Astronomy and Astrophysics*, 1964.

Problems

11.1. If θ is the angle between a normal drawn to the surface of the sun and the lines of force emerging from a sunspot, a fair approximation for θ is found to be $\theta = (\pi \cdot r)/(2a)$, where a is the radius of the spot to the outer edge of its penumbra and r is the distance from the center. Another valid approximation concerns the strength H of the spot field; it is that $H = H_0(1 - r^2/a^2)$.
   (a) Find the emergent flux $F = H \cos \theta$.
   (b) F is related to the "pole strength" m of the spot by $F = 2\pi m$; calculate in terms of the earth's dipole moment, the magnetic dipole moment of a

bipolar spot that consists of two spots separated by 175,000 km. The radius of each is $a = 20,000$ km, and the field strength is $H_0 = 3900$ gauss.

11.2. Under the conditions of the preceding problem, calculate the force of mutual attraction between the two spots.

11.3. The amount of photospheric radiation at the wavelength of Hα ($\lambda = 6563$ Å) is ~2×10^7 ergs cm^{-2} Å^{-1} second^{-1} from the quiet sun. Consider a 3b flare that has an area of 2×10^{-3} of the visible disc; Hα is observed to double in brightness over this area, in a wavelength interval of 2 Å. The flare lasts for 10^4 seconds. Measurements at the earth reveal that about 10^{30} ergs were emitted by the solar flare in the form of particles; compare this with the total energy radiated at Hα.

11.4. The total half-width of a solar spectral line $\delta\lambda$ is given by $2(\lambda - \lambda_0) = 2([2\lambda_0^2 kT \ln 2]/c^2 M)^{1/2}$ for a molecule of mass M at a temperature T, radiating at a central wavelength λ_0. Consider the Fe^{56} XIV $\lambda 5303$ line in the solar corona and assume the gas kinetic temperature to be 2×10^6°K. Find $\delta\lambda$.

11.5. The companion of Sirius has a mass equal to $0.98M_\odot$ and a radius $R = 0.02R_\odot$. Find the average pressure in this star.

11.6. A gaseous sphere of the same mass and radius as the sun has a constant temperature of 3×10^6°K. Calculate the following:
(a) The total energy stored in radiation.
(b) The total energy stored in gas kinetic energy, assuming the mass to be composed of hydrogen.
(c) The total energy stored as energy of ionization.

11.7. Two stars are similar in their mass density distributions, with ρ in both varying as r^{-1} from the center. If the first star has 10 times the mass and 10 times the radius of the second, what is the ratio of central densities of the two stars?

11.8. Show that the "ratio of specific heats" for a pure photon "gas" (radiation) isotropically distributed in an enclosure is 4/3.

11.9. Show that the mean molecular weight μ of completely ionized material inside a star is

$$\mu = \frac{1}{1.5X + 0.25Y + 0.5},$$

where we assume that the hydrogen content X and the helium content Y are known. (Thus μ may be replaced in our equations by the observables X and Y.)

11.10. What is the mean molecular weight of the material inside a star if ionization is complete and if the hydrogen content is 80% and the helium content is 15%?

11.11. Express the luminosity of a shell of stellar material in radiative equilibrium as a function of P, T, and the composition of the material, if the material contains only hydrogen and helium.

11.12. The radiation pressure in the photosphere of a star is 25.1 ergs cm^{-3}.
   (a) What is the temperature in this region?
   (b) If this star has the same absolute luminosity as the sun, what is its radius in $R_\odot$?

11.13. (a) What fraction of calcium atoms are in the singly-ionized condition (known as CaII) in the atmosphere of Sirius if $T = 10,000°$K and $P_e = 300$ dynes cm^{-2}? For calcium $I = 6.11$ eV and $\log 2\, u_1(T)/u_0(T) = 0.18$.
   (b) The second ionization potential of calcium is 11.87 eV and $u_0 = 1.0$ for doubly-ionized calcium, leading to $\log 2\, u_2(T)/u_1(T) = -0.25$. Is the ionized calcium predominately CaII or CaIII?

11.14. The effective temperatures of the sun, Sirius, and Rigel are 5700°K, 10,000°K, and 15,000°K. If the electron concentrations N_e are 4×10^{13}, 2.8×10^{24}, and 3×10^{15} cm^{-3}, respectively, compare the number of ionized hydrogen atoms relative to neutral hydrogen atoms in the three stellar atmospheres.

11.15. Assume that both the guillotine and Gaunt factors are equal to unity at a point inside a star where the density $\rho = 100$ gm cm^{-3}. If the temperature T at this point is $15 \times 10^{6°}$K and the hydrogen content $X = 85\%$ while $Y = 12\%$, what is the opacity K at this point?

11.16. Show that in the convective core of a star the differential equation $dP/P = \gamma/(\gamma - 1)\, dT/T$ is valid.

11.17. (a) Compute the energy carried to the surface of the sun by a granule of diameter $d = 10^8$ cm, vertical velocity 0.5 km per second, and density $\rho = 10^{-7}$ gm cm^{-3}.
   (b) If at any time there are 10^6 granules on the solar surface, with an average lifetime of 300 seconds for each, what is the rate of energy transport by the granules?
   (c) The corona and chromosphere together apparently radiate some 10^{27} ergs per second. Is there sufficient energy transported to the base of the chromosphere to supply this loss?

11.18. Suppose that at a point $r = 10^{10}$ cm from the center of a star the pressure P is known to be 10^{12} dynes cm^{-2} and the temperature T is known to be $10^{6°}$K, while $L(r) = 10^{33}$ ergs per second there. Assume that the mean molecular weight $= 0.5$ and that the opacity at this depth is 4 and independent of depth in this region. How much different is the temperature $\Delta r = 100$ km farther out toward the surface? (It is clear that to obtain most accurate results in calculations like this, Δr should be made very small, leading to extensive calculations.)

11.19. Is the star in the previous problem in convective or radiative equilibrium at $r = 10^{10}$ cm? Assume $\gamma = 5/3$.

11.20. Compare the rates of energy generation by the carbon cycle and the proton-proton chain at a point inside a star like the sun where the temperature is 1.5×10^{7}°K, the hydrogen content $X = 0.85$, and the heavy element content Z is 0.03.

11.21. Show, from the expressions for ϵ_{CN} and ϵ_{PP}, that the CNO cycle takes place in a core close to the center of a star, whereas the proton-proton chain operates at much greater distances from the center.

11.22. Assume that the proportion of carbon plus nitrogen to the total amount of the heavier constituents of a stellar interior is 12%. For $X = 0.60$, $Y = 0.35$, $T = 20 \times 10^{6}$°K, and $\rho = 100$, calculate the relative energy outputs by the carbon cycle and by the proton-proton reaction.

11.23. At what temperature, for the preceding values of ρ, X, and Y, will the energy output by the two mechanisms be equal?

11.24. (a) Why is one more likely to find a convective core in a star where the carbon cycle is dominant than in one where the proton-proton chain dominates?

(b) Be^8 is formed in the triple-alpha process, but since the sum of the masses of 2 He^4 nuclei is smaller than the mass of Be^8, the latter decays spontaneously into the former. Why then should there be a sufficient amount of Be^8 present to allow the formation of C^{12} in some stars? (Assume the primordial concentration of Be^8 to be zero.)

11.25. Assume that a star in radiative equilibrium has only a radioactive energy source, so that the energy production per unit mass is a constant ϵ_0. Also assume that the opacity (due to electron scattering), k, is a constant k_0.
(a) Find the relation between L_r and M_r.
(b) Find the relationship between P and T.

11.26. A star whose distance is calculated by heliocentric parallax is found to have a luminosity $L = 10^4 L_s$, where L_s is the total solar luminosity. Assume that the measured energy, distributed as black body radiation, is emitted from a region where the density $\rho/\mu = 10^{-2}$ g cm^{-3}, and that the (photon) radiation pressure is equal to 8.2×10^{-11} of the (perfect) gas pressure in this region. Find the following:
(a) The effective temperature of the emitting region.
(b) The star's effective radius in terms of a solar radius.

11.27. Consider a star whose density profile is described by the equations

$$\rho(r) = \rho_0 \left(\frac{1 - r^2}{R^2} \right), \qquad 0 \leqslant r \leqslant \frac{R}{2}$$

and

$$\rho(r) = \frac{3}{2} \rho_0 \left(\frac{1 - r}{R} \right), \qquad \frac{R}{2} < r \leqslant R,$$

where ρ_0 is the central density and R is the photospheric radius of the star.
(a) What is the total mass of the star?
(b) What is the total gravitational energy of the star?
(c) What is the central pressure of the star?
(d) If the stellar material is assumed to be a perfect gas, what is the temperature profile of the star?
(Give all answers in terms of the constants R, ρ_0, G, k [Boltzmann gas constant], and so on.)

11.28. Assume that a stable star in static equilibrium can be described by an inner spherical core generating nuclear energy at a total rate Q (ergs per second) [i.e., $Q = \int_0^{R_0} 4\pi r^2 \epsilon(r) \rho(r)\, dr$, where R_0 is the radius of the core, $\epsilon(r)$ is the energy generation rate per gram, and $\rho(r)$ is the density]. Assume a perfect gas equation of state.
(a) If the star's photosphere has an effective temperature T_0, what is the radius of the star in units of R_0?
(b) If Q, R_0, and T_0 are all doubled, what happens to the radius?

Chapter 12

THE EVOLUTION OF STARS

The sun and the other stars seem to be unchanging. Yet stars shine; they produce energy and must therefore be steadily consuming their fuel supply. We therefore conclude that stars *must* change with time, even though the times involved are huge on our time scale. The study of how stars form and how they evolve after they have begun shining is a most active area of research in space science, the study of stellar evolution.

12.1 Stellar Formation

We believe that what we call a star starts off as a cold cloud of gas atoms and dust. Many such clouds may be seen in our galaxy alone; they make themselves known by obscuring the stars that lie behind them (see Figure 12.1.) In particular, dark globules (small patches of dark nebulous material) are observed in and around clouds of gas glowing by fluorescence. Once thought to be holes in the sky, these globules presumably represent the "prestar" stage of evolution; they are formed by self-gravity and radiation pressure by the surrounding gas. Examples are shown in Figure 12.2. Once formed, the globular mass increases rapidly through accretion; it may double its mass in only 3×10^7 years.

The dust particles in a gas cloud absorb radiation from various directions. They are therefore driven together by radiation pressure into a globule that grows with time.

A cloud of gas will condense under self-gravity, if the temperature is sufficiently low. As the condensation proceeds, the temperature of the particles steadily rises, because gravitational potential energy is made available. Eventually the temperature rises to the point ($\sim 10^{7 \circ}$K) where nuclear reactions set in; the constituents (principally hydrogen) begin to burn. As we outlined in Chapter 11, the proton-proton chain is the first to set in. If carbon-12 is available, the CNO cycle will become significant at higher temperatures.

When the nuclear reactions start, the contraction ceases; sufficient pressure is built up by the evolved nuclear energy to balance the weight of the overlying material. A star has been born. It seems likely that the so-called Herbig-Haro objects (see Figure 12.3) are representative of this stage. One of these has actually been observed to "turn on"; FU Orionis increased in brightness

Figure 12.1   Horsehead nebula in Orion. The obscuration is caused by interstellar dust. Courtesy Hale Observatories, California Institute of Technology.

by a factor of ~ 100 in 120 days, presumably while it collapsed. It then stabilized in brightness.

There is little doubt that star formation is taking place today, as well as having occurred in the past. The objects called *T-Tauri* stars are found in association with dark nebulae in the galaxy. (Also found around galactic dust clouds are the intensely hot O and B stars [see Section 12.4], which are— as we shall see—more massive than the sun and presumably much younger; they are only $\sim 10^7$ years old.)

The T-Tauri stars are irregular variables that lie in spectral classes G, K, and M (see Table 12.1); rapid and erratic brightness fluctuations of a few magnitudes are observed in the initial contraction phase that we have been discussing. The association of T-Tauri objects with young stars is taken to mean that they too are "new"; presumably the differences are caused by mass differences, with a typical T-Tauri mass being about 1 $M_\odot$. One remarkable feature is that they are hardly in hydrostatic equilibrium; mass ejection has been detected. The detection technique is interesting.

One observes spectral lines in these objects that are split; one component is

Figure 12.2 Section of the nebula NGC 2237, photographed with the 48-inch Schmidt telescopy in red light. Notice the many dark globules which may be stars beginning to form. Courtesy Hale Observatories, California Institute of Technology.

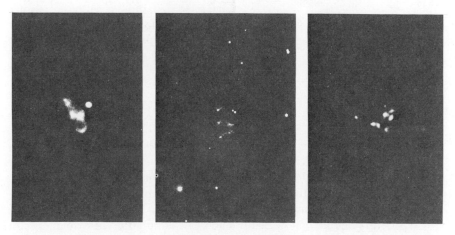

Figure 12.3 Three Herbig-Haro objects, photographed with the 120-inch telescope, may be stars in the process of becoming visible. Lick Observatory photograph.

TABLE 12.1. SPECTRAL SEQUENCE

Spectral Class	Color	Approximate Temperature (°K)	Principal Features
O	Blue	>25,000	Relatively few absorption lines in observable spectrum. Lines of ionized helium, doubly ionized nitrogen, triply ionized silicon, and other lines of highly ionized atoms. Hydrogen lines appear only weakly
B	Blue	11,000–25,000	Lines of neutral helium, singly- and doubly-ionized silicon, singly-ionized oxygen, and magnesium. Hydrogen lines more pronounced than in O-type stars
A	Blue	7,500–11,000	Strong lines of hydrogen. Also lines of singly-ionized magnesium, silicon, iron, titanium, calcium, and others. Lines of some neutral metals show weakly
F	Blue to white	6,000–7,500	Hydrogen lines are weaker than in A-type stars but are still conspicuous. Lines of singly-ionized calcium, iron, and chromium, and also lines of neutral iron and chromium are present, as are lines of other neutral metals
G	White to yellow	5,000–6,000	Lines of ionized calcium are the most conspicuous spectral features. Many lines of ionized and neutral metals are present. Hydrogen lines are weaker even than in F-type stars. Bands of CH, the hydrocarbon radical, are strong
K	Orange to red	3,500–5,000	Lines of neutral metals predominate. The CH bands are still present
M	Red	<3,500	Strong lines of neutral metals and molecular bands of titanium oxide dominate

at the laboratory wavelength for the particular element, but the other component is blue-shifted. This shift is interpreted as being due to the Doppler effect operating on the light from a mass of gas that has been ejected from the star and is moving toward the observer. Perhaps 10^{-8} $M_\odot$ is ejected per year from a T-Tauri star.

Considerable effort is now being made to examine the ultraviolet spectra of objects near dust clouds in the galaxy. Ultraviolet light is not very penetrating, so that the observations must be conducted above the earth's atmosphere, with rockets and satellites.

One expects that in general young stars will be hot and luminous, other things being equal. A hot star will probably emit most of its light in the ultraviolet, according to the Planck law. Thus extensive observations conducted in the ultraviolet should reveal features of the stars in the universe as they form.

12.2 Stellar Magnitudes

The history of a star must be described in terms of observable quantities. The first thing we notice about a star is its apparent brightness. The brightness of a star may be measured in terms of its *apparent magnitude* (optical brightness at the earth) and its distance from the earth. The apparent magnitude of a star, m, may be defined in terms of the magnitude m_2 of another celestial object through the relation

$$m_1 - m_2 = 2.5 \log \left(\frac{B_2}{B_1} \right), \tag{12.1}$$

where B_1 and B_2 are the respective brightnesses. A *factor* of 100 in brightness corresponds to a magnitude *difference* of 5. Astronomers prefer to talk in terms of magnitudes rather than brightness, because B varies tremendously (by many orders of magnitude) for the various stars. The *brighter a star, the less its magnitude*. A star that has $m = +1.0$ is exactly 100 times as bright as a star that has $m = +6.0$.

The naked eye can just perceive stars of magnitude $+6$; these stars are the last to become visible as night comes on. The first to become visible at dusk are the 100-fold brighter first-magnitude objects. Since this approximate scale of magnitudes was set up, however, it has become apparent that there are objects it does not describe. The planet Venus, for example, is 100 times brighter than a first-magnitude star. Hence Venus has been assigned a negative figure, an apparent magnitude of -4.0. The sun has an apparent magnitude of -26.72, while the full moon's brightness is characterized by $m = -12$.

A quantity related to m is the *absolute magnitude M*. This is the apparent magnitude that a star would have if it were at a distance of 10 parsecs (1 psc = 3.26 light-years). The distance to a star must be known if its absolute magnitude is to be measured. Since the sun has an apparent magnitude of -26.72 at $r = 1$ A.U., and since brightness varies as r^{-2}, the absolute magnitude of the sun is $+4.7$. The sun would be dimly visible if it were at a distance $d = 32.6$ light-years from the earth, since $m - M = 5 + 5 \log d$. The *luminosity* of a star is intimately related to its absolute magnitude. In units of solar luminosity $L_{\odot}$, for example, stars with the same M have equal L-values.

This last statement does not take into account the spectrum of the star; differences in stellar spectra require that corrections for them be made. Such a correction is called the *bolometric correction*. The human eye (for that matter, any optical instrument), doesn't perceive all wavelengths equally well. Each light detector has some spectral response factor that must be taken into account; this factor is the bolometric correction. A star that emits at the same wavelengths most easily perceived by the human eye has a bolometric correction of zero applied to it; the visual magnitude of the sun is placed equal to its bolometric magnitude. However, a very blue (very hot) star might have a calculated bolometric correction of as much as −8 magnitudes; red stars are assigned positive bolometric corrections.

The *difference* between the photographic magnitude (i.e., that recorded with a blue-sensitive photographic plate) and the photovisual magnitude (recorded with yellow-sensitive film) of a star is called its *color index*, CI; $CI = m_{pg} - m_{pv}$. Since the development in the 1950s of the more precise methods of photoelectric photometry, it has become customary to measure the light of stars in three or more spectral regions instead of only two. The so-called UBV system is now in general use. The UBV system utilizes filters in the ultraviolet (U), blue (B), and green-yellow or visual (V) spectral regions. It provides two independent color indices, blue-visual and ultraviolet-blue. The absolute calibration of the system is shown in Table 15.2. The colors of stars may be used to correct our estimates of stellar distances. Interstellar dust causes *reddening* of the spectrum, through Rayleigh scattering. (This is the same effect that causes the sky to appear blue and the sun seem to be redder at dawn and sunset). If we "know" the "true" luminosity of a star—through its spectrum, say—and we can measure how much redder its apparent spectrum is, we can deduce how much brighter the star actually is than the observed magnitude. This in turn enables us to correct the initial distance estimates.

12.3   Stellar Surface Temperatures

Another observable characteristic of a star is its color. The color of the visible surface may be related to the surface temperature, T_e (often called the *effective* temperature). We saw in Section 4.2 how the effective temperature of the sun might be measured, through measurements of the solar constant, the diameter of the sun, and the astronomical unit.

Stars other than the sun are so far away that it is not practical to attempt a measurement of the "stellar constant." Instead one may measure the spectrum of the continuum radiation from the star and find that wavelength λ_{max} where the emission is a maximum. If one differentiates Planck's expression for the radiant energy per unit volume at wavelengths between λ

and $\lambda + d\lambda$, $\psi(\lambda)\, d\lambda$, where

$$\psi(\lambda)\, d\lambda = 8\pi \frac{d\lambda}{\lambda^4} \frac{1}{e^{h\nu/kT} - 1} \tag{12.2}$$

with respect to λ, and sets the derivative equal to zero, the *Wien law* results. The Wien law is

$$\lambda_{max} T = \text{constant.} \tag{12.3}$$

In cgs units, where λ is in centimeters and T in degrees Kelvin, (12.3) reads

$$\lambda_{max} T_e = 0.2898. \tag{12.3a}$$

A measurement of λ_{max} yields the surface temperature T_e of the star.

Stars are classified by astronomers on the basis of their surface temperatures. In the optical region of wavelengths there are seven such "spectral classes", labeled O, B, A, F, G, K, and M ("Oh, be a fine girl, kiss me," is one way of remembering them). Class O stars are the hottest; they shine blue-white because $T_e \sim 50,000°K$. Class M stars, on the other hand, are the coolest stars. They are so cool ($T_e \approx 3000°K$) that they appear to be dim red, if of the same size. The sun ($T_e = 5700°K$) appears yellow; it is in class G.

Like all classification schemes, the spectral classification turns out to require subgroups. The letter classification is followed by a number that ranges from 0 to 10; the number indicates how close a star is to the next cooler letter classification. A K5 star is halfway between K0 and M0. The sun is a G2 star.

The most accurate means of assigning spectral classes to stars is on the basis of the spectral lines observed in their light. Table 12.1 provides a summary of the spectral sequence, together with some illustrative examples. The lines that we refer to here are the dark lines due to absorption of the (continuous-spectrum) light emitted from regions of the star that are lower down (i.e., closer to the stellar center).

12.4   The Hertzsprung-Russell Diagram

The Hertzsprung-Russell (H-R) diagram is extremely important for astrophysics. It is the chief source of information on stellar evolution.

The H-R diagram is a plot of brightness versus color for the various stars. *Each star represents one point on the diagram.* Alternatively, if the various stellar distances are known, the H-R diagram may be put in terms of $(L/L_\odot)$ versus T_e for the various stars. For the sun $L/L_\odot = 1$; the quantity $\log (L/L_\odot) = 0$ for the sun. Also, since $T_\odot = 5770°K$, $\log T_e = 3.5$ for the solar point. A typical H-R diagram for the stars near the sun is shown in Figure 12.4.

The most remarkable feature of the H-R diagram is that the points are *not* randomly distributed over the entire diagram. Instead almost 90% of the

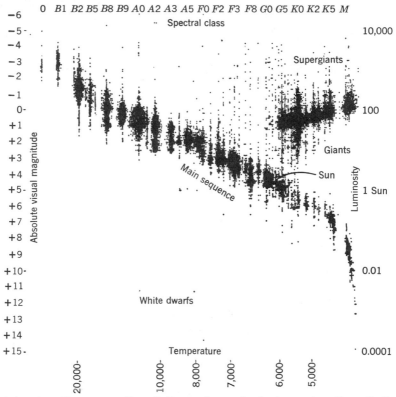

Figure 12.4   Hertzsprung-Russell diagram for randomly chosen stars. From *Stellar Evolution* (O. Struve, Princeton, N.J.), Princeton University Press, 1950.

points fall along a nearly straight line that runs from the upper left to the lower right corner of the diagram. The curve that is the best fit to these points is called the "main sequence." The sun is a main sequence star, and it apparently falls nearly exactly in the middle of the main sequence. This is the basis for our earlier assertion that the sun is a very typical star.

Stellar luminosities are proportional to the surface areas of the stars and to the fourth powers of T_e. That is,

$$L_* = 4\pi R_*^2 \sigma T_e^4, \tag{12.4}$$

where R_* is the stellar radius. Thus a star that has $T_e \approx T_\odot$ but is much brighter than the sun is presumably much larger than our star; the star is referred to as a "giant."

In addition to the main sequence stars there is another stellar grouping evident on the H-R diagram: the red giants. The diameter of Betelgeuse is ~500 times that of the sun. Betelgeuse is an extremely tenuous globe of gas;

a density of only 10^{-7} of the solar density has been inferred. A few stars lie between the main sequence and the red giants; these lie on what is called the "horizontal branch" of the H-R diagram.

There are many other stellar groupings, to be sure. For example, the "supergiants" are poorly understood stars that are positioned on the diagram all across the top of the plot. They are about 10,000 times as luminous as the sun (log $[L/L_{\odot}] \approx 4$), and—even more odd—they have surface temperatures that range from ~3000°K to beyond 30,000°K.

A numerous and fascinating group is found in the lower left-hand corner. These are extremely hot stars that are intrinsically quite faint. We deduce from (12.4) that their radii are quite small, perhaps only about $10^{-3}R_{\odot}$, so that these stars—which presumably have masses almost as large as that of the sun—have diameters about that of the earth. These are the famous "white dwarfs." The smallest star known is a 16th-magnitude white dwarf that has a diameter intermediate between those of the moon and Mercury. The material in white dwarf stars is presumably at densities of the order of 10^5–10^6 gm cm^{-3}, more than 10,000 times the density of lead.

There are also, as one might guess, "red dwarfs" in evidence. Indeed, red dwarfs are the most plentiful stars in the solar neighborhood of the galaxy. They are like the white dwarfs in that L is 0.1–1 % of the solar luminosity, but the surface temperature of the red dwarfs is ~3000°K. Red dwarfs such as Proxima and Krüger 60 are about the size of the larger planets of the solar system; they are a little larger than the white dwarfs.

12.5   Main Sequence Stars

The sun burns hydrogen to form helium. Since the sun is apparently typical of the stars on the main sequence, we conclude that they too are hydrogen burners. Some 90% of the stars in the galaxy (and perhaps beyond) are presently deriving their energy from the conversion of hydrogen to helium.

The beginnings of an "evolutionary track" may now be seen. The cloud of gas gravitationally contracts until hydrogen-burning nuclear reactions set in. On the H-R diagram this corresponds to the point moving from right to left. The point probably moves somewhat upward as well, since the temperature of the gases is increasing. The luminosity increases slightly, as the nuclear "fires" start to "burn." There should, however, be a very slight drop in luminosity once gravitational contraction ceases, because gravitational potential energy is no longer released.

Once the nuclear reactions start, the star is "on" the main sequence. It does not move (to a first approximation) *on* the main sequence while it is burning hydrogen. The vast majority of a star's life is spent on the main sequence. Since stars are found all along the sequence, we ask what decides the point on the main sequence at which the star is to be found.

The answer lies in the *mass* of the gas cloud, which is assumed here, for the sake of simplicity, to be composed of 100% hydrogen. A cloud that is very massive ($M > M_\odot$) will have much more hydrogen to burn per second and will therefore be much more luminous than a cloud of small mass. Hence we expect that the stars in the upper left-hand portion of the main sequence will be very massive stars, while those at the lower right-hand side will have stellar masses smaller than that of the sun. A cloud that is more massive than the sun will contract leftward to a higher position (than the sun) on the main sequence.

These ideas are supported by the few "direct" measurements of stellar masses that have been made. If we plot the luminosity of a star as a function of its mass, we find that we may approximate the result as

$$\frac{L}{L} = \left(\frac{M}{M_\odot}\right)^n, \tag{12.5}$$

where the exponent n is in itself rather mass-dependent. For $M \approx M_\odot$, $n = 5$; a star only 20% more massive than the sun will emit 2.3 times as much energy per unit time as does the sun. Observation indicates that $n = 3$ when $M \approx 9M_\odot$. A common approximation is to set $n = 3.5$ in (12.5) over a large range of stellar masses; stellar luminosities are (nearly) dependent on the masses to the fourth power.

The measurement of stellar masses presents a most vexing and important problem. Equation 12.5 is crucial to our understanding of astrophysics and yet is based on only about 50 measured stellar masses! And even these are only the masses of components of binary systems (so that Kepler's third law may be employed); a worrisome thought is that these may not be truly representative of all stellar masses. It is to be hoped that more mass determinations will be made in the future.

The existence of a mass-luminosity relationship like (12.5) is the reason for the "linear" appearance of the main sequence. That is, massive stars are hot stars (they evolve energy rapidly) and they are also bright. The main sequence is interpreted as the locus of points on the H-R diagram representing stars of essentially similar chemical composition but different mass. The observed scatter about the main sequence presumably is due to slight differences in the chemical compositions of actual stars.

12.6 Lifetimes

Now we may form an estimate of how long a star will remain on the main sequence. This amounts to asking how long hydrogen burning in the stellar core will be the main energy source. If the energy liberated on the main sequence is E ergs, then

$$E = F \times M \times 6.4 \times 10^{18}, \tag{12.6}$$

where M is the stellar mass in grams and, as we saw in the preceding chapter, 6.4×10^{18} ergs are liberated per gram. If we again assume that the star originally consisted of 100% hydrogen, the quantity F is the fraction of the mass of hydrogen that is consumed before gravitational contraction sets in. According to detailed numerical computations of stellar models, the fraction is somewhat mass-dependent, but roughly 10% of the stellar mass is consumed before contraction becomes serious; $F \approx 0.10$.

Now we also have that

$$E = L \times T_{\text{M-S}}, \tag{12.7}$$

where, as before, L is the luminosity of the star in ergs per second, while $T_{\text{M-S}}$ is the lifetime on the main sequence. We may equate the right-hand sides of (12.6) and (12.7) and employ (12.5) to eliminate the luminosity. The result is

$$T_{\text{M-S}} = 11 \left(\frac{M}{M_\odot} \right)^{-3}, \tag{12.8}$$

where we have taken $n = 4$ in (12.5) and $T_{\text{M-S}}$ is expressed in billions of years. The sun, according to (12.8), will spend about 11 billion years on the main sequence. Since it is only 4.6 billion years old, it has a long time to go. This is the same result—somewhat generalized—that we arrived at in Section 11.17.

It is useful to rewrite (12.8) in terms of other parameters. Since the luminosity is sometimes easier to observe than the mass of a star, we may rewrite (12.8) as

$$T_{\text{M-S}} = 11 \left(\frac{L}{L} \right)^{-3/4}. \tag{12.9}$$

Relation 12.9 indicates that the extremely hot stars that we describe as types O and B are very young. The relation indicates that their ages cannot be appreciably greater than approximately 10 million years.

Since there is more hydrogen than anything else in stars, we expect the lifetime on the main sequence to be considerably greater than any other period in the history of a star. Thus a star spends the vast majority of its total life on the main sequence, burning hydrogen.

12.7 Nuclear Reactions on the Main Sequence

We saw in Chapter 11 that two sets of hydrogen-burning nuclear reactions were possible. They were the CNO cycle (Section 11.18) and the P-P cycle (Section 11.19). We seek now to determine which set predominates for a particular star.

We saw in Section 11.18 that the CNO cycle will be predominant in the hotter stars. In the present context, therefore, the CNO cycle will be the major energy source in stars more massive than the sun. Detailed calculations indicate that the two sets liberate equal amounts of energy when $M = 1.3 M_\odot$; only 1 % of the sun's energy output is derived from the CNO cycle. About 99 % of the solar output is generated by the P-P cycle. Equation 12.8 tells us that the more massive stars have shorter lifetimes; at any given time there should be fewer of them. Hence most of the stars in the galaxy derive their energy from the P-P cycle.

These ideas gain support from the observation that there are apparently fewer stars in the upper left-hand end of the main sequence than there are at the lower right. This, however, is not completely definite, for the intrinsically fainter stars are harder to observe; it is just this factor that leads astronomers to believe that there are more faint stars than bright ones.

12.8   Evolution on the Main Sequence

Equation 12.5 should not be interpreted as meaning that the luminosity of a main sequence star is strictly constant. Stellar evolution along the main sequence depends on the value of the mean molecular weight μ and on the degree of internal mixing. As we have seen, the proton-proton chain becomes operative as the star enters the main sequence; the central temperature is then of the order of $10^{7\circ}$K.

But as this thermonuclear process goes on, the mean molecular weight of the star increases, since more and more hydrogen is transformed into helium. An analysis of the equations of stellar structure will show that an increase in μ results in an increase in the interior temperature of the star. This increase in T accelerates the proton-proton reaction rate; helium is formed even faster, so that the evolutionary rate tends to increase. It follows from this that as long as the chemical composition remains uniform through-out the star (because of mixing), the star moves up along the main sequence. This luminosity increase occurs very slowly at first, but more and more rapidly as the hydrogen is used up.

If we assume that a thoroughly mixed star consisted initially of hydrogen only, it cannot have evolved far beyond the sun's position on the main sequence, since under such conditions only the proton-proton chain can operate and this cannot release energy fast enough to account for the luminous B and O stars. The reason for this is that there is no carbon present and therefore no way in which the CNO cycle can proceed. However, if a star starts out with some heavy matter (e.g., carbon, nitrogen, and some metals), it may evolve past the sun on the main sequence, assuming that it remains thoroughly mixed.

It now appears that the great majority of stars are chemically inhomo-geneous. The sun is believed to be one such star. If the sun was initially homogeneous and consisted primarily of hydrogen, the variation in temperature throughout the star has caused hydrogen burning to occur at different rates in the star. Consequently there now exists a gradation in the hydrogen content, ranging from zero in the central core (a helium core) to its original value near the surface.

If one takes the spectroscopic data on the sun ($X = 0.80$, $Y = 0.18$, $Z = 0.02$) to be accurate and representative of the original composition, a solar model may be constructed and its evolution in time followed by a computer. We find from the evolutionary study that the present solar luminosity is about 1.6 times the initial luminosity (the luminosity 5×10^9 years ago), the radius is about 1.04 times the initial radius, and T_e now is about 1.1 times the original T_e.

Calculations like these provide the basis for our earlier assertion that, to a first approximation, the point representing the star does not move during its life on the main sequence. We see that the solar temperature and luminosity only varied by 10% and 60%, respectively, during 5 billion years. These small changes become quite tiny on logarithmic scales, such as those employed in the H-R diagram, but it is nevertheless true that one expects the whole main sequence to shift upward a trifle as time progresses.

One conclusion that has been drawn from the calculations on the sun is that in the early Precambrian era the sun was a smaller, cooler, less luminous and redder star. The average temperature on earth may have been close to the freezing point of water then, all other parameters such as the albedo being equal. (See, however, Section 16.10.)

The evolutionary study also tells us about the present-day makeup of the sun. We find that there is a convective zone extending from the photosphere down to where the temperature is $\sim 1 \times 10^{5\circ}$K. This occurs at $r = 0.8R_\odot$. The central density and temperature are also model-dependent, although values of ~ 130 gm cm^{-3} and $15 \times 10^{6\circ}$K are commonly quoted.

12.9 The Red Giants

When approximately 10% of the hydrogen in the star is consumed, the nuclear reactions that characterize the main sequence (hydrogen to helium transmutation) cease for lack of fuel, and the core of the star gravitationally contracts. During this contraction phase, energy is now available from two sources: the continued burning of hydrogen in a shell surrounding the core, and also the release of gravitational potential energy. The time spent in this contraction phase is mass-dependent, but is presumably short compared to T_{M-S}. In general, the greater the mass, the shorter the evolutionary time. Chandrasekhar has shown that the inert helium core can support the weight of the material lying above it as well as its own weight if the total mass of the star, when it consists entirely of helium, is not more than $1.44M_\odot$. If the total mass is much in excess of this *Chandrasekhar limit*, the helium core is unable to support the overlying weight of the star and contraction proceeds fairly rapidly in the core. Smaller-mass stars leave the main sequence more slowly.

Now the gravitational potential energy that is released from the contracting core regions forces the outer parts of the star to distend themselves greatly, as does the heating from the nuclear burning that takes place in the shell outside the core. The expansion of the outer layers (which consist of unburned hydrogen) causes them to cool; the star becomes red because the released energy is not only radiated but also "used" by the outer layers in the expansion process. In other words, the stars move off the main sequence and become red giants. As we have seen, the contraction will halt when helium burning sets in through the triple-alpha process. A red giant, unlike a main sequence star, is one that burns helium to supply its energy, again assuming that thermonuclear fusion really does power the stars.

If the sun follows a typical evolutionary track, we may expect it to expand to about the orbit of the earth when it becomes a red giant, some 6.4×10^9 years from now. Mercury, Venus, and the earth may vaporize and disappear into the star when it expands. We may expect that the climatological conditions on planets such as Mars will undergo remarkable changes.

The size of this future red giant will thus be about $200R_\odot$. Although the stellar atmosphere will thus envelope the earth, the pressure will only be about 10^{-3} mm at the orbit of the earth, the vacuum achieved by a "roughing pump" in present-day laboratories. It is not clear how long a star of mass $M = 1M_\odot$ will remain a red giant, but estimates are that it might spend some 6×10^8 years burning helium.

If the transition from the main sequence to the red giant region of the H-R diagram occurs along the horizontal branch, we have some support for the calculated result that the time spent in the contraction phase is relatively short. This is because if the contraction is rapid there will be relatively few stars in this stage at any given instant. Thus there should be only a few in the horizontal branch.

We now see how our evolutionary track of Section 12.5 may be extended. A star contracts onto the main sequence from right to left in the H-R diagram. It leaves the main sequence from left to right and stops in the red giant region. The two horizontal portions of the track, however, are not quite coincident because of the slight change in luminosity while on the main sequence

12.10   Post-Red Giant Phases

Next to the main sequence the most numerous group of stars is the white dwarfs. The density of a white dwarf is so high that electron degeneracy provides most of the interior support pressure. While the mass of a white dwarf may be as large as $\sim M_\odot$, typical radii of these stars are only $\sim 10^{-2} R_\odot$. The radius of a white dwarf remains essentially constant as it evolves; the radius is practically independent of the interior temperature distribution. The primary source of escaping energy is the thermal energy of nondegenerate heavy nuclei. The white dwarf very slowly cools and becomes less luminous as time passes; a white dwarf may remain as such for $\sim 10^{12}$ years, so that there are as yet few if any "black dwarfs" (the ultimate end) in our galaxy.

Since there are so many white dwarfs, we conclude that at least a large fraction of all stars pass through the white dwarf phase before becoming completely inactive. The fact that most white dwarfs are characterized by masses less than 1 $M_\odot$ suggests that many stars lose mass prior to becoming white dwarfs. The mechanism of mass ejection is not known. The evolutionary track from the red giant phase to the white dwarf region of the H-R diagram has not been worked out, however. Since mass ejection apparently is involved, this portion of the track is likely to be complex.

Chandrasekhar has shown that a star can reach the white dwarf stage and remain stable as a white dwarf only if its mass is less than a critical value M_c. The mass of the star after it has become a white dwarf must be less than $M_c = 5.75 M_\odot / (\rho / n_e M_p)^2$, where ρ is the density of the stellar matter. n_e is the number density of free electrons, and M_p is the mass of the proton.

To see that an expression of this form is valid, consider the degenerate situation in which the separation δx between electrons is related to the electron density n_e by $n_e \sim (1/\delta x)^3$. The kinetic energy of these electrons may be written as $p^2/2m$ and set equal to (GMM_p/R) for a star of radius R. An expression for the limiting mass follows when our previous results for n_e are employed.

The denominator of the expression for M_c is essentially the mean molecular weight that must be assigned to the material in the white dwarf. Nuclei are not included in the molecular weight, because they move very slowly compared with the free electrons and hence contribute very little to the pressure exerted by the gaseous material in white dwarfs. As long as $\rho \leqslant 10^8$ gm cm^{-3}, the nuclei retain their separate identity and the degenerate electron gas maintains the star in its state of equilibrium; we have a white dwarf.

Stars that have masses greater than M_c can reach the white dwarf stage only by ejecting matter. It is thought that white dwarfs are the result of an ejection of material, possibly an explosive ejection. One example of such an explosive ejection is provided by novae. A typical nova ejects $10^{-5} M_\odot$. If novae are in fact responsible, one is led to believe that novae are *recurrent*. About 25 occur annually in our galaxy; only if each explodes ~ 1000 times can we equate novae with the 10^8 white dwarfs apparently present (see Figures 12.5 and 12.6).

Figure 12.5 Photographs of Nova Herculis taken with the 120-inch reflector. The picture on the left was taken 10 March 1935; that on the right, 6 May 1935. Lick Observatory photograph.

12.11 Neutron Stars and Ultimate Evolution

If the pressure of the degenerate electron gas (i.e., the electron gas moving through the closely packed heavy nuclei found within white dwarfs) deep within a star is not sufficient to compensate for the gravitational force, the core of this massive star becomes unstable and contracts until there are more than 10^{32} electrons cm^{-3}. Many electrons will then be moving at enormous velocities because of the exclusion principle. Indeed, some velocities will be near the speed of light; relativistic electrons are produced by the contraction.

Under these conditions the electrons having very high energies ($E \geqslant 800$ keV) begin to combine with protons inside nuclei. The *inverse beta decay process* occurs; the reaction is

$$p + e \rightarrow n + \nu_e;$$ (12.10)

so that protons are transformed into neutrons. But this process cannot continue indefinitely; nuclear instability results if there are too many neutrons and the unstable heavy nuclei decay into lighter nuclei (with the emission of individual nucleons).

This process leads to stars with core densities much greater than 10^8 gm cm^{-3}, since this decay happens to many nuclei at the same time. These

Figure 12.6   The expanding nebulosity around Nova Persei 1901. Photographed with the 200-inch telescope. Courtesy Hale Observatories, California Institute of Technology.

objects then consist of a mixture of three degenerate gases: neutron, proton, and electron. This is present only in the core of the star (where ρ is very great); the outer layers consist of ordinary nuclei and degenerate electrons, and the material near the surface itself contains unionized atoms.

If continued contraction is caused by the large stellar mass, values of n_e ranging from 10^{34} to 10^{38} electrons cm^{-3} are finally reached in the core, and most of the nuclei have been disrupted by the inverse beta decay process. There are now thousands of times more neutrons than protons present (notice that the number of protons always equals the number of electrons; electrical neutrality is thought always to exist), and the core becomes what is called a *neutron star*. Oppenheimer and Volkoff showed that a neutron star can exist (if it exists as anything other than a logical possibility) in equilibrium if its mass lies between 0.3 and $0.7 M_{\odot}$. The radius of a neutron star is only ~ 20 km; its surface gravity is 2×10^{11} g and the initial surface temperature some 10×10^{6}°K; the central temperature is two orders of magnitude greater.

It is possible that pulsars are associated with rotating neutron stars and might in fact be observational evidence for the actual existence of neutron stars in the universe. The small size of neutron stars would make them difficult to detect directly.

The energy released in the contraction process, if released quickly so that the material would be heated, would provide a truly titanic explosion. Hence many investigators are of the opinion that supernovae "result in" the formation of neutron stars.

It is interesting to speculate about what would happen if there were further contraction. When ρ reaches 10^{15} gm cm^{-3}, the Pauli principle leads us to expect that high-energy neutrons will be present and the normal decay of hyperons (particles that exist momentarily in normal nuclei) into nucleons will be inhibited; more and more neutrons would be changed into hyperons until there were equal numbers of hyperons and neutrons. Free mesons are *stable* (because the exclusion principle prevents their decay into electrons) in hyperon stars, which are hypothetical stars that have masses about that of the solar mass, but have radii of only a few kilometers. Such an object would have a dense hyperon core and a surrounding layer composed primarily of degenerate neutrons (no hyperons) with a few electrons and protons. The outer portions of the star would resemble the material of a white dwarf; they would include a layer of degenerate nuclei and electrons and an outermost layer (probably only a few meters thick) that contained ordinary atoms.

The Pauli exclusion principle is the underlying reason why black dwarfs are the ultimate stage of a star's evolution, assuming, as we do throughout the volume, that the physics operative elsewhere in the cosmos is the same as found valid on earth. If we picture a star as contracting to its ultimate stage, its entire matter will be completely degenerate and therefore unable to emit any energy. For if any particle (e.g., electron, hyperon) were to emit any energy, it would have to fall into a lower state. But this is impossible according to the Pauli principle; there is no room in any lower energy state for the particle to fall into after it emits the energy. These black dwarfs can be quite hot and yet not visible.

We may summarize the ideas on the structure and evolution of stars by saying that the structure is determined by the balance of inward-directed gravitational forces and outward-directed forces exerted by internal pressures. There are at least three stable, long-lived stellar states; these occur when the internal pressures are supplied by (a) gas pressure (normal stars), (b) electron degeneracy (white dwarfs), and (c) nucleon degeneracy (neutron stars). All other stellar evolutionary phases represent transitions between the stable states.

12.12   Cluster Stars

Stellar groupings such as the Pleiades and the Hyades are visible to the naked eye. When examined through telescopes, they are found to consist of

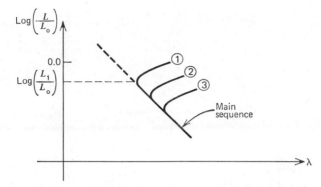

Figure 12.7 Schematic illustration of the H-R diagrams for several open clusters. Cluster 1 is probably the youngest of those shown because the lifetimes of the stars within it are the shortest on the average.

clusters of 20–100 stars each, with some nebulosity in between. They are within our galaxy, since virtually all such open clusters have low galactic latitudes; they are in or near the Milky Way.

Since the stars in a cluster are all at about the same distance from the earth, the variations in apparent magnitude among them are due to variations in their (intrinsic) luminosity. Thus we may draw some conclusions concerning stellar evolution without any nagging doubts concerning the accuracy of distance determinations, by preparing an H-R diagram for the stars in a cluster.

When an H-R diagram is plotted for the cluster stars, only the main sequence and the horizontal branch are seen. The interesting thing is that the only difference between clusters is in the upper left-hand edge of the main sequence, where different "turn-offs" are observed. That is, the main sequence for one cluster will extend to lower values of L than for another cluster. This suggests that the *ages* of the clusters are different. Equation 12.9 will then— at least roughly—tell us the age in years, where the luminosity to be used is the luminosity of the turn-off, L_i (see Figure 12.7). The luminosity L_i is assumed to be related to that time when contraction to the helium-burning phase sets in.

There are other clusters that are called the *globular clusters* because of their distinctive spherical shape. The most famous of these is in the constellation Hercules; it is visible to the naked eye and rather resembles a white billiard ball seen through a pair of binoculars (see Figure 12.8). A globular cluster may contain 10^5 stars, so tightly packed that the average spacing among them is only light-months rather than light-years (the average spacing in the solar neighborhood).

Figure 12.8   The globular star cluster in the constellation Hercules photographed with the 200-inch Mount Palomar telescope. Also known as M13 and NGC 6205, it is about 0.1° in angular diameter and has a mass of 3×10^5 M⊙. Courtesy Hale Observatories, California Institute of Technology.

Globular clusters are distributed, it seems, more or less spherically about the plane of the galaxy; these stars are not bound to the spiral arms that lie in the galactic plane. Lifetimes computed from relation (12.9) for the globular clusters give $T \approx 11$ billion years. Thus the stars comprising the globular clusters are the oldest known members of the galaxy.

One remarkable feature has been noted for the globular clusters. The stars in these old objects are three orders of magnitude poorer in the metals than are stars such as the sun. (Those with sunlike metal concentrations are called population I by astronomers; the more numerous metal-deficient stars are denoted population II.) This therefore suggests that heavy elements (iron, say) have only appeared within the past 11 billion years in our galaxy. Since the sun is "rich" in the metals—some 2% of the solar mass is in the form of $Z \geqslant 2$ elements—the inference is drawn that the sun is at least a second-generation star. That is, the spiral arm gas from which the sun coalesced was already enriched in the metals. If supernovae are indeed the site of heavy element nucleosynthesis (see Section 11.22), the heavy elements on and within the earth may be the result of some ancient stellar explosion.

There is additional evidence for present-day nucleosynthesis, other than the calculations presented in this and the preceding chapter. Merrill detected the presence of technetium in the spectra of some variables and cool carbon stars in 1952. The half-life of the longest-lived isotope of technetium is only about 2×10^5 years, so that this element must have been synthesized recently. If element synthesis occurs in stars, we have seen how the various elements can be cooked up if an initial concentration of hydrogen is provided and Tc stars characteristically show overabundant heavy elements near the neutron magic number 50, which is taken to be evidence for s-process nucleosynthesis.

It is a remarkable fact that no matter where we look in the universe, we see light from the same elements. Thus the same kinds of atoms exist in pieces of matter too widely separated possibly to affect each other. We therefore conclude that a common process operative throughout the universe produces these atoms.

12.13 The Helium Problem

The calculation of an evolutionary track for a star, assuming the ideas of stellar nucleosynthesis, depends on the initial concentration of helium that is chosen. We have seen in Section 12.6 how a lifetime may be computed if one naively assumes the initial composition to be 100% hydrogen. We have also seen that the hydrogen is converted into helium during this lifetime, and the star then burns the helium during subsequent phases of its life.

These calculated times will obviously be altered if helium was present initially. It turns out that if one chooses an initial helium abundance of 10%, a much more heavily populated horizontal branch results than if one chooses the initial helium abundance to be $\sim$35% of the stellar mass. The latter figure gives results that are in accord with observation. A choice of the lower value cannot be fitted to observation unless one assumes that 30–40% of the stellar mass is lost between the main sequence phase and the onset of the horizontal branch phase. Such a mass loss might occur through strong "stellar winds" or through violent mass expulsion, but no observational evidence exists for either possibility.

The difficulty is that stars near the left (i.e., blue) end of the horizontal branch appear to have a very low helium surface abundance. If this is really representative of most stars, the foregoing theory is in considerable trouble. It may be, however, that the surface abundance of helium is low because of gravity; the heavier helium atoms sink and are lost to view. Very few helium abundance measurements have been made.

While it is true that the existence of helium was first detected in 1868 in lines of the solar spectrum, the relative abundance of helium in the solar surface is not determinable spectroscopically. The energy levels of the helium atom lie so high that great temperatures are required to excite the atoms. The reason helium was discovered in the sun is the high temperature of the corona, about 2,000,000°K. Thus the solar helium lines are coronal lines; they are not emitted by atoms in the stellar surface.

The surface temperature is too low to excite spectral helium lines. All that we really know is that hydrogen is the most abundant species in the surface. The surface is of interest, for its composition is presumably typical of the solar material as it was before any nuclear burning took place; it may therefore be used as a guide to the composition of metal-rich stars.

If one performs a numerical calculation on a computer in which a star of mass $= 1 M_\odot$ is allowed to evolve for 4.6 billion years with a given (assumed) initial composition that best fits the present (observed) luminosity and surface temperature of the sun, a model that assumes an initial abundance of about 70% hydrogen and 28% helium (by mass) seems to give the best fit. Unfortunately even this cannot be done for population II stars. No mass determinations yet exist for metal-poor stars, so that we cannot assign even a *total* mass for a calculation.

In elucidating this "cosmic abundances" problem, the abundances derived from measurements of the cosmic ray charge spectrum (see Chapter 13) are clearly of interest. These indicate an abundance of nearly 50% helium (by mass), so that a high value is easy to come by.

However, it is by no means clear that the charge spectrum of extremely high energy particles is the same as that present in stellar envelopes. Particles of different Z are conceivably differently accelerated. Also, it is quite possible that fragmentation of the cosmic ray particles occurs while the particles are in space. Thus the charge spectrum observed at the earth would not be representative of that found at the injection sites of the cosmic radiation.

Indeed, the observation that the abundances of the light nuclei (lithium, beryllium, and boron) in cosmic rays are much higher than those found in the solar surface makes one suspect that such fragmentation has occurred. This fragmentation may be due to an interaction of the cosmic ray particles with those particles that comprise the interstellar medium.

It is possible that direct solar wind measurements made on deep space probes may provide the most valuable means of settling the question of how much helium is present in the sun. A sensitive mass spectrometer on an interplanetary spacecraft that can make direct abundance measurements of helium, C^{12}, N^{14}, and O^{16} is highly desirable.

The recent discovery of the "primordial fireball" radiation at microwave frequencies has bearing on the helium problem. This radiation has been interpreted as evidence for a very high temperature state of the universe in the remote past. High temperatures provide a means for cooking up helium quickly, so that this interpretation favors the higher helium concentration required by our present ideas of stellar evolution. (Indeed, a major difficulty has been to learn where the large required amounts of primordial helium came from; a high-temperature phase at some ancient time is necessary, if we hold to the view that all the elements were developed from hydrogen.) We shall postpone a further discussion of the fireball radiation, however, until the chapter on radio astronomy (Section 14.17).

Problems

12.1. A spherical silvered balloon (like one of the Echo satellites) of radius R is orbiting the earth at an altitude of h kilometers. It is illuminated by the sun, which has an apparent visual magnitude of -26.72.
   (a) Give the stellar magnitude of this satellite as a function of distance from an observer on the earth.
   (b) If $R = 50$ feet and $h = 1000$ miles, what is the apparent magnitude m?

12.2. If the entrance pupil of the human eye is about $\frac{1}{4}$ inch in diameter, what is the limiting magnitude visible to the Mount Palomar telescope? (Assume that the eye of the human in question can detect objects as faint as $m = +6.0$.)

12.3. If a star is observed to have a surface temperature of $11,550°K$ and an absolute magnitude of $+9.7$, what is its radius, in solar radii?

12.4. A star with a luminosity 16 times that of the sun is of spectral type A2, so that its effective temperature is about $10,000°K$.
   (a) What is the radius of the star?
   (b) How long will the star exist on the main sequence?

12.5. Derive Wien's law from the Planck law.

12.6. Find an expression for the upper limit to the mass of a white dwarf in terms of the electron density n_e.

12.7. The equations of stellar structure tell us that the mean pressure $\bar{P}$ of a star of mass M and radius R varies as (M^2/R^4) and that the mean temperature $\bar{T}$ varies as (M/R). Using the fact that the mean density $\bar{\rho}$ varies as (M/R^3) and assuming the opacity to be a constant, show that the luminosity L of the star varies as M^3. This represents a crude derivation of the mass-luminosity relationship.

12.8. A star in its initial hydrogen-burning stage has an absolute luminosity of $10L_\odot$.
   (a) What is the mass of the star, in $M_\odot$?
   (b) How long will this star stay on the main sequence?

12.9. A certain star on the main sequence has a mass of $10M_\odot$. What fraction of its mass is lost per year in electromagnetic radiation?

12.10. Compare the lifetime of a hypothetical O-type star of pure hydrogen with the lifetime of the sun (assuming that the sun is also composed only of hydrogen).

12.11. (a) Estimate the "stellar birth rate" ψ (i.e., the number of star births per cubic parsec per year per magnitude interval) in terms of $T_{\text{M-S}}$.

(b) Use ψ to find what fraction of the interstellar gas that goes into star formation is used to form bright stars (brighter, say, than $+3$ in visual absolute magnitude).

12.12. A star is observed spectroscopically to have a surface temperature of 10,000°K. Assume there is no required color correction to its visual magnitude. If the star is observed to have an apparent magnitude of $+17$ and a parallax of 0.01 seconds of arc, what type of star (main sequence, giant, supergiant, white dwarf, neutron star, etc.) is it, and how far along in stellar evolution is it?

12.13. If a star had a mass of 2×10^{34} g and a radius of 1.5×10^6 cm, what would its luminosity be? Explain your answer in terms of stellar evolutionary models, and show a calculation to indicate how you obtained the answer for the luminosity. (*Hint*: Compare the radius and mass to a solar radius and mass.)

Chapter 13

COSMIC RAYS

It was found early that the electrostatically charged leaves in laboratory electroscopes would eventually lose their charge, regardless of how well insulated they were. It therefore seemed that the gas between the leaves was somehow being ionized. The most likely candidate was radiation from the slightly radioactive surroundings. Therefore this effect could be eliminated by shielding the instruments with lead. This shielding, when emplaced, did in fact help; it markedly reduced the rate at which the charge was neutralized. Nevertheless the neutralization rate could not be reduced to zero; no matter how much shielding was put around the electroscope, some slight ionization, about 10 pairs cm^{-3} second^{-1} at standard temperature and pressure, always seemed to be present. The presence of ionization led to the speculation that high-energy radiation of fantastic penetrating power was present—radiation that could, for example, penetrate several meters of solid lead.

The study of this radiation now called cosmic rays has proved to be of great benefit to many other areas of space science. For example, the study of meteoritics involves cosmic rays; the dating and sizing of meteorites becomes possible when the cosmic radiation effects are examined. The distribution of the geomagnetic field far from the earth may be inferred from the distribution of cosmic ray flux over the surface of the earth. Those cosmic rays emanating from the sun reveal information on the composition of the outer layers of that star, and the radiation emanating from beyond the solar system constitutes our only direct sample of the chemical composition of the matter beyond our planetary system. This latter radiation also provides data on interstellar magnetic fields and studies of the temporal variations of the solar component have helped to reveal the spiral structure of the interplanetary field.

Cosmic rays are also of interest in areas of study not ordinarily included in space science, for example, experiments in very-high-energy physics can only be done with cosmic rays; they are our only source of particles with energies in excess of 10^{11} eV. Then, too, many of the particles and processes of nuclear physics were first discovered in cosmic ray experiments. Archaeology has also benefited from this research: cosmic ray (secondary) neutrons produce C^{14}

in the atmosphere, and studies of this isotope permit calculations to be made of the length of time organic material has been dead.

13.1   Discovery and Definition of Cosmic Rays

The pre-World War I balloon flights of V. F. Hess showed that the ionization produced in a gas by this mysterious radiation decreased with altitude, at least up to altitudes of $\sim$2000 feet above the surface. This was to be expected, if the radiation really did emanate from the ground, rocks, and so on. The flights also showed, however, that the radiation intensity *increased* at altitudes greater than this; a steady increase to perhaps 50 times the sea level value was noted, up to the maximum altitudes then achievable ($\sim$30,000 feet). Hess therefore proposed that there existed a very penetrating ionizing radiation that entered the earth's atmosphere from outer space. This radiation would have to be very penetrating indeed; it could pierce the entire thickness of the planetary atmosphere. Subsequent experiments confirmed the existence of such a radiation, and R. A. Millikan proposed the name "cosmic radiation" in the mid-1920's. Cosmic ray research developed rapidly during the 1930s but was interrupted by World War II. The "heroic era" of cosmic ray study was the late 1940s, when a large number of significant developments occurred, but the subject still forms a most active research area at the present time.

Nowadays we define the term cosmic rays to include those charged particles that reach the earth's magnetosphere from interplanetary space with velocities greater than the solar wind velocity. Frequently one also speaks of "cosmic ray" neutrinos and neutrons, but usually in the context of neutral radiation that is secondary to the (primary) charged cosmic rays. It will be noted that our definition of cosmic radiation excludes electromagnetic photons of extraterrestrial origin, such as radio, optical, x-ray, and gamma ray quanta; these certainly exist and may be related to the source or sources of cosmic rays, but are of interest in their own right and are usually treated separately.

13.2   Galactic and Solar Cosmic Rays

Galactic cosmic rays are those that reach the earth from outside the solar system. Indeed, it is possible that some of them originate from beyond our galaxy. The flux of galactic cosmic rays is thought to be isotropic and constant in time over periods of at least tens of thousands of years.

The particles comprising the galactic cosmic radiation are nuclei that have energies from $\sim 10^7$ eV up to at least 10^{19} eV, a truly staggering amount of energy for a single particle. In approaching the earth's orbit, the flux of those with energies less than $\sim$15 BeV is reduced by the solar wind and the interplanetary magnetic field; the more active the sun, the fewer of these "low-energy" particles at the earth. Since they are nuclei, the vast majority of the cosmic rays are *positively charged*.

The flux of higher-energy particles at the earth remains constant; it is approximately one particle per cm²-second at midlatitudes. The maximum flux of galactic cosmic rays observed at the earth (which occurs when the sun is in the quiet part of its 11-year cycle and the sunspot number is near a relative minimum) may equal the interstellar flux. However, it is at least possible that the number of lower-energy particles is always depressed at the earth's orbit, since there is always some solar wind. In subsequent sections we describe the galactic radiation as it is observed at the earth and then consider the consequences of supposing that the interstellar flux equals the maximum flux observed.

As the name implies, solar cosmic rays are produced by the sun. Like the galactic rays they are nuclei, but unlike their galactic counterparts their production is only sporadic; it apparently happens only occasionally in those violent outbursts that also produce radio noise, optical flares, and x-radiation. The particles reach the earth a few minutes to an hour after the solar eruption. The more energetic particles arrive first, because they are faster. Thus it is clear that major differences exist between the solar and galactic components of the cosmic radiation.

The solar cosmic rays are in general of lower energy and occur only occasionally. The galactic flux is (we think) isotropic and is sensibly constant in time. The galactic cosmic ray energy spectrum extends to very high energies; it is this radiation that was responsible for the ionization detected by the turn-of-the-century electroscopes.

13.3 Energy Spectrum

Figure 13.1 shows the integral energy spectrum of cosmic rays observed at the earth. Above $E = 2 \times 10^9$ eV the spectrum is well represented by a power law; the functional dependence of the inbound flux J on kinetic energy T is given by

$$J(>T) = \frac{K}{(T + 0.931)^{\gamma-1}}, \tag{13.1}$$

where $J(>T)$ represents the flux of particles with kinetic energy greater than T; K is a constant. The values of γ may be read from Figure 13.1, and T is the kinetic energy per nucleon, expressed in billion-electron volts (1 BeV = 1×10^9 eV).

It is seen that the spectrum obeys the same power law over an amazingly broad energy range; $\gamma \sim 2.6$ from 10^9 eV to the highest energies yet measured ($\sim 10^{20}$ eV). There is, however, some evidence for an interesting break at $\sim 10^{16}$ eV, since it seems that γ becomes closer to 3 at greater energies. If it is real, this break has significant consequences for theories of the origin of the radiation, as we shall see later.

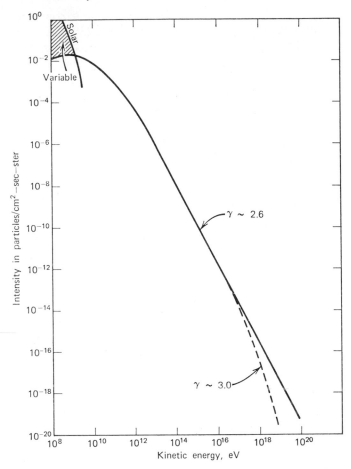

Figure 13.1   The energy spectrum of the cosmic radiation.

From (13.1), the differential energy spectrum is

$$-\frac{d}{dT}J(T) = \frac{K(\gamma - 1)}{(T + 0.931)^{\gamma}} \text{ particles/cm}^2\text{-sec-ster-MeV,} \qquad (13.2)$$

where K is such that the flux of all nuclei with kinetic energies in excess of 500 MeV is $J(>500 \text{ MeV}) \sim 0.25$ per cm²-second-steradian.

The relationships (13.1) and (13.2) are useful for the galactic radiation, but the energy spectrum of solar cosmic rays varies from event to event. It also varies during an event. The integral energy spectrum of the solar radiation may be written as a power law for $T \gtrsim 1$ BeV;

$$J(>T) = K/(T + 0.931)^{\beta - 1},$$

where β may be anywhere from 4 to 10. Below 1 BeV, however, this representation fails and an expression such as $J(>T) = AT^{-m}$ is more accurate, with $m \sim 4\text{-}10$. Thus the integral energy spectrum rises sharply with decreasing energy at least down to $T = 10$ MeV, and possibly lower. The peak intensity ranges from 1 to 10^4 particles per cm²-second-steradian, when all particle types are considered and for energies in excess of 10 MeV.

13.4 Composition

Although the vast majority of the galactic radiation consists of protons, heavier nuclei are also observed. Indeed, there is evidence that nuclei as heavy as uranium ($Z = 92$) have been detected. (Interestingly, the energy spectra of all of the different nuclei appears to be approximately the same, at least up to $\sim 10^{11}$ eV per nucleon, which is the highest energy for which reliable data on the heavier nuclei presently exist.) The charge spectrum (i.e., the relative abundances of the different nuclei) is not well known above $\sim 10^{11}$ eV per nucleon. It varies significantly from event to event in the solar component.

Table 13.1 lists the measured fluxes of galactic cosmic rays with energies in excess of 2.4 BeV per nucleon. The measurements may be made with nuclear emulsions, which are thick pieces of photographic film. The passage of a charged particle through such an emulsion causes individual grains to be developed; the width of the resulting track depends on the atomic number of of the charged particle (see Figures 13.2, 13.3, and 13.4).

The abundances in the sun and in the solar cosmic radiation are also shown, for comparison purposes in table 13.1, where the solar radiation has energies in the range 50–200 MeV per nucleon. All these particles, of course, are positively charged. The number of relativistic ($T > 1$ MeV) electrons among galactic cosmic rays amounts to a small percentage of the proton flux, and it appears that there are more electrons than positrons at the earth.

We should note here an interesting feature of the (galactic) charge spectrum at high energies. Nuclei with $Z = 2$ up to about $Z = 10$ are more abundant in cosmic rays, compared to heavier nuclei, than is the case with the "universal" abundance. This is particularly so with Li, Be, and B, which are rather rare in the universe. This relative enhancement might mean either that the sources of the galactic radiation are enriched in light nuclei or, as seems more likely, that heavy nuclei such as iron are originally injected and accelerated, along with protons and alpha particles. The "heavies" are then fragmented into lighter nuclei as the cosmic rays pass through interstellar matter. We shall return to the interesting consequences of this fragmentation hypothesis in Section 13.12.

TABLE 13.1. ABUNDANCE OF ELEMENTS

Element	Z	Relative to $O^{16} = 1.0$[a]		Universal Abundance	Galactic Cosmic Rays: Flux[b] (particles/m² sec-ster)
		Solar Cosmic Rays	Sun		
H	1	Variable		1950	1510 ± 150
He	2	107 ± 14		150	89 ± 3
Li	3		$<10^{-5}$	$<10^{-5}$	
Be, B	4, 5	<0.02	$<10^{-5}$	$<10^{-5}$	2.0 ± 0.2
C	6	0.59 ± 0.07	0.6 ± 0.1	0.26	
N	7	0.19 ± 0.04	0.15 ± 0.05	0.20	
O	8	1.0 (definition)	1.00	1.0	5.6 ± 0.2
F	9	<0.03	0.001 (?)	$<10^{-4}$	
Ne	10	0.13 ± 0.02	0.1 (?)	0.36	
Na	11		0.002	0.002	
Mg	12	0.043 ± 0.011	0.027	0.040	
Al	13		0.002	0.004	1.88 ± 0.3
Si	14	0.033 ± 0.011	0.045	0.045	
	15–21	0.057 ± 0.017	0.032	0.024	
	22–28	≤0.02	0.006	0.033	0.69 ± 0.16
					200 (electrons, 3–8 MeV)

[a] After C. Biswas and C. Fichtel, *Space Science Reviews*, 1965.
[b] After J. Waddington, *Progress in Nuclear Physics*, 1960.

Residual Range = 24·0gms/cm²
Kinetic Energy ∼ 1,000 Mev/n

Residual Range = 12·0 gms/cm²
Kinetic Energy ∼ 600 Mev/n

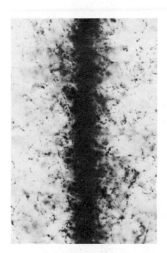

Residual Range = 7·5gms/cm²
Kinetic Energy ∼ 440 Mev/n

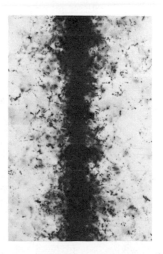

Residual Range = 3·0 gms/cm²
Kinetic Energy ∼ 240 Mev/n

├─── 50 μ ───┤

Figure 13.2 Sections of track of slowing nucleus $Z \sim 56$ in a nuclear emulsion carried into the stratosphere over Palestine, Texas, May 1967. Courtesy Dr. Peter Fowler.

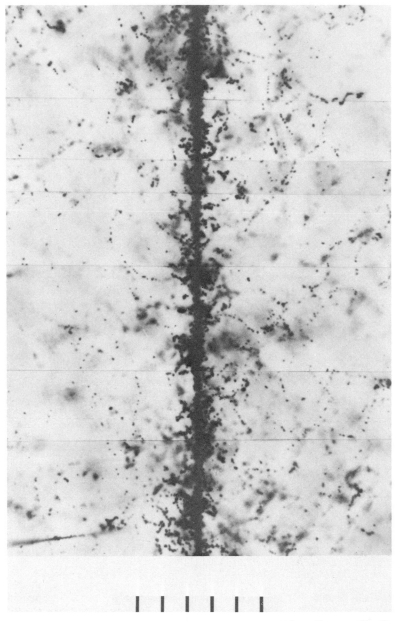

Figure 13.3　A $Z \sim 40$ cosmic ray track in nuclear emulsion. Courtesy Dr. P. Fowler.

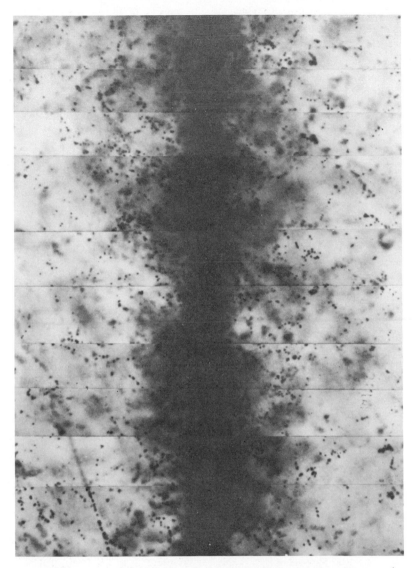

Figure 13.4   A $Z \sim 90$ cosmic ray traversing a nuclear emulsion. Contrast the width of the track with that in Fig. 13.3. Courtesy Dr. P. Fowler.

13.5 Interactions of Cosmic Rays with the Atmosphere

The cosmic radiation so far discussed is known as the *primary* radiation; it is that energetic charged radiation that is incident on the earth's magnetosphere from beyond. The discussion contained in Appendix C, however, shows that a very complex chain of interactions is to be expected when a high-energy cosmic ray nucleus enters the atmosphere (or other bulk matter) and collides with the atoms and nuclei comprising the latter.

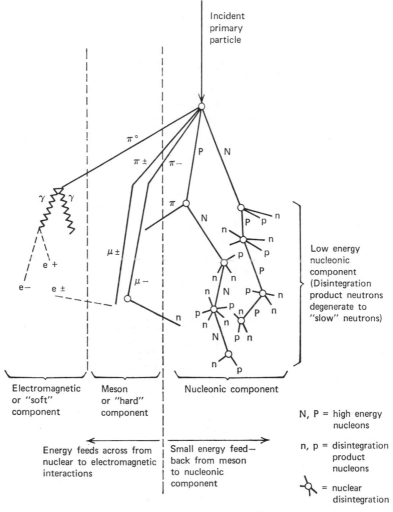

Figure 13.5 Schematic representation of the development of secondary cosmic radiation within the atmosphere as a result of the interaction of an incident primary particle with the nucleus of a high-altitude air nucleus. From J. A. Simpson, W. Fonger, and S. B. Treiman, *Physical Review*, 1953.

About 1 eV cm^{-3} is brought in by the primary radiation. At midlatitudes about 64% of this is lost to ionization in the atmosphere by the secondary radiation. Some 24% is given up to neutrinos, and about 0.8% is expended in overcoming the binding energy of nuclei. Only 4% is absorbed by the solid earth.

Figure 13.5 shows the atmospheric interactions schematically. One sees that many *secondary* particles are created, some of which are unstable and may decay in flight. It is the secondary radiation that is actually detected at sea level; the primary particles themselves are for all practical purposes completely absorbed by the atmosphere.

The intensity and energy spectrum of each type of secondary radiation depends on several factors, such as the zenith angle of the direction of motion, the atmospheric depth, and the geomagnetic latitude. For example, the total number of particles actually *increases* with increasing depth X up to $\sim X = 100$ gm cm^{-2} and then decreases more or less exponentially at greater depths where absorption predominates over multiplication. Thus the maximum flux of particles occurs at ~ 100 gm cm^{-2}; this maximum is called the *Pfotzer maximum*. The exact depth of the Pfotzer maximum depends somewhat on particle type and geomagnetic latitude (Figures 13.6, 13.7).

It is possible to calculate these curves of flux versus depth, and they fit the experimental data reasonably well. The nucleonic component (nucleons with $E \geq 1$ BeV) may be thought of as the source of the other radiations. It interacts with a mean free path that is essentially geometric (~ 65 gm cm^{-2}). In the following, we discuss three members of the nucleonic component (protons, neutrons, and pions) separately, as well as muons, which result from pion decay.

Protons

An average of 2.7 shower particles is produced in each interaction with the atmosphere; of these 20% are fast protons and the remainder are pions. Hence the change in the flux N of the proton component in passing through an atmospheric thickness dx is

$$dN = N \, dx \left(-\frac{1}{L} + \frac{2.7 \times 0.2}{L} \right),$$

or

$$N = N_0 e^{-x/L_a} \tag{13.3}$$

where L_a is the absorption mean free path.

We observe that the proton component's mean free path for absorption at depths greater than the Pfotzer maximum is ~ 125–140 gm cm^{-2}, not the 65 gm cm^{-2} expected from nuclear geometry. This merely means that proton production as well as absorption occurs throughout the atmosphere; the

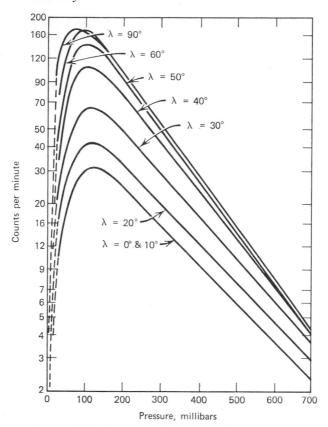

Figure 13.6 Intensity of secondary slow neutrons versus atmospheric pressure for various geomagnetic latitudes. These data were obtained with balloon-borne BF$_3$-filled proportional counters, 30 cm long × 5 cm diameter; the filling pressure was 20 cm Hg and the background due to other radiation types has been subtracted. Note the pronounced Pfotzer maximum near 100 mb; it is deepest in the atmosphere at the latitudes at which the primary spectrum is hardest. From R. K. Soberman and S. A. Korff, *Physical Review*, 1956.

decrease in proton flux is not nearly as steep as predicted by absorption alone. Another way of saying this is that catastrophic absorption is not the only operative process.

Neutrons

Neutrons are also produced in these interactions. The total number of neutrons born as a result is about equal to the proton production rate. Most of the neutrons emerge from the struck nuclei with several million-electron

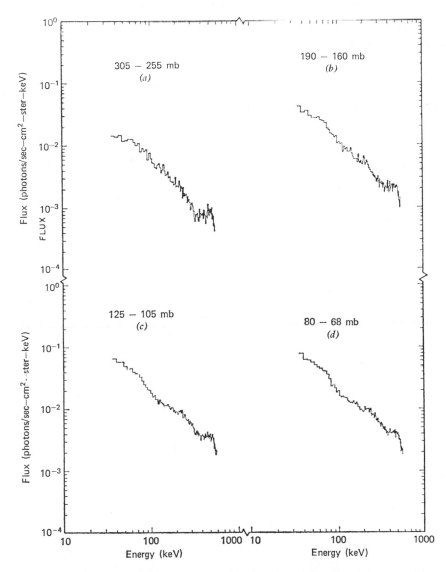

Figure 13.7 Spectra of secondary low-energy gamma radiation within the earth's atmosphere. The left-most group of plots shows atmospheric spectra derived for relatively large pressures (measured in millibars) or depths; the other groups describe the situation that existed over Texas at progressively higher altitudes during 1967. Note that the Pfotzer maximum is in the 80–68 mb range. From R. C. Haymes, S. W. Glenn, G. J. Fishman, and F. R. Harnden, Jr., *Journal of Geophysical Research*, 1969.

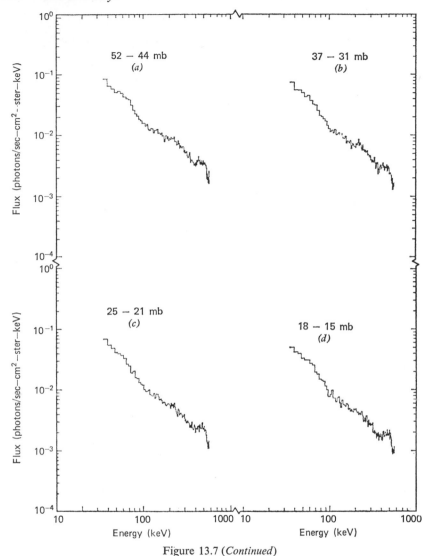

Figure 13.7 (*Continued*)

volts of energy, typically 1–2 MeV in the center of mass system. Neutrons of this type are called "evaporation neutrons"; it is as though the incident particle "heats up" the target nucleus and the components of the nucleus then "boil off" (with several million-electron volts of energy each), so that the energy imparted to the nucleus is shared more or less evenly by all the nuclear components.

There are also higher-energy neutrons. This latter variety has come to be

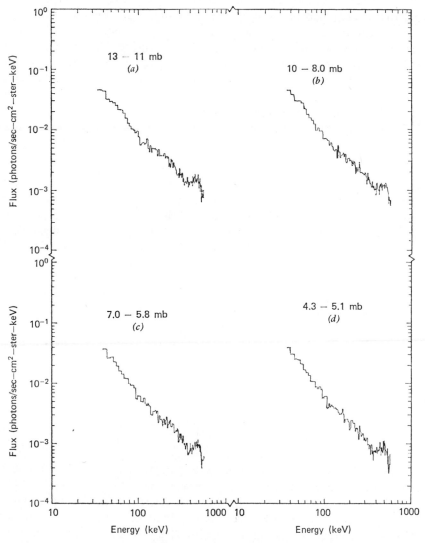

Figure 13.7 (*Continued*)

called "knock-on" particles. In this picture the incident particle communicates its energy to just a few of the nuclear components; they are knocked out of the nucleus with a great deal of energy. The knock-on process, however, turns out to be only ~15–20% as probable as the evaporative mechanism.

Regardless of how they are born, the neutrons are scattered and slowed in the atmosphere before they "die" (i.e., are absorbed). Thus a continuous spectrum of neutrons is present in the atmosphere that generally rises with decreasing energy down to about $2kT$. Neutron absorption by nitrogen becomes important at lesser energies; a peak is present in the differential neutron energy spectrum at roughly 0.05 eV, at atmospheric depths such that absorption occurs before either escape to space (the so-called neutron albedo) or capture by the ground may occur. This region of the atmosphere, sometimes called the equilibrium region, is found between 150 and 750 gm cm⁻². The neutron production rate equals the absorption rate in the equilibrium region.

Pions

Since the rest lifetime of a charged π-meson is only about 0.02 microseconds, in-flight decay may be expected to be an important factor in the altitude distribution of this component of the secondary radiation. Pions produced in nuclear collisions decay after traveling an average distance l_D of only ~200 meters. At sea level this corresponds to a thickness of only 26 gm cm⁻². While pions interact with nuclei just as nucleons do, it is clear that this short mean free path for decay makes decay more probable than interaction. Pion production is proportional to the nucleonic flux N and to the atmospheric density ρ; it is therefore also inversely proportional to the nucleonic mean free path for absorption (interaction), L. The pion flux itself will be proportional to the product of the pion production and l_D. Hence in the region of the atmosphere where decay is the predominant loss mechanism for pions, we may estimate the pion flux π. This region is in the depth range 50–500 gm cm⁻², and the appropriate relation is

$$\pi = Nl_D\rho\left(\frac{2.2}{L}\right), \tag{13.4}$$

where $\rho = X/H$ in terms of the atmospheric depth X gm cm⁻² and the scale height H kilometers. Relation (13.4) may be rewritten as $\pi = NX/1200$, which indicates a flux that increases with depth. It should be recalled, however, that this increase will be offset at least partially by the increased probability at great depths that the pion will undergo a nuclear interaction before it decays.

Muons

In contrast with charged pions, the average μ-meson travels an appreciable fraction of the atmosphere before decay; the muon flux decreases more gradually than the pion flux as the depth increases. This is because the rest lifetime of the muon is two orders of magnitude greater than that of the charged π-meson.

Muon decay is a source of fast electrons. Indeed, it is the dominant electron source at depths greater than $\sim$600 gm cm^{-2}. At lesser depths (higher altitudes) electrons from the decay of neutral pions predominate. Neutral π-mesons decay with a rest lifetime of 10^{-16} second into two gamma photons; the photons may then create electron-positron pairs, which generate more photons through bremmstrahlung and annihilation radiation; these photons in turn produce even more electrons in a cascade.

Muons are the principal constituents of what is called the "hard component" at sea level. The hard component is that radiation that can penetrate at least 10 cm of lead; the "soft component" of the radiation is defined to be that which is adequately shielded by 10 cm of lead. The total sea level cosmic ray flux at midlatitudes is about 1 particle per cm^2 per minute, integrated over zenith angle and energy. The hard component, which accounts for about 70% of this flux, depends on zenith angle θ as $\cos^2 \theta$; the soft component is isotropic.

The vertical intensity at sea level has been measured many times. The midlatitude flux in the vertical direction of muons is 0.008 per cm^2-second-steradian, and the average nucleon flux is only 0.0001 per cm^2-second-steradian. The terrain type sharply influences this last factor; water (the oceans) is a good neutron absorber. By contrast, the vertical electron flux is 0.003 per cm^2-second-steradian.

The muon flux depends on the atmospheric temperature. This is because muons decay, and the higher the temperature the greater the spatial extent of the atmosphere, so that the decay probability is increased. A sea level temperature coefficient of -0.18% per degree Kelvin has been measured for the muon component.

This sea level flux also depends on the total mass of air overhead; the atmosphere is an absorber. This holds even more strongly for the nucleonic component. The fluxes thus depend on barometric pressure. For muons, a pressure coefficient of -3.45% per cm of mercury is found, while -9.8% per cm of mercury is the corresponding nucleonic coefficient. These figures are somewhat latitude-dependent, since the average energy of the particles depends on the effects of the geomagnetic field.

13.6 Splash Albedo

"Splash albedo" is the term applied to the flux of secondary particles projected upward from the top of the atmosphere. It is of special interest for balloon-borne experiments, for which it often constitutes an unwanted background. Another type of albedo is the *re-entrant albedo;* it is that charged splash albedo (with magnetic rigidity less than the magnetic cut-off rigidity at the point it leaves the atmosphere) that is returned by the geomagnetic field to the same latitude on one hemisphere or the other. The following fluxes have been measured at $\lambda = 41°$ for vertical splash albedo: protons 90–350 MeV = 5–15% of primary flux ($= \sim 0.02$ per cm²-second-steradian); electrons 10–500 MeV = primary flux ($= \sim 0.2$ per cm²-second-steradian). The re-entrant albedo was measured to have a flux equal to the splash albedo flux.

A neutron albedo also exists; it is composed of those neutrons scattered by atmospheric nuclei (upward) away from the earth. At $\lambda = 41°$ the fast neutron albedo = 0.24 per cm²-second (1 MeV $\leqslant T \leqslant 15$ MeV), while the thermal neutron albedo has been measured to have a density of $\leqslant 10^{-7}$ cm⁻³.

13.7 Geomagnetic Effects

It is not true that a cosmic ray near the earth may move freely in any direction. When the complete trajectories of these charged high-energy particles are considered, it is found that some of the trajectories are bound in the geomagnetic field in such a way that they are inaccessible to particles starting at infinite distances from the earth. In other instances, although the trajectory may be traced back to infinity, it has already passed through the volume of the solid earth before reaching the point of observation. The corresponding direction of observation is then said to lie within the earth's *shadow;* such a direction is forbidden.

An extremely simple example of these trajectories is provided by particles of very low momentum. As we saw in Chapter 7, the trajectories of such particles are helices about the field lines; it is easily seen that they can only carry such cosmic ray particles as they arrive at the poles.

At a given point on the earth (neglecting atmospheric absorption and scattering) and for some given particle momentum, cosmic ray particles can only be observed in certain directions, called the "allowed cones." The other directions are "forbidden." The allowed cones result from detailed numerical calculations on large numbers of trajectories.

Detailed calculations of these so-called Störmer cones are found in Appendix D. It suffices for our present purposes, however, to examine a particularly simple case physically and see how a direction can be classified as forbidden or allowed.

Figure 13.8 illustrates this case. It is that of a charged particle moving in the earth's equatorial plane. The total energy of the particle is not altered by the magnetic field (assumed dipolar). Thus the radius of curvature of its path as a function of distance r from the center of the earth will be proportional to the inverse of the magnetic field; it will be proportional to r^3. The path of a particle that arrives at some given point with a momentum p and a direction φ (measured from the eastern horizon) will be the same as that of a particle of opposite sign ejected from the earth's surface along the same line of motion, and with the same momentum.

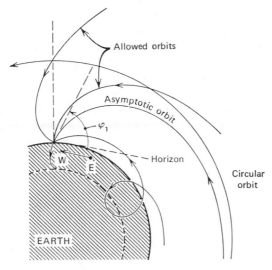

Figure 13.8 The motion of positively charged particles in the equatorial plane of the earth. From J. E. Hooper and M. Scharff, *The Cosmic Radiation*, (Methuen) Wiley, New York, 1958.

When p is small, a negative particle sent off with a direction that lies close to the eastern horizon (small φ) will be bent back toward the earth again, so that (neglecting the fact that the earth is solid) it will remain in a bound orbit that always stays close to the earth. Such a particle cannot reach infinity; no cosmic ray particles (positive particles) that have this value of p can reach the earth in the chosen direction. As φ is increased, the maximum distance from the planetary center that the particle achieves before being bent back increases. This increase occurs until, for a certain value $\varphi = \varphi_1$, which depends on p, the path asymptotically approaches a circular orbit that is concentric with the earth and whose radius r' is related to p and the dipole moment M of the earth by

$$p = \frac{ZeM}{c} (r')^{-2}, \tag{13.5}$$

where c, as usual, is the speed of light. It may be shown that the circular orbit can be approached by a particle of momentum p either from the inside, as discussed above, or from the outside.

If the angle of emission φ is increased still further, beyond φ_1, the particle will be able to cross the circular orbit and escape to infinity. Thus *positive* particles with momentum p can only approach the earth in its equatorial plane at angles greater than φ_1. (If the approaching particles had been negatively charged, the only difference would be that the angle φ would have to be measured from the western horizon.)

For a given value of p, all orbits that reach the surface of the earth with an angle $< \varphi_1$ are bounded, and, as seen above, no cosmic ray primaries may arrive from this portion of the sky. For $\varphi > \varphi_1$, however, all directions are allowed; these orbits are unbounded.

Equation 13.5 permits an approximate calculation of the smallest momentum at which all directions are allowed, for the condition that the primaries can reach the earth horizontally from the eastern direction corresponds to the condition that the radius of the earth R_E is equal to the radius of the circular orbit discussed above. Thus the minimum value of momentum, p_{min}, required to penetrate from the east is

$$p_{min} = \frac{ZeM}{cR_E{}^2},$$

which tells us that that $p_{min} = 59.6$ BeV/c; a proton will need nearly 60 BeV of energy to strike the earth at its equator, if it is incident from the eastern horizon.

Positive particles such as cosmic rays can be detected on the earth with lower momenta than p_{min} if they come from the west. The geomagnetic field is sufficiently strong on the equator, however, that they need a good deal of energy, even if from the west ($p_{min} = 0$ only at the geomagnetic poles in this theory). It turns out that cosmic rays coming from the west can get in as far as $(\sqrt{2} - 1)R_E = 0.414R_E$ from the dipole axis. Since (13.5) shows us that the radius of the orbit is inversely proportional to $p^{\frac{1}{2}}$, the minimum particle momentum required to penetrate from the west is given by $59.6 \times (0.414)^2 = 10$ BeV/c. Particles with momenta less than 10 BeV/c will not reach the earth at all at the equator, and at just 10 BeV/c they will arrive from the western horizon if negatively charged. At an intermediate momentum, which turns out to be 15 BeV/c, positively charged particles can arrive at any angle between the western horizon and the zenith; the minimum momentum required for vertical penetration at the equator is 15 BeV/c.

The kinetic energy T of a particle with momentum p is $T = (p^2c^2 + m_0{}^2c^4)^{\frac{1}{2}} - m_0c^2$, where m_0 is the rest mass. If $W_0 = m_0c^2$ in million-electron volts and $p_c = p$ in units of MeV/c, $T = (p_c{}^2 + W_0{}^2)^{\frac{1}{2}} - W_0$, in million-electron volts. If $T \gg m_0c^2$ (0.5 MeV for electrons, 0.93 BeV for protons) $T \simeq p_c$. Hence the kinetic energy in million-electron volts is numerically equal to the momentum in MeV/c. Thus protons require kinetic energies of at least 15 BeV to penetrate the geomagnetic field to the earth at the equator from the vertical direction.

It is clear that the case discussed above will lead to an *east-west asymmetry* if the incoming particles are predominantly of one sign. Such is the case for cosmic rays, even at sea level. The asymmetry becomes greater at high altitudes, but is confused there because of the effects of *re-entrant albedo* particles (see Section 13.6).

13.8   The Latitude Variation

It is shown in Appendix D that there exists a relationship between the arrival direction φ and the geomagnetic latitude λ; it is

$$\cos \varphi = \frac{2\gamma}{r \cos \lambda} + \frac{r \cos \lambda}{r^3}, \tag{13.6}$$

where $|2\gamma|$ is equal to the impact parameter, the perpendicular distance between the asymptotic velocity vector and the dipole axis. In the case of radiation that is isotropic and homogeneous at infinity, such as cosmic radiation, orbits with all values of γ occur. The problem of finding the cutoff momentum—the minimum momentum that can penetrate the field to the earth—at a given λ for a given φ is therefore reduced to finding the γ for which orbits reach a minimum distance from the dipole under these given conditions.

The minimum value of r is found from (13.6) when $\gamma = -1$; the result is

$$\frac{1}{r_{\min}} = \frac{1}{\cos^2 \lambda} (1 \pm \sqrt{1 + \cos^3 \lambda \cos \varphi}). \tag{13.7}$$

The positive root must be taken in (13.7) to make r a minimum. Equation 13.5 tells us what momentum p is to be associated with $r_{\min}$.

An interesting conclusion that emerges from a study of (13.5) is that the effective radius of the earth is proportional to $p^{1/2}$. That is, for other momenta the curvature of the paths drawn in Fig. 13.3 will be altered, but it may be shown that this can be considered merely a matter of scale. From this we find that the cutoff momentum p is given by

$$p = \frac{59.6 \cos^4 \lambda}{(1 + \sqrt{1 + \cos^3 \lambda \cos \varphi})^2} \left[\frac{\text{BeV}}{c}\right]; \tag{13.8}$$

the cutoff momentum decreases rapidly with latitude λ as $\lambda \to 90°$. At every λ the cut-off momentum has a maximum for particles arriving from the east, where $\varphi = \pi$; it has a minimum for those coming from the west ($\varphi = 0$). The cutoff for vertical arrival is

$$p_v = \frac{59.6}{4} \cos^4 \lambda, \tag{13.9a}$$

or

$$p_v = 15 \cos^4 \lambda \left[\frac{\text{BeV}}{c}\right], \tag{13.9b}$$

so that, as before, vertically incident protons require 15 BeV of kinetic energy to arrive at the earth's surface from infinity, at the equator.

We see that the geomagnetic field makes a good and useful charged particle energy analyzer. The flux depends sharply on the latitude, according to (13.9), since the energy spectrum is a decreasing function of the energy. In fact, while the pole to equator flux ratio is only $\sim 14\%$ at sea level, it approaches a *factor* of 13 at small atmospheric depths upon the energies measured. Thus flux measurements made simultaneously at different latitudes with similar detectors permit good integral energy spectrum measurements to be made in the otherwise difficult 10^{10}-eV energy range. One particularly interesting result of such an energy analysis has been that the spectrum inside the magnetosphere is dependent on the 11-year solar cycle; at times

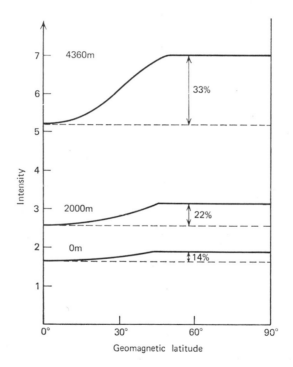

Figure 13.9   The cosmic-ray geomagnetic latitude effect, observed with ionization chambers at three different altitudes ranging from sea level to mountain heights. The effect becomes even more pronounced at greater altitudes than those shown. From J. E. Hooper and M. Scharff, *The Cosmic Radiation*, Wiley-Methuen, New York, 1958.

near maximum solar activity there is a "knee" in the measured latitude variation near $\lambda \sim 50°$, and this knee disappears at solar minimum, when there is less plasma emanating from the sun. Some of these effects are summarized in Figure 13.9.

13.9 Temporal Variations of the Galactic Radiation

When the sun is inactive there are two categories of temporal variation of the galactic radiation to be considered. One is diurnal and the other depends on the 11-year solar cycle. Solar activity adds another type of variation to this list; the cosmic ray intensity is depressed during geomagnetic storms, and this variation, which may amount to as much as 20% for the sea level nucleonic component, is over 10 times the amplitude of the other two at sea level. The greatest fluctuations, of course, are associated with the solar cosmic radiation; solar flares have caused sea level increases of as much as a factor of 5 in an hour or so.

The very existence of the diurnal variation was long in doubt, but now seems well established (see Figure 13.10). The amplitude is between 0.5% and 1.0% for rigidities between 1 and 200 Bv. The maximum flux appears to come from a direction in the ecliptic plane $\sim 85°$ east of the sun, on the

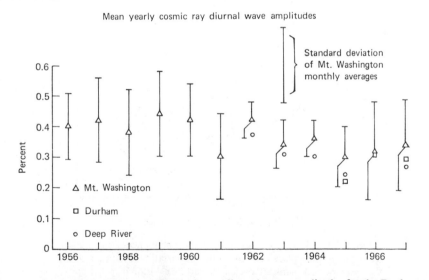

Figure 13.10 Mean yearly cosmic-ray diurnal wave amplitudes for the Durham, New Hampshire, Mt. Washington, and Deep River neutron monitors during their periods of operation through 1967. From R. W. Jenkins and J. A. Lockwood, *Journal of Geophysical Research*, 1970.

average. The dependence of the directional flux j on direction may be written as

$$j(R) = j_0(R)[1 + 4 \times 10^{-3} \cos b \cos (l - 85°)], \qquad (13.10)$$

where R signifies magnetic rigidity and b is the ecliptic latitude. The quantity l in (13.10) is the longitude measured from the solar direction. It should be noted that the direction of maximum intensity sometimes varies by as much as 90° in a matter of days.

The angular anisotropy produces a diurnal variation in the radiation observed at a point fixed on the earth. The local time of maximum flux depends on the rigidity observed and hence on the latitude.

It is believed that in interstellar space, beyond the shock boundary between the solar wind and interstellar medium, cosmic rays are homogeneous,

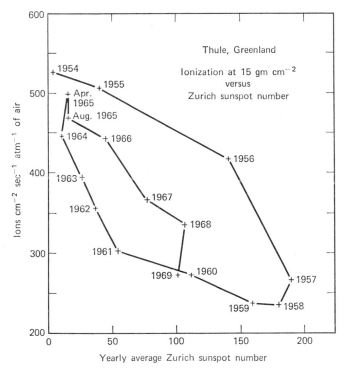

Figure 13.11   Ionization in the stratosphere near the north geomagnetic pole observed with similar ionization chambers over a 15-year period (sunspot cycles 19 and 20). Minima in the spot numbers are seen to have occurred in 1954 and late in 1964. The "hysteresis effect" seems to be larger for cycle 19, the largest amplitude cycle since the year 1700, than it is for cycle 20. From H. V. Neher, *Journal of Geophysical Research*, 1970.

isotropic, and constant in time over many thousands of years. The anisotropy at the earth is caused by the streaming of the galactic particles inward through the more or less regular spiral of the interplanetary magnetic field (see Chapter 9).

No sidereal time variation as large as 0.1% has ever been detected. One might expect such a variation to exist, either because of streaming along the (spiral) magnetic field lines in the galaxy or because of galactic discrete sources of the radiation.

The 11-year dependence of the flux is also well established. The minimum intensity occurs approximately when the sunspot number and solar radio activity are at their highest. There is a phase lag of 6–12 months, however; the lag is greatest when solar activity is decreasing rather than increasing (see Figure 13.11). Lower-energy particles are admitted to the earth's atmosphere during solar minimum; these observations are summarized for a low-cut-off latitude in Figure 13.12. Another way of putting this is that low-energy particles are excluded from the planetary system when the sun is active; the interplanetary flux is depressed below the interstellar flux by the sun, presumably through the interplanetary magnetic field that is transported by the solar wind.

The depression of the flux inside the solar system below the interstellar level implies that there may exist a time-dependent gradient of intensity with radial distance from the sun. Experiments conducted with interplanetary spacecraft have found such a gradient; $\sim 10\%$ per A.U. has been measured over several years between 0.75 A.U. (Venus) and 1.50 A.U. (Mars). The "true" interstellar flux has of course never been measured, but it may be approximately like that at the earth near solar minimum.

Forbush decreases (see Figure 13.13) are those decreases observed to occur in the cosmic ray flux up to rigidities of ~ 100 Bv, when a geomagnetic storm occurs. They are of the largest amplitude when observed with instruments that primarily sense the lower-energy portion of the cosmic ray beam, such as neutron monitors. Meson monitors and ionization-chamber Forbush decreases are typically only 1–5% in amplitude.

It now appears that the explanation of a Forbush decrease is that it occurs when a stream of fast solar plasma (the agency causing magnetic storms) overtakes the earth. As such a stream expands, the cosmic ray density in it is temporarily reduced below that in surrounding regions; recovery occurs as cosmic rays diffuse back in.

13.10 Solar Cosmic Rays and Riometers

Solar cosmic rays may be detected at the earth by counters above the atmosphere at high latitudes. Occasionally, when very high energy particles

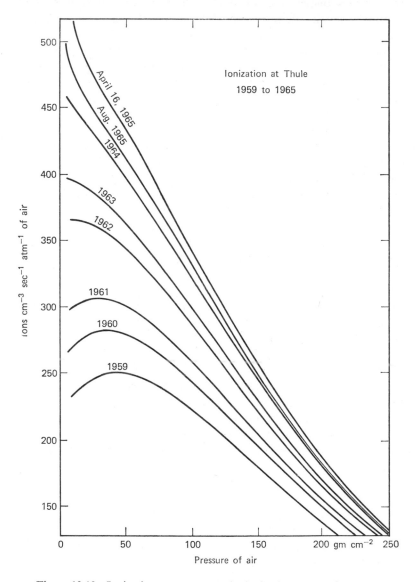

Figure 13.12 Ionization versus atmospheric depth at a very low geomagnetic-cutoff site. Note the disappearance of the Pfotzer maximum after 1961; only slightly penetrating (i.e., low-energy) particles are present at depths less than ~50 gm cm⁻² after that year and they persist through the 1964 minimum in the sunspot number. From H. V. Neher, *Journal of Geophysical Research*, 1970.

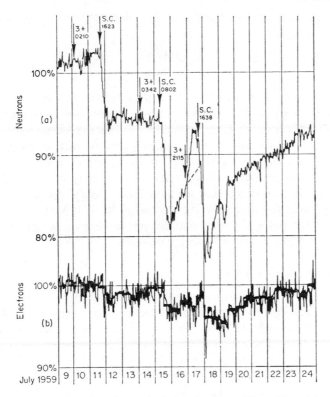

Figure 13.13 Hourly totals from the Deep River (Canada) neutron monitor are shown (*upper curve*) with the times of sudden commencements (S.C.) of geomagnetic storms and the times of large solar flares (3+). The combined Forbush decreases total more than 20% in amplitude for the neutron monitor but only ~5% for the nearby ionization chamber (*lower curve*). From H. Carmichael and J. F. Steljes, *Physical Review Letters*, 1959.

occur, they may be detected by particle monitors on the surface, such as neutron monitors and ionization chambers.

A powerful technique uses the *riometer*, a receiver on the ground that records cosmic radio noise (the superposition of the radio energy emitted by the various discrete galactic and extragalactic sources as well as diffuse sources) in the 1–50 MHz range of frequencies. When particles from the sun strike the atmosphere so that ionization occurs, the electron density n is given approximately by

$$\frac{dn}{dt} = kJ - \alpha n^2, \tag{13.11}$$

where k and α are constants and J is the solar particle flux. At equilibrium $dn/dt = 0$, so that

$$n = \sqrt{\frac{kJ}{\alpha}}.$$ (13.12)

The particle flux J is related to the electron density n, which in turn determines the absorption of the cosmic noise. Absorption measurements of J are surprisingly sensitive; fluxes as small as ~ 1 per cm²-second-steradian may be detected with riometers. Also, since particles with energies down to ~ 10 MeV are effective in producing atmospheric ionization down to ionospheric altitudes, particle with energies this low may be detected from the ground. This in turn means that the technique is useful primarily at latitudes higher than that for which the cut-off energy is ~ 10 MeV. Thus riometers are used in the polar regions; solar cosmic ray events are called polar cap absorption (PCA) events by riometer workers.

13.11   Solar Particle Fluxes

During a large solar flare electromagnetic radiation is observed promptly at the earth and charged particles are detected after a delay of 10–60 minutes. Fast particles arrive first; the slower ones do not reach a peak intensity until many hours later. The composition of the radiation varies; proton to alpha ratios have been measured that vary from 1 to 30, but the charge spectrum above $Z = 2$ is about that of the solar abundance.

The fluxes from various flares decay with a time t dependence that varies from $t^{-3/2}$ to e^{-t/t_0}. The former describes diffusion into an infinite diffusing medium, and the latter, leakage from a trapping mechanism. The behavior of the radiation depends on the condition of the interplanetary medium. Sometimes the increase may be observed again 27 days later, as the rotation of the sun carries the solar region back into a direction favorable for beaming particles to the earth.

13.12   Origins of the Galactic Radiation

The origin of the galactic radiation is not known. There have been many attempts to explain it over the years, but only limited success has been achieved.

The principal features of the radiation that must be accounted for are the following.

The Approximate Constancy of the Radiation, as Deduced from C^{14} and Meteorite Data. The source must be either continual or impulsive; if it is impulsive it must have occurred either long ago or else be repetitive and distant.

The Isotropy of the Radiation, up to the Highest Energies Observed. This seems to require that galactic *and* intergalactic space are uniformly filled with the radiation, or that the galactic magnetic field is stronger than the several microgauss inferred from the polarization of starlight. That there is no appearance of an enhanced flux in the direction of the galactic plane also appears to require an equal flux in the halo, but there is a containment problem for the highest-energy particles even there. A 5-microgauss field produces a 15-kpc turning radius for 7×10^{18}-eV protons (the radius of the halo, from the radio data, is about 15 kpc). Thus it is by no means clear that the "galactic" radiation is restricted to our own galaxy. This leads to another feature.

The Energy Density of Cosmic Rays. Their energy density at the earth, ω_{CR}, is ~ 1 eV cm^{-3}, which is about equal to the energy density of the interstellar magnetic field and also equal to the energy density of starlight. The total rate of energy loss by cosmic rays (if confined to our galaxy alone) is Q_{CR}, where

$$Q_{CR} \sim \frac{\omega_{CR} \times \text{volume of galaxy}}{\tau_{CR}},$$

and where Q_{CR} is also the rate of energy input to cosmic rays, assuming that equilibrium exists.

The quantity τ_{CR} is the lifetime against leakage from the galaxy, again assuming that the particles are all somehow accelerated within our galaxy. This may be estimated from the fragmentation hypothesis mentioned in Section 13.4; passage through ~ 5 gm cm^{-2} of matter, with interstellar gas densities that vary from $\sim 10^{-26}$ to 10^{-24} gm cm^{-3}, leads to path lengths that range from 5×10^{6} to 10^{9} light-years. This means lifetimes of 5×10^{6} to 1 billion years. This nuclear lifetime must also approximate τ_{CR} for the fragmentation hypothesis to be tenable. Hence $\tau_{CR} \sim 10^{14}$–10^{16} seconds; $Q_{CR} = 10^{42}$–10^{40} ergs per second, at least two orders of magnitude greater than the galactic nonthermal radio emission.

The flight of Gemini 11 in August 1966 provided an opportunity to measure the chemical composition of the incoming cosmic ray beam beyond the atmosphere, where the additional fragmentation had introduced troublesome corrections. Astronauts Conrad and Gordon recovered the nuclear emulsions from the outer surface of the spacecraft (where they had been covered by only 0.07 gm cm^{-2} of aluminum) after an 18-hour exposure. The orbit ranged between the latitudes of 29° N and 29° S. Table 13.2 shows the measured relative abundances of the elements at energies $\geqslant 1$ BeV per nucleon. Also C. Fichtel, M. Shapiro, and their colleagues deduced the chemical composition of the "source" of cosmic rays, by allowing for passage through 4 gm cm^{-2} of interstellar matter. These results are listed in the second

TABLE 13.2. RELATIVE ABUNDANCES OF ELEMENTS (NORMALIZED TO OXYGEN)

	Cosmic Rays[a] (Gemini 11)	Cosmic Rays at Source[b]	"Universe"	Solar Energetic Particles	Sun
Be	0.06 ± 0.03	0^c	2.9×10^{-8}	<0.02	$<10^{-5}$
B	0.28 ± 0.05	0^c	2.6×10^{-7}	<0.02	$<10^{-5}$
C	1.09 ± 0.08	0.89	0.57	0.59 ± 0.07	0.60 ± 0.10
N	0.25 ± 0.04	0.03	0.10	$0.19 + 0.04 - 0.07$	0.15 ± 0.05
O	1.00 ± 0.08	1.00	1.00	1.00	1.00
F	$\leqslant 0.04$	0^c	1.5×10^{-4}	<0.03	0.001 (?)
Ne	0.24 ± 0.04	0.19	0.10	0.13 ± 0.02	0.1 (?)
Na	$\leqslant 0.03$	0	2.7×10^{-3}	?	0.002
Mg	0.26 ± 0.04	0.25	0.04	0.042 ± 0.011	0.051 ± 0.015
Al	$\leqslant 0.06$	$\leqslant 0.03$	3.6×10^{-3}	?	0.002
Si	0.24 ± 0.04	0.33	0.04	0.09	0.045
P-K	0.10 ± 0.03	0.04	0.03	$\leqslant 0.02$	0.04^d
Ca-Ni	0.18 ± 0.03	0.24	0.04	0.011 ± 0.003	0.01 (?)

[a] Compared at the same energies per nucleon.
[b] Extrapolated through 4 gm cm^{-2} to source. The errors associated with the numbers in this column are typically $\approx 20\%$. However, nitrogen and the P-K group are uncertain by a factor of ≈ 2.
[c] Assumed.
[d] Assuming Ar to be about half as abundant as S, and the abundance of Cl to be small compared to S.

column of Table 13.2; it is seen that there are interesting differences between the composition of the source and the universal and solar abundances. There appears to be a relative enrichment of the heavier nuclei, and carbon is "overabundant" in the source, while nitrogen seems relatively scarce.

13.13 The Fermi Mechanism

One way of accelerating particles to cosmic ray energies was proposed by Fermi; his method is statistical and produces a $T^{-\gamma}$ energy spectrum. According to it, the moving interstellar gas clouds contain regions of higher than average magnetic field, which reflect cosmic rays. The latter gain or lose energy, depending on whether the collisions are head-on or overtaking. Per unit time, the former are more probable; there is an average acceleration.

In other words, speaking loosely, equipartition of energy occurs. There are two "gases" in contact. One consists of the large, moving HII regions. The particles of the other "gas" are the cosmic ray particles. If the particles of the latter are accelerated to the average kinetic energies of the former, which are huge because of the relatively great mass, very high cosmic ray energies will result.

If v is the cosmic ray velocity (nonrelativistic case), V is the gas cloud velocity, λ is the mean free path between collisions with clouds, and τ is the mean life of a cosmic ray, then dT/dt, the average rate of energy gain by a cosmic ray, is

$$\frac{dT}{dt} \sim \frac{4TV^2}{c\lambda} \frac{v}{c}.$$

This tells us that $T = T_0 e^{\alpha t}$, where $\alpha = 2V^2 v/c^2 \lambda$. Now if N is the number of cosmic rays, their lifetimes are

$$\frac{dN}{N} = \frac{dt}{\tau};$$

$$N(>t) = N_0 e^{-t/\tau}.$$

We may combine these two time-dependent expressions. The result is

$$N(>E) = N_0 \left(\frac{E_0}{E}\right)^{1/\alpha t},$$

the required power law energy spectrum. It now appears, however, that the Fermi mechanism is not the correct one to explain cosmic ray acceleration, because the charge spectrum is found experimentally to be energy-independent. Fermi's theory says that the slope of the energy spectrum is greater for smaller τ; it should therefore be greater for heavier nuclei, and this is not observed.

13.14   The Supernova Mechanism

Only supernovae seem to produce enough energy in particles to supply Q_{CR}. Solar flares, for example, typically generate only 10^{23} ergs per second; the 10^{11} such stars in the galaxy, therefore, could only generate $\sim 10^{34}$ ergs per second in the form of particles, which is far short of the 10^{40} ergs per second required.

Supernovae seem to occur once per century in an average galaxy, although this number is quite uncertain (recent estimates are that the true frequency may be about 3 times this figure). Some 10^{48} ergs go into relativistic electrons, from the observed synchrotron radiation. If we assume that ~ 100 times as much goes into relativistic nuclei, since they are 10^3 times as massive as electrons, we find that

$$Q_{CR} \sim \frac{10^{48} \text{ ergs} \times 10^2}{100 \text{ years}} = 10^{40} \text{ ergs per second},$$

about the required amount. Protons of up to 10^{18} eV can be contained in the dimensions of supernova remnants such as the Crab nebula, where fields of up to 1 milligauss are present. At the present time it appears that supernovae are the most likely sources of cosmic rays.

Problems

13.1. (a) What is the total electric current flow into the earth's atmosphere due to cosmic ray bombardment?
(b) How is this current neutralized or the circuit completed?

13.2. How long a time will a 10^{19}-eV neutron take to traverse the 10^5 light-year diameter of our galaxy, in the neutron's frame of reference? Compare this time to the half-life of the neutron.

13.3. A certain cosmic ray secondary particle is produced in an interaction at a height of 100 km, and it moves vertically downward. The particle has no electric charge and is unstable, decaying at rest after 100 microseconds. What must its velocity be in order to reach the earth's surface?

13.4. Assume that the maximum magnetic fields in the interstellar gas and dust are of the order of 20 gammas, and the maximum random velocity of the clouds is of order 1000 km per second. What could you say about the importance of the statistical Fermi acceleration mechanism for cosmic rays, if these gas clouds were as numerous as the stars in the galaxy?

13.5. Taking the average density of interstellar hydrogen to be 1 cm^{-3}, calculate the average time of travel between nuclear collisions for a relativistic cosmic ray iron nucleus.

13.6. Suppose that a vertically incident, very energetic primary cosmic ray interacts in the earth's atmosphere at an altitude of 30 km over the geomagnetic equator and produces a positive and negative π-meson, which then travel vertically down. Making any necessary assumptions,
   (a) Calculate how far from each other will be the two points where these mesons hit the earth's surface (assuming that they do so).
   (b) What will be the magnetic bearing of the positive meson's impact point from that of the negative meson?
   (c) Are the two likely actually to hit the earth? Why?

13.7. The number of C^{14} nuclei produced per second per cm^2 of the earth's surface is approximately equal to the energy flux brought into the atmosphere by cosmic rays with kinetic energy $\geqslant 1$ BeV, where we assume that all cosmic rays are protons.
   (a) If the earth's magnetic dipole moment were double its present value, what would the global average production rate of C^{14} be?
   (b) Suppose that until 5800 years ago the magnetic dipole moment was twice its present value. An object has been dated by the C^{14} method on the assumption that C^{14} has been produced at a constant rate, and the "C^{14} age" is 3000 years. What is its chronological age, if the magnetic field really changed as described?

13.8. Iron atoms ($z = 26$, $A = 56$) have a photoelectric cross-section of $\sigma = 1.2 \times 10^{-23}$ cm^2 per atom for 100-keV photons. How many grams per square centimeter of iron shielding are required to reduce a flux of such photons by a factor e^2?

13.9. Cosmic rays are detected on the upper and lower sides of a barrier of mass density 10 gm cm^{-3} and thickness 10 cm. The material has an average number density of 10^{26} atoms cm^{-3}. If the ratio of fluxes on the upper and lower sides is 148.41:1, what is the cross-section of a barrier atom?

Chapter 14

RADIO ASTRONOMY

Information about the universe flows to us on the earth's surface through several "channels." The corpuscular radiation called cosmic rays and the study of neutrino astronomy provide two such channels. Another channel is optical astronomy; great amounts of information have been gathered through the detection and analysis of electromagnetic radiation with wavelengths between 4×10^{-5} and 8×10^{-5} cm. There is also the *radio* channel, which is occupied by electromagnetic waves that have lengths between ~ 1 cm and perhaps 30 meters. The radio channel of information provides the data for the subject known as radio astronomy. Radio astronomy has added greatly to our knowledge of the cosmos, particularly within the past 20 years or so.

14.1 Radio Wave Absorption

The great value of the radio channel lies in the fact that radio waves can penetrate most of the matter in the galaxy, whereas optical photons are absorbed. This is because the radio waves are much longer than the typical sizes of dust particles found in the galaxy, while optical wavelengths are comparable to, or shorter than, the dimensions of cosmic dust particles. In the galactic plane we can see optically for perhaps 10,000 light-years. In the radio, however, we can see for a whole galactic diameter ($\sim 100{,}000$ light-years). One of the major triumphs of radio astronomy has been the elucidation of the structure of our galaxy.

The short-wave limit of the radio channel is caused by absorption by molecules of oxygen and water vapor in the terrestrial atmosphere. The long-wave limit, on the other hand, is caused by the ionosphere; electromagnetic waves longer than ~ 30 meters cannot readily penetrate the ionosphere. These limits will presumably be eliminated with the advent of radio astronomy satellites in orbit about the earth. The limits then will be primarily technological in nature; sensitive receivers of extremely high frequencies and huge antennae are equally difficult to construct.

14.2 Radio Telescopes

Because radio waves are so much longer than optical waves, the collecting areas must be greater. The critical parameter is the collecting area, expressed in square wavelengths.

402

Present-day steerable antennas range in size up to the 250-foot-diameter parabolic "dish" at Jodrell Bank in England. The largest nonsteerable instrument is the 1000-foot spherical reflector located at Arecibo, Puerto Rico. The rotation of the earth permits a nonsteerable antenna to scan the sky.

It is an engineering feat to build a large antenna. The surface must be mechanically true—even in strong winds—to about 0.1 wavelength. Hence the surface must be accurate to perhaps $\frac{3}{8}$ inch over its entire surface. Particularly at the shorter wavelengths, the beamwidth and shape or profile of the angular-response pattern may be distorted by the small variations in deflection due to gravity (even in wind-calm conditions) that take place as the antenna of a steerable instrument moves between the zenith and the horizon.

If a "feed" (signal source) is placed at the focal point, an antenna like this may be used for *radar* astronomy. As we have seen, radar has been extremely helpful in planetary studies.

14.3   Radio Sources

Radio waves have been detected from the planets and from the sun. These studies have led to the discovery of magnetically trapped radiation around Jupiter and also to an understanding of some of the processes operative in the solar corona.

Emissions have also been detected from beyond the solar system. A listing of the types of discrete sources include (a) the galactic arms and center, (b) ionized hydrogen clouds, (c) neutral hydrogen regions, (d) supernova remnants, (e) radiogalaxies outside our own galaxy, (f) quasars, and (g) quasistellar galaxies. We discuss each of these sources later.

It will be noted that stars are not on this list. No radio waves have ever been detected from stars other than the sun (except for a few that seem to be ejecting a considerable amount of mass). This is presumably due to the sensitivity limits of the present instruments. Present radio telescopes can detect discrete sources down to $\sim 2 \times 10^{-26}$ watt meter^{-2} Hz^{-1} at a few meters' wavelength. Radio astronomers refer to this signal strength as a flux density of two "units."

A large outburst from our sun produces something like 10^{-15} watt meter^{-2} Hz^{-1} at these frequencies, which is roughly 10^{10} times the minimum detectable signal. We could therefore detect a comparable flare on another star out to about $(10^{10})^{1/2} = 10^5$ times the earth-sun distance. Stellar flares could therefore be detected to a distance of 10^5 A.U. $= 1.5$ light-years from the earth. The star closest to the sun is over 4 light-years away from us.

It is to be hoped that as new telescopes become operational other stars may be detected through their presumed radio flares. Proxima Centauri, the star closest to the sun, is optically a flare star. It is possible that the detection of flares on this and other stars will increase our understanding of the flare mechanism.

There are also emissions found from nondiscrete sources. Radio energy has been detected that emanates from a roughly spherical region that surrounds our galaxy. This region is called the galactic "halo."

Recently, isotropic microwave background radiation has also been detected; it is not (apparently) associated with our galaxy or any other discrete object in the skies. This radiation has received the name "primordial fireball radiation," but it is by no means clear that it really is "primordial." We shall return to these distributed sources later.

14.4   Emission Mechanisms

We know of primarily three ways by which cosmic radio energy may be emitted. These are the thermal mechanism, atomic emissions from transitions between hyperfine states of atoms, and synchrotron emission. As we shall see, the first two are independent of magnetic fields, while synchrotron radiation requires that a magnetic field be present. If we can identify the precise mechanism responsible for a given radio signal, we may be able to learn a great deal about the nature of the emitting region.

There are two principal ways of identifying the cause of a given signal. One is to examine its frequency spectrum. An analysis of the received power versus frequency may be made and compared with that predicted by the source mechanisms.

Another interesting test of a signal is to examine the *polarization* of the radio waves. The synchrotron mechanism, for example, results in plane-polarized electromagnetic energy if ordered magnetic fields are present in the source region.

The polarization test, however, is much more difficult to perform than a measurement of the frequency spectrum. This is due to the effects of magnetic fields between the source and the observer; the intervening fields may change the polarization of the emitted waves through the *Faraday effect*.

In 1845 Michael Faraday discovered that when a block of glass is subjected to a strong magnetic field, the glass affects the propagation of light passing through it. In particular, when plane-polarized light is sent through glass in a direction parallel to the applied magnetic field, the plane of the electric vector is rotated. The amount of rotation is proportional to the field strength and to the path length through the medium; the constant of proportionality, it was learned, is a function of the medium and its density.

A classical explanation of the effect is as follows. Suppose that the electric field of a plane-polarized wave extends vertically upward from the paper and that the magnetic field has a component in the plane of the paper. The electric field will tend to make the electrons of the medium oscillate up and down.

The magnetic field, however forces them to move also in the plane of the paper, as they start to execute this motion. The wave thereby gains an electric field component in the plane of the paper; the wave becomes elliptically polarized. The wide variety of possible different wave configurations allows two to occur simultaneously (for a given frequency of oscillation) that have different wavelengths and velocities. This is the double refraction effect; the two waves are usually referred to as the ordinary (O) wave and the extra-ordinary (E) wave.

Some of the properties of the O- and E-waves may be understood in terms of the relatively simple case of propagation along the magnetic field. Both waves are then circularly polarized (one clockwise, the other counter-clockwise). We normally write the index of refraction of a medium, k, as $k^2 = 1 - \omega_p^2/\omega^2$ (see Section 3.19) for radiation of angular frequency ω. Here, however, the appropriate expression is

$$k^2 = 1 - \omega_p^2(\omega^2 \pm \omega\omega_e)^{-1}, \tag{14.1}$$

where the upper sign refers to the O-wave and the lower to the E-wave. In (14.1) the quantity ω_e is just the cyclotron frequency; $\omega_e = eB/m$.

Now consider (14.1) when $\omega_e \ll \omega$. If the effect of ω_e is vanishingly small, we have two circularly polarized waves traveling with the same velocity. It is easily shown that the combination of a right- and a left-handed rotating vector gives a fixed vector that corresponds to a plane-polarized wave. However, ω_e does have a slight effect. It causes the two waves to travel with slightly different speeds. The result is that the plane-polarized (combination) wave rotates its plane of polarization slowly as it travels forward through the medium. This is the basis of the Faraday effect. If random magnetic fields exist between a strongly polarized source and the observer, the radiation detected by the observer will exhibit a much smaller degree of polarization than it initially possessed.

14.5   The Thermal Emission Mechanism

A hot object radiates energy, as we have seen many times in this volume. Thus any object that has an absolute temperature greater than zero should produce some thermal radio emission. We may use the black body equations to estimate how much energy is emitted.

At wavelengths that are large compared to that of the maximum in the Planck radiation law, we may employ the Rayleigh-Jeans approximation to the Planck distribution. Radio wavelengths satisfy this condition for most sources met in nature.

Therefore the *radio brightness* b of a given area of the radiation is

$$b = \frac{2kT_b}{\lambda^2}, \tag{14.2a}$$

where b is expressed in units of watts per square meter per steradian per hertz. Hence we may rewrite the expression for b at a wavelength λ meters as

$$b = 2.8 \times 10^{-23} \frac{T_b}{\lambda^2} \text{ [watts m}^{-2} \text{ ster}^{-1} \text{ Hz}^{-1}]. \tag{14.2b}$$

The quantity T_b is the "equivalent brightness temperature"; astronomers find it a convenient way of characterizing sources. The brightness is closely related to the radio flux S that we introduced in Section 4.8. The flux density S is just $S = b\Omega$, where Ω is the solid angle in steradians subtended by the source.

The relations (14.2) are only accurate, however, if the source is dense. That is, the source is assumed to be many wavelengths in thickness. For a nearly transparent ("thin") source, the brightness temperature expression must be modified. We may write that

$$T_b = T\{1 - e^{-Kx}\}, \tag{14.3}$$

where T is the "true" equivalent temperature. The absorption coefficient K is related to the opacity; as we have seen previously, the opacity is a complicated function of the wavelength λ. The path length of the radiation through the source is denoted by x in (14.3).

Therefore the radio brightness becomes

$$b = \frac{2kT}{\lambda^2} \{1 - e^{-Kx}\} \tag{14.4}$$

for a thin source. This, of course, also applies for a thin *absorber*, such as a nearly transparent cloud between the observer and a source. An absorber is also an emitter, and the absorption process is sometimes easier to describe. We see that if $Kx \gg 1$, (14.4) reduces to our black body limit for b, namely, $2kT/\lambda^2$. The black body laws are strictly applicable only for thick radiators (or absorbers).

However, if $Kx \ll 1$, we have a very nearly transparent cloud. In this case (14.4) becomes

$$b = \frac{2kTKx}{\lambda^2} \; ; \tag{14.5}$$

the brightness of a thin source depends on the thickness of the source. The opacity depends in a complicated way on the ion and electron densities and on the temperature, as well as on the wavelength. Thus Shklovsky has shown that the absorption coefficient K_ν for diffuse ionized interstellar matter may be written as

$$K_\nu = 9.8 \times 10^{-3} \frac{N_i N_e}{T^{3/2} \nu^2} \left[19.8 + \ln \left(\frac{T^{3/2}}{\nu} \right) \right] \tag{14.6}$$

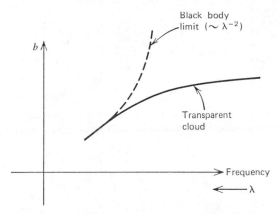

Figure 14.1 Frequency spectrum of purely thermal radio emission.

for radiation of frequency ν that passes through a region where the ion and electron number densities are N_i and N_e, respectively. The temperature in (14.6) is the kinetic temperature of the absorbing cloud; a Maxwellian velocity distribution was used in the derivation of (14.6).

Now we usually have that $N_i \approx N_e = N$. We may substitute (14.6) (with this approximation) for K into (14.5). The result is that

$$b = \text{constant} \times \frac{N^2 x}{T^{1/2}}. \tag{14.7}$$

For fully ionized hydrogen the value of the constant is about $0.2k$. Thus if the cloud thickness is small compared to a wavelength, b is *independent of the wavelength*. (Alternatively, the brightness temperature is proportional to λ^2.)

We may put all of the foregoing together and form a picture of the radio frequency spectrum expected from thermal emission (or absorption). The result is shown schematically in Figure 14.1. Thus if we detect a spectral distribution of this shape, we may more or less confidently ascribe it to radio emission (or absorption) by a plasma at a temperature T.

14.6 Thermal Radiation and the H II Regions

When radio astronomers conduct a sky survey, the first problem encountered in analyzing the results is to separate the thermal and nonthermal components of the emission. From the preceding section, the former component would give us the distribution of ionized matter (mostly hydrogen) in the galaxy. One of the great achievements of radio astronomy has been the understanding of the structure of our galaxy that it has brought to us.

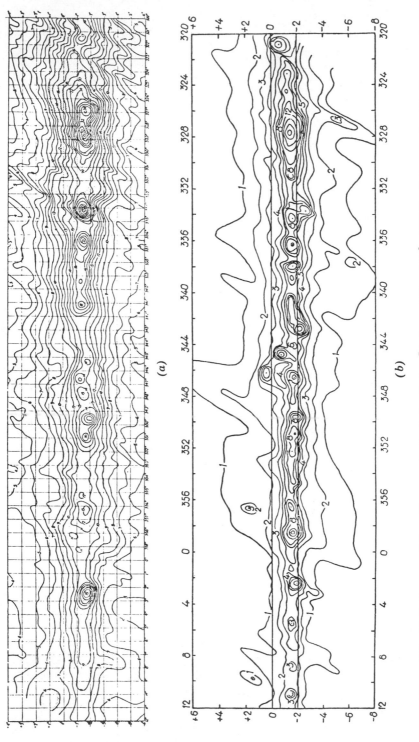

Figure 14.2 (a) Radio map of the milky way at galactic longitudes l^I between 320 and 12° at 85 Mhz (3.5 m wavelength). The numerals refer to the brightness temperature in units of 1000°K. Data obtained by E. R. Hill, B. Slee, and B. Y. Mills in 1958. (b) Same sky region as in (a) ($+6° \geqslant b^I \geqslant -6°$), with an equal beamwidth (0.6–0.8°) but at a wavelength of 22 cm. The numerals here refer to the brightness temperature in units of 3.25°K. Data obtained by G. Westerhout. Curves (a) and (b) taken from J. L. Steinberg and J. Lequeux, *Radio Astronomy*, Dunod, 1960. The Galactic center is apparent at both wavelengths, in the region surrounding $l^I = 327°$, $b^I = -1°$. Other discrete sources are also apparent.

Contours of thermal radio brightness are shown in Figure 14.2. We see the galactic center and also several radio sources that are spread along the Milky Way. The thermal sources are identified as large, bright (at short wavelengths of ∼22 cm) clouds of ionized hydrogen that astronomers call H II regions. In this nomenclature neutral hydrogen atoms are referred to as H I; triply-ionized oxygen atoms are designated O II.

A typical H II region may have dimensions of the order of 10 light-years. H II regions are sometimes visible to the naked eye. The most noteworthy (and among the most beautiful) is the Orion nebula, which is barely visible to the naked eye but easily seen with binoculars. Plasma clouds like this glow by fluorescence; a hot O or B star nearby photoionizes the gas with its ultraviolet radiation, and the gas then emits in the visible region as the electrons recombine with the protons and cascade down through the various energy levels of the hydrogen atom. The size of an H II region is determined by the temperature of the nearby stars. An O6 star on the main sequence can photoionize a region ∼400 light-years in diameter if the gas density is ∼1 atom cm^{-3}; the H II region caused by an A0 star is only ∼4 light-years across.

Temperatures within an H II region are typically $10,000°K$; the H I regions that bound them are quite cold. This is because a hot star radiates primarily in the ultraviolet (according to the Planck law); photons of ultraviolet wavelengths can photoionize hydrogen, since the binding energy of the electron in the hydrogen atom corresponds to a wavelength of 912 Å.

At long wavelengths the radio picture reverses from the picture seen at short wavelengths. Thus at $\lambda = 15$ meters the formerly bright regions become dark ones on a lighter background. The background is due to the nonthermal component. The H II clouds have brightness temperatures of only about $10^{4}°K$ thus their emission is negligible compared to the $5 \times 10^{5}°K$ brightness of the nonthermal background at $\lambda = 15$ meters. They *absorb*, however, the nonthermal radiation coming from sources behind them.

It is found that the total mass of the ionized hydrogen in the spiral arms of the galaxy is less than $6 \times 10^{7} M_{\odot}$, or only about 0.06% of the total galactic mass. Even though it is a small fraction of the total mass, ionized hydrogen is an extremely important part of the galaxy. As we shall see in the next section, the ionized gas holds the galactic magnetic field in place; there would be no "general field" without it, and it is possible that the structure of the galaxy would be greatly modified by the absence of such a field.

Other galaxies are also sources of radio energy. The nearest spiral galaxy to ours is known variously as M31, and as the Great Galaxy in Andromeda. Figure 14.3 illustrates its appearance at radio wavelengths.

The *motion* of H II regions may also be studied through the Doppler effect. It is found that shifts in the absorption frequency of a line may be interpreted

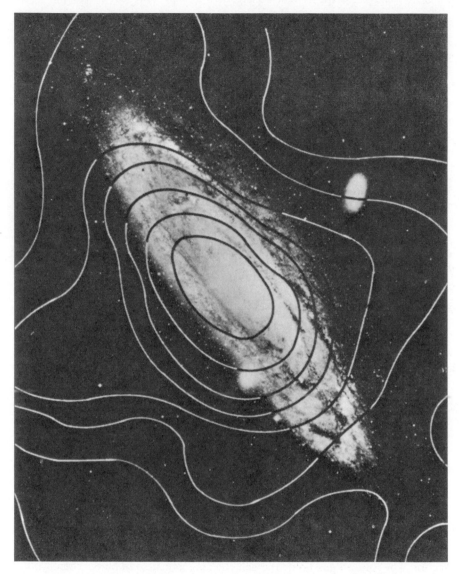

Figure 14.3 The Great Nebula in Andromeda (M31). A smoothed radio image at a wavelength of 75 cm is superimposed on an optical photo obtained with the 200-inch telescope on Mount Palomar. Courtesy Hale Observatories, California Institute of Technology. The radio data were obtained in 1957 with a 2.0° beamwidth, using a 25-m paraboloid. The interval between contours is 0.8°K and the integrated 75 cm flux is 78 flux units. Notice that the radio isophotes are much less flattened than is the optical brightness distribution; the radio galactic corona is several times the size of the visible galaxy, which is about 5° in diameter. Adapted from C. L. Seeger, G. Westerhout, and R. G. Conway, *The Astrophysical Journal*, 1957.

as cloud speeds that range up to several hundreds of kilometers per second. A typical value is 5–10 km per second.

14.7 Interstellar Magnetic Fields

We expect that interstellar magnetic fields exist. The interstellar plasma consists of ionized gas particles (mainly hydrogen), to a first approximation. If there are any temperature gradients or pressure gradients in the medium, currents will flow, giving rise to the magnetic fields.

Any turbulence in the ionized medium will have the effect of amplifying these initial fields. That is, random twisting of the frozen-in field lines will increase the density of the lines. Hence the field strength increases.

The amplification will increase the field until the condition

$$\frac{B^2}{2\mu} = \tfrac{1}{2}\rho v^2 \tag{14.8}$$

is obeyed, where B is the field strength at equilibrium, ρ is the density of the medium, and v is the mean speed of the randomly moving clouds of ionized hydrogen. (As we have seen, radio astronomy can measure the speeds of the clouds.) Physically, we may say that this "equipartition of energy" between the fields and plasma arises because the field amplifies until it is strong enough to oppose further deformation (twisting) of the lines of force.

The amplification through twisting occurs because the "diffusion time" τ is *very* long; the field lines are frozen into the medium. In fact, if we assume a size L of 15 light-years for such a cloud of plasma and a conductivity of $\sim 6 \times 10^{-9}$ emu (the value of σ appropriate for a density of 1 cm^{-3} at a temperature of $\sim 10^{4}$°K), we find from (10.40) that

$$\tau = \mu\sigma L^2 \approx 10^{22} \text{ years (!)}$$

for the relaxation time in an H II cloud.

Now if we take $\rho \approx 10^{-24}$ gm cm^{-3} for the density of a cloud (about 1 hydrogen ion per cm^3) and a mean speed v of about 10 km per second, we find from (14.8) that $B \approx 1$ gamma in interstellar space. That is, we expect field strengths of roughly 10^{-5} gauss to be found in interstellar space within our galaxy.

Optical observation confirms this; fields of this strength are in fact present. It is found that the light from distant stars is partially polarized, with the electric vector generally in the plane of the galaxy. The polarization is weak— only about 5%—but definitely present.

The polarization is believed to be caused by reflection of the starlight by nonspherical paramagnetic dust grains that are aligned by the interstellar magnetic fields. The alignment is such that their long axes are parallel to the lines of force. (One speculation is that the particles are composed of ice.) An interesting result of these studies is that the magnetic field lines seem to occur along the spiral arms of the galaxy.

The presence of field lines along the spiral arms may be the reason that the arms do not contract gravitationally into thin filaments. The stars and gas in the arms should cause such a gravitational collapse of the arms. Presumably, therefore, the lateral magnetic pressure prevents the collapse; it appears at this writing that a field strength of the order of 1 gamma would suffice to prevent the collapse.

14.8 Emission from Neutral Hydrogen

Radio emission may be observed from *cold neutral hydrogen atoms*. This may seem at first glance to be paradoxical; no thermal motion or ionization is present, and yet electromagnetic energy is available.

The hydrogen atom exhibits hyperfine splitting of even the ground state. This is due to the spin of the nucleus (a proton). The spin of the orbital electron may be either parallel or antiparallel to the spin of the proton. There is consequently a slight energy difference E between the two states. This energy difference corresponds to a frequency f, where $f = E/h$. It is found that $f = 1420.4$ MHz, so that hydrogen *line emission* is observed at a wavelength λ of 21.1 cm.

We may calculate from quantum mechanics the lifetime of the excited state against spontaneous emission. The result is 11 million years; the transition probability and hence the intensity of the spontaneously emitted 21-cm radiation are quite small. This is because the transition probabilities depend (see Section 5.5) on ν^3, and this frequency is quite low. It is also because the transition is a forbidden one; the observed radiation is magnetic-dipole in character, which means that we expect it to be $\sim 10^5$ times weaker than electric dipole transitions. In the galaxy, however, each hydrogen atom suffers occasional collisions with its neighbors. Some of the collisions *induce* transitions, which have the effect of lowering the lifetime of the excited state to ~ 100 years. This radiation is detectable by modern receivers (even in spite of the 100-year lifetime) because there are so many hydrogen atoms along the line of sight in the galaxy.

The detectability of 21-cm radiation was first pointed out by van de Hulst in 1944; it is of vast importance to astronomy, because the presence of cold hydrogen atoms in space could not otherwise be detected, even from space. This is because the Lyman series of hydrogen is in the ultraviolet region, which is completely absorbed by the earth's atmosphere. Even if we do attempt ultraviolet measurements from space vehicles outside our atmosphere, cold atoms will practically never be even in the first excited state, so that even Lyman alpha emissions (1216 Å) will be faint.

While it is possible to detect the radiation, it is perhaps instructive to see just how weak it is. The power received by an antenna of effective area A and gain g and operating over bandwidth ΔF hertz is just $(4kTA\,\Delta F)/g\lambda^2$, where T is the brightness temperature of the source. Let us take the whole sky as the source, and assume that the radiation from it is all due to cold neutral hydrogen atoms. The average brightness temperature at $\lambda = 21$ cm for the whole sky is $\sim 3°$K. If we assume that $\Delta F = 100$ kHz and that A, for the sake of discussion, is the surface area of the earth, we find that the received power from the whole *universe* is only ~ 0.5 watt. That radiation this weak may be detected must be considered a tribute to technology.

This branch of radio astronomy—the detection of the 21-cm line—has made it possible to map our galaxy. We are now able to determine where hydrogen, its predominant component, is concentrated in space.

Another feature emerges from studies of the 21-cm line. We may measure the radial component of the neutral hydrogen clouds with respect to the earth, through measurements of the Doppler shift of the line. Hence we may determine the motions of the clouds with respect to each other.

14.9   Structure of the Galaxy

It has been learned from studies of the types described above that ionized hydrogen is found mainly in clouds that occupy a flat disc in the galactic plane. The clouds are regions within the much larger amount of un-ionized, cold hydrogen atoms.

In order to make some generalizations about the distribution and motions of neutral hydrogen in the galaxy, some assumptions are necessary; a model of the galaxy must be devised. We assume that the entire galaxy is rotating about its center. We take it that the innermost parts rotate fastest and the rate of rotation gradually decreases with increasing distance from the galactic center. It is further assumed that all the gas at a given distance from the center rotates at the same (circular) velocity and that all of the galactic gas is at the same temperature.

Because of the differential rotation, any point in the galaxy will have a particular velocity toward or away from the earth. If we look in a given direction and measure the Doppler shift, the distance of the radiating region may be found from the model. The observations are consistent with all the above assumptions.

Most of the neutral hydrogen (see Figure 14.4) is confined to a flat disc of diameter $\sim$80,000 light-years. The thickness is only about 800 light-years. The mass of interstellar neutral hydrogen is perhaps 2% of the galactic mass. Most of the mass of the galaxy is presumably in the stars; our galaxy (from a Keplerian analysis of the sun's rotation about the galactic center) contains a

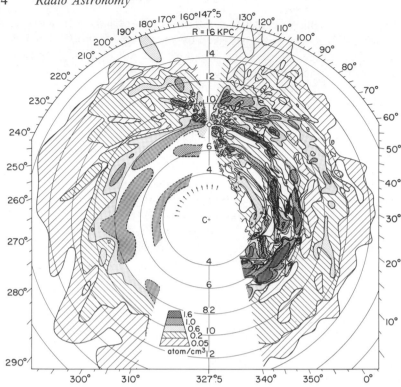

Figure 14.4   Radio map of our galaxy at the wavelength of the neutral-hydrogen-atom line, 21 cm. Contours of equal number-density of hydrogen atoms are shown as a function of the old system of galactic longitudes, l^I. At distances R less than $\sim$8 kps four or five spiral arms are apparent. From J. H. Oort, F. T. Kerr, and G. Westerhout, *Monthly Notices of the Royal Society*, 1958.

mass of $1 \times 10^{11}M_\odot$, so that we conclude that there are about 100 billion stars in the Milky Way.

Within the disc the gas is arranged in arms that show a spiral form. There seem to be at least four distinct arms, called the Sagittarius, Cygnus, Perseus, and Norma arms. The sun apparently lies within the Cygnus arm.

Stars are associated with the gas, so that the whole galaxy has a spiral structure (see Figure 14.5). Within 20,000 light-years of its center (i.e., $R/2$) the spiral arms define a plane to 1 part in 10^3. It is curious that the outermost parts of the galaxy are seriously distorted; the plane seems to be "twisted." Seen edge-on, the Milky Way would apparently resemble an integral sign lying down. The reason for this is unknown, but may be connected with motion of the galaxy through the plasma and magnetic fields of the intergalactic medium.

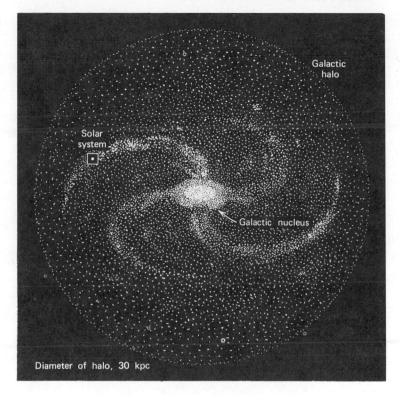

Figure 14.5   Schematic view of our galaxy.

At $R/2$ the density of neutral hydrogen reaches a maximum; it is about 1 atom cm^{-3} there. At $r = R$ some 40,000 light-years from the center, the density is down to about one-tenth its maximum value.

There is no information on the distribution of hydrogen radially between the earth and the galactic center. This is because of the model described above. A Doppler measurement of the speed should yield the distance in all directions except toward the center, where the motion should all be transverse to the line of sight. At the distance of the earth, for example ($r \sim 25,000$ light-years), the rotational speed of the galaxy is about 200 km per second. The speed is nearly constant between this distance and about $r = 9000$ light-years, where it is approximately 175 km per second.

A surprising feature becomes evident at distances of less than 9000 light-years. A weak wing is observed in the 21-cm emission from these regions; a Doppler shift corresponding to about 50 km per second is observed at $r \sim 9000$ light-years, and speeds of recession from the galactic center as high as 200 km per second have been measured nearer the center. It is as though some gas is moving radially toward us in the form of a shell that has now reached the 9000-light-year point. This raises the fascinating possibility that the center of our galaxy has undergone an explosion; the expelled gases are moving outward at about 50 km per second. Of course, it is entirely possible that the speed of the radial motion is much greater than 50 km per second in directions perpendicular to the galactic plane; collisions with interstellar matter will be less frequent in such a direction, and the inhibiting effects of the galactic magnetic field will be reduced. The explosion hypothesis is supported somewhat, in that a strong nonthermal source has been detected near the thermal center of the galaxy, called Sagittarius A. The nonthermal region is consistent with $2.5 \times 10^5 M_\odot$ of ionized hydrogen at a density of about 85 cm^{-3}; this gas is apparently being formed at the rate of $\approx 1 M_\odot$ per year. We shall return to the subject of galactic explosions in Section 14.14.

14.10   Interstellar Molecules and Chemistry

At this writing, radio lines from over 40 molecules have been found to be emanating from regions in galactic space. The first discovery, which took place in the early 1960s, was of four closely spaced lines near 1666 Mhz that are emitted by the OH^- radical. A brightness temperature of $10^{12}°K$ is observed for this quadruplet.

The lines are all quite sharp. When closely spaced in frequency as in a multiplet, they frequently have a nonthermal intensity distribution, as does a maser in the laboratory.

Many of the molecules detected to date are organic. The most complex thus far is cyanoacetylene, $CN-C{=}C-H$. Molecular hydrogen, H_2, has also been found, but not through radio techniques, because of its spectrum; a rocket-borne ultraviolet spectrometer detected the presence of H_2.

Much of the molecular emission originates in small, dense clouds within our galaxy. Some of the clouds are smaller than the solar system.

In these clouds, the molecular lifetime against photodissociation by ultraviolet is less than a thousand years. This short a time raises questions concerning the mechanisms by which the molecules originate. Those suggested thus far include two-body radiative association (the densities are too low to make three-body processes likely) and molecular formation on grain surfaces.

Interstellar Chemistry, a branch of nonequilibrium chemistry, attempts to describe the formation of such molecules. It seems likely that interstellar chemistry and the study of lines in radio astronomy are destined to become active and important areas of space science.

14.11   Synchrotron Emission

Synchrotron emission is sometimes called "magnetic bremsstrahlung." It is the electromagnetic radiation produced by ultrarelativistic charged particles moving in a magnetic field.

In the nonrelativistic case we have the well-known "cyclotron" radiation. An electron, say, radiates isotropically like a dipole and at only one frequency F_L, where

$$F_L = \frac{eB}{2\pi m_0}.$$ (14.9)

This frequency is known as the "cyclotron frequency" or the "Larmor frequency." It amounts to 28 Hz per gamma of field strength.

When the total energy E of the particle, however, is much greater than m_0c^2, the emitted radiation is not isotropic. Also, a quasi-continuum of frequencies is emitted by even a single electron. This is called synchrotron radiation. First observed around accelerating machines, synchrotron radiation was worked out theoretically by Schwinger in 1949.

Let us first discuss the single-electron case. If the relativistic electron is circling about a field line, it radiates in a little cone of directions (of angle θ) centered about the direction of the instantaneous velocity, where

$$\theta \approx \frac{m_0c^2}{E},$$ (14.10)

and where m_0 is the rest mass of the electron. (For example, if $E = 1$ BeV, $\theta = 1.5$ minutes of arc.)

That the radiation is not isotropic in the laboratory frame of reference follows from considering the radiation in the frame of reference moving with the electron. The radiation is like that of a dipole in the electron's frame, but transfer to the laboratory frame strongly concentrates the radiated energy into a forward cone of directions.

It is useful to digress for the moment and consider the Doppler effect at very high energies. In the ultrarelativistic case we may write

$$E = \frac{m_0c^2}{(1 - v^2/c^2)^{1/2}},$$ (14.11)

so that

$$\left(\frac{E}{m_0c^2}\right)^2 = \left(1 - \frac{v^2}{c^2}\right)^{-1}.$$ (14.12)

Now we may also write

$$\left(1 - \frac{v^2}{c^2}\right) = \left(1 + \frac{v}{c}\right)\left(1 - \frac{v}{c}\right).$$ (14.13)

But $v/c \approx 1$, so that (14.13) becomes

$$\left(1 - \frac{v^2}{c^2}\right) \approx 2\left(1 - \frac{v}{c}\right).$$ (14.14)

The Doppler effect says that

$$\tau = \tau'\left(1 - \frac{v}{c}\right),$$ (14.15)

where τ' is time as measured by an observer moving with the electron. Therefore, we may write that

$$\left(1 - \frac{v^2}{c^2}\right) = \frac{2\tau}{\tau}. \tag{14.16}$$

Hence (14.12) becomes

$$\left(\frac{E}{m_0 c^2}\right) = \frac{\tau'}{2\tau}, \tag{14.17}$$

which may be regarded as the Doppler effect at very high energies.

We now return to our high-energy electron. If the electron is moving in a circular orbit of radius R at a velocity very nearly that of the velocity of light, we have that

$$R \approx \frac{c}{2\pi F_L}. \tag{14.18}$$

The observer sees pulses of radiation only when the electron comes toward him. The duration of each pulse is given by $(R\theta/c)(m_0 c^2/E)$ and is just τ. Therefore the duration of the pulse τ (seen by the observer) is $(\theta/2\pi F_L) \times (m_0 c^2/E)$; it is

$$\tau = \frac{1}{2} \frac{2\pi m_0}{eB} \left(\frac{m_0 c^2}{E}\right)^2. \tag{14.19}$$

These pulses recur with a frequency F_L. This "interrupted" radiation may now be regarded as the sum of harmonics of a fundamental frequency F_L. There is a maximum power output of the radiation at a frequency F_m, where

$$F_m \approx \frac{1}{\tau}. \tag{14.20}$$

This tells us that

$$F_m = \frac{F_L}{2} \left(\frac{E}{m_0 c^2}\right)^2 \tag{14.21}$$

is the frequency where most of the emission is to be found. We noted that a nonrelativistic electron radiates at 28 Hz in a 1-gamma field; a 1-BeV electron, on the other hand, radiates predominantly at $F_m = 54$ MHz!

This megahertz emission may be thought of as the 1.9-millionth harmonic of the fundamental 28 Hz. The electron is "actually" radiating a series of lines 28 Hz apart, but this is so closely spaced that it may be thought of as a continuum that peaks at 54 MHz.

Now 28-Hz emission is not observable from the ground. But 54 MHz certainly is detectable by ground-based instruments, since it represents a wavelength of 5.6 meters.

We may construct a sketch of the spectrum emitted by our single electron. The emission is broadly peaked about F_m and falls off at higher frequencies (see Figure 14.6). This is a characteristically different shape from our earlier sketch of the frequency spectrum of thermal emission.

Synchrotron emission is strongly polarized. The electric vector of the emission lies primarily in the plane of the electron's orbit.

If we now have a distribution in energy of *many* electrons, we must take the distribution into account. The radio brightness must be integrated over the electron energy spectrum and over all the electrons along the line of sight

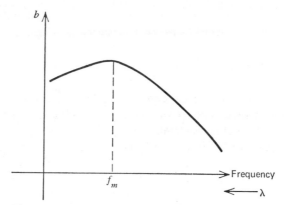

Figure 14.6 The synchrotron frequency spectrum emitted by a single relativistic electron.

to the observer. Therefore for randomly oriented magnetic fields the brightness b is given by

$$b = \frac{1}{4\pi} \iint P(F, E) N(E, R) \, dE \, dR, \tag{14.22}$$

where $N(E, R)$ is the number density of electrons, which depends on the energy E and the distance R between the source and the observer. The $(4\pi)^{-1}$ factor arises from the random orientation of the magnetic fields; the synchrotron radiation is therefore isotropic. The factor $P(F, E)$ represents the power output radiated at a frequency F for an electron of energy E.

If we now assume that the relativistic electrons in the source have a differential unchanging power law energy spectrum, we may write that

$$N(E) \, dE = K E^{-\gamma} \, dE, \tag{14.23}$$

where K and γ are constants to be determined. We should remark that this assumption has some support, in that cosmic ray electrons do seem to have a power law spectrum. It may well be, however, that the electrons in a radio source have some other spectrum; this will affect the conclusions that follow.

If we insert (14.23) into (14.22), we may perform the required integration. The result is

$$b(F) = \frac{3}{\pi} (2\pi)^{(1-\gamma)/2} \frac{e^3 B}{m_0 c^2} \left(\frac{2eB}{m_0^3 c^5} \right)^{(\gamma-1)/2} U(\gamma) R K F^{(1-\gamma)/2} \tag{14.24}$$

for the dependence of the radio brightness upon frequency, where $U(\gamma)$ is a slowly varying function given by

$$U(\gamma) = \frac{\sqrt{2\pi}}{16} \int_0^\infty U^{1/4} \exp\left[-\tfrac{4}{3}U^{3/2}\right]U^{(3\gamma-5)/4} \, dU$$

when we are dealing with frequencies greater than F_m, and

$$U(\gamma) = 0.256 \int_0^\infty U^{(3\gamma-5)/4} \, dU$$

when $F \ll F_m$. A typical numerical value of U is approximately 0.1 (at $\gamma = 1.9$).

Notice that the brightness depends inversely on the fourth power of the mass m_0 of the particle. This is why electrons, rather than protons, are the primary source of synchrotron radiation.

We may rewrite (14.24) in terms of the wavelength λ. The result is

$$b(\lambda) = 1.3 \times 10^{-33}(2.8 \times 10^8)^{(\gamma-1)/2}KB^{(\gamma-1)/2}R\lambda^{(\gamma-1)/2}\left[\frac{\text{watts}}{\text{m}^2\text{-Hz-ster}}\right], \quad (14.25)$$

where we have taken $U = 0.1$. The brightness b may, of course, be expressed in terms of the brightness temperature T_b, through $b = 2kT_b\lambda^{-2}$.

We may also find the median energy of the electrons that produce a given F_m. If we call the median energy E_m and express it in million-electron volts,

$$\frac{E_m}{m_0c^2} = \left(\frac{F_m}{1.26B_\perp}\right)^{1/2}, \quad (14.26)$$

where F_m is expressed in megahertz and $B_\perp$, the component of the field strength perpendicular to the orbital plane of the electrons, is in gauss. Thus, if $B_\perp = 10^{-4}$ gauss, the electrons causing 100-MHz radio emission have energies of about 46 MeV; those causing optical radiation at 5000 Å have about 110 BeV of energy, while x-rays of 20-keV energy would result from electrons with 10^{13}-eV energies.

14.12   The Galactic Halo

The galaxy appears, at meter wavelengths, to be roughly spherical rather than disc-shaped. That is, radio energy comes to us from outside the galactic plane as well as from within it. The region outside the plane is frequently referred to as the "halo."

It is interesting to inquire into the source of the radio energy coming to us from outside the plane of the galaxy; it has a nonthermal spectrum and we therefore suspect that the synchrotron mechanism is responsible. The observation is that $T_b \approx 450°\text{K}$ near the galactic poles at $\lambda = 3$ meters. It is also found that T_b is proportional to $\lambda = 2.8$ at wavelengths less than 30 meters, from high galactic latitudes. If we assume that the radiogalaxy is spherical, the distance to the poles, R, may be taken as roughly 25,000 light-years if the radius of the sphere is 40,000 light-years. All this tells us that if the synchrotron mechanism is indeed responsible for the emission, then

$$\frac{\gamma - 1}{2} = 0.8,$$

so that $\gamma = 2.6$, very much like the spectrum of the cosmic radiation. We may also determine how many relativistic electrons are required for the observation.

If we assume that $B = 1$ gamma in the vicinity of the galactic poles, we find from (14.25) that $K = 260$ eV$^{1.6}$ cm^{-3}. The density of electrons with energy greater than E_0 electron volts is given by

$$N(E > E_0) = \int_{E_0}^{\infty} N(E)\, dE = \frac{260}{1.6 E_0^{-1.6}}. \tag{14.27}$$

Therefore, if E_0 is 10^9 eV, the density of electrons with energies greater than 1 BeV is only $\sim 4 \times 10^{-13}$ cm^{-3}. Thus only a very small flux (about 0.01 cm^{-2} second^{-1}) of relativistic electrons is required to yield the observed intensity and spectrum of radio waves coming from the galactic poles. This required flux is about equal to the flux of heavy primaries in the cosmic radiation. The observations conducted to date of the energetic electron component of the cosmic radiation indicate that the flux is indeed 10^{-2} cm^{-2} second^{-1}.

One might ask where these high-energy electrons come from. One source is the interaction of cosmic rays with galactic hydrogen. These interactions produce pions and neutrons, which subsequently decay into high-energy electrons. It seems doubtful, however, that this process will yield a sufficiently large number of electrons, and it now appears that we must search for additional sources.

We noted previously that a good test of the synchrotron origin of the radiation was to determine the degree of polarization. Synchrotron radiation is polarized in the plane of the electron's orbit. For a power law electron spectrum it turns out that the degree of polarization π is

$$\pi = \frac{\gamma + 1}{\gamma + \frac{7}{3}}. \tag{14.28}$$

This amounts to 73% if $\gamma = 2.6$. Nothing like this high a degree of polarization is observed for the halo radiation; some 2–3% is actually found. It is presently believed that the reduction in the polarization is due to Faraday rotation in the intervening space by the randomly oriented magnetic fields.

We may put all our results on galactic structure in a different way. It is found that $T_b = 1 \times 10^{6}$°K for the quiet sun at $\lambda = 15$ meters. Since stars occupy only about 10^{-8} of the total area of the sky and since the sun is believed to be a typical star, we might expect the average brightness temperature of the sky to be only about 0.01°K.

This is not the case; 10^{6}°K is observed in the galactic plane at a 15-meter wavelength. Thus thermal and synchrotron emissions are observed from the whole sky, and in particular from interstellar space. We infer from this that a magnetic field and a halo of relativistic electrons pervade and surround the galaxy of stars.

14.13   Supernova Remnants

Many discrete sources of radio emission have been found within our galaxy. Many of those optically identified are associated with the remnants of supernovae, and some astronomers believe that all nonthermal galactic radio sources are supernova remnants.

The most celebrated case is that of the Crab nebula, visually a tenth-magnitude amorphous mass of rapidly expanding, glowing gas found where Chinese astronomers reported the observation of a "guest star" in the year A.D. 1054. This "new star" became visible in broad daylight and achieved a peak brilliance about equal to that of Venus.

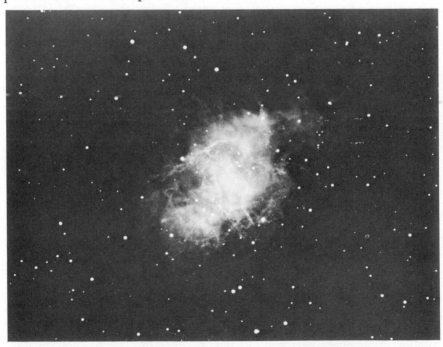

Figure 14.7   The Crab nebula in the constellation Taurus, photographed with the 200-inch telescope on Mount Palomar. Now approximately 0.1° in angular extent, it is expanding at about 1300 km/sec and is believed to be the remnants of the 1054 A.D. supernova. Courtesy Hale Observatories, California Institute of Technology.

The Crab (see Figure 14.7) is about 3500 light-years distant from the earth and is now about 5 light-years across its largest dimension, so that it subtends ~0.1° at the earth. The Crab bears the label M1; it was the first item in C. Messier's 1771 catalog of peculiar objects or nebulae. The speed of expansion, ~1300 km per second, is sufficient to account for an explosion around A.D. 1054, if it has been slightly accelerated in the time since the explosion by an amount of ~10^{-3} cm second^{-2}.

T_b is found to be about $4 \times 10^{6}°$K at $\lambda = 3$ meters for the Crab. This is far too hot, because of the known optical brightness. If the emission really were thermal, the Crab would have to be some 400 times as bright optically as is observed. Also, the radio brightness is found to decrease gradually at higher frequencies; as we have seen, this is not in agreement with the spectrum expected for thermal emission.

Shklovsky predicted in 1953 that synchrotron radiation was present and that both the radio and optical radiation should accordingly be polarized. Observation confirmed this prediction, and magnetic "maps" have been

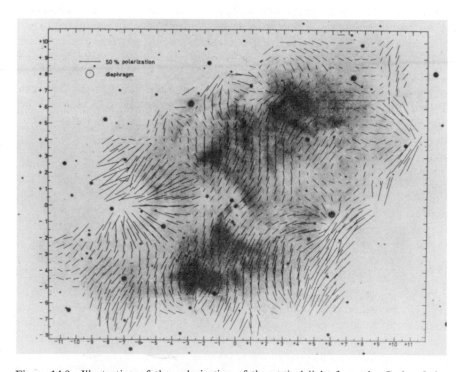

Figure 14.8 Illustration of the polarization of the optical light from the Crab nebula. The electric vector lies along each of the short lines drawn. Notice the distinctive "bays" near the boundaries of the nebula. Courtesy *Bulletin of the Astronomical Institutes of the Netherlands*, 1957.

constructed that show the presence of relatively strong ordered fields within the nebula (see Figure 14.8).

Measurements of the radio spectrum of the Crab reveal that $\gamma = 1.5$. This value of the exponent is considerably different from that found for the cosmic radiation; the spectrum of electrons in a supernova remnant is much harder than the spectrum of energetic particles observed at the earth.

It is also interesting to estimate the strength of the magnetic field present within the nebula. If the relativistic electrons are to be confined to the nebular volume, they are presumably held there by magnetic pressure of a highly irregular field, since a regular field would not cause confinement. A minimum-energy situation results when the pressure of the relativistic electrons is just balanced by the magnetic pressure. Recalling that particle pressure is given by one-third the particle energy density for isotropically distributed radiation, we may write that

$$\frac{1}{3} \int EN(E)\, dE = \frac{B^2}{2\mu}, \tag{14.29}$$

where B is a typical magnetic field strength for the nebula. Thus the constant K in the electron energy spectrum (which gives the number of relativistic electrons) is not independent of the magnetic field strength; the magnetic field strength controls the population of relativistic electrons. We need some additional data in order to obtain B uniquely (and to also find K).

This is provided by the observation of internal motions in the Crab. Typical speeds are ~ 300 km per second. The density ρ of the moving material may be 10^{-24} gm cm^{-3} (we can see stars through the nebula), and we may resort to the formula

$$\tfrac{1}{2}\rho v^2 = \frac{B^2}{2\mu}. \tag{14.30}$$

Equation 14.30 tells us that B is somewhere between 10 and 50 gammas, about an order of magnitude stronger than the general interstellar field.

If we compute the product $B^2/2\mu \times$ the volume of the nebula, we find that the energy of the magnetic field is about 10^{48} ergs. The Crab is expanding at about 1300 km per second, from measurements of the Doppler shift of the optical lines. (Both the blue *and* the red components are seen; the nebula is so rarefied that we can see through it to the gas rushing *away* from us, a result in accordance with the fact that background stars may be seen through the Crab.) If the mass of the object is $1 M_\odot$, the kinetic energy of the expanding gas is also about 10^{48} ergs. *There is at least as much energy in the fields and energetic particles as there is in the expanding gas.*

For purposes of comparison, our sun—a normal star—would only emit some 10^{43} ergs during the nine-century lifetime of the Crab. The Crab has produced over 100,000 times as much energy as that radiated by a normal star in the same period.

It is interesting that the magnetic field must have been created after the explosion. Field energies this large, if compressed to stellar dimensions, would disrupt any normal star because of magnetic pressure.

If, as seems to be the case, the optical radiation is also due to the synchrotron process, then electrons with at least 300 BeV of energy must exist there. We shall see in the next chapter that x- and gamma ray astronomy have shown that electrons 1000 times as energetic as this may also be present in the Crab.

Measurements of the optical spectrum imply that $\gamma = 3$ at these wavelengths. Some 2.4×10^{-9} electron per cm^3 with energies in excess of 300 BeV is required to produce the observed optical spectrum, on the basis of the synchrotron model.

The energy source of the Crab is unknown. About 10^{37} ergs per second are now being radiated. Some investigators believe that the decay of radioisotopes formed in the explosion is the energy source, while others think that a large initial pulse of ultraviolet radiation is causing the neighboring interstellar medium to fluoresce.

Quite recently the Crab has been found to contain two radio pulsars (Section 14.16) and, in addition, an optical pulsar, the period of which equals that of one of the radio pulsars (about 0.03 second). The optical pulsar is coincident with one of the two "stars" near the center of the nebula. Most investigators feel that pulsars are rapidly rotating neutron stars; the period of rotation equals that of the pulsar in this view, and one possibility is that the angular spin energy of the neutron star supplies the present luminosity of the Crab.

We close this discussion of the Crab with a remark on the variability of the total luminosity. Fluctuations are occasionally observed in the Crab. These ripples or "wisps" change the total optical luminosity by 0.01%, and they appear to move in the nebula with speeds up to about $c/10$. The lifetime of a typical wisp is of the order of weeks.

The wisps—on the basis of the synchrotron model—are thought to be due to fluctuations in the Crab's magnetic field. They remind one of the motions observed in terrestrial aurorae. It is not known whether the radio or x-radiation also detected from the Crab fluctuates in a similar manner.

14.14 Radiogalaxies

Extragalactic discrete sources of radio energy have been detected. The radiogalaxies represent one class of these sources; they are galaxies of stars that emit as much as 1000 times as much radio energy as does the Milky Way.

The strongest of these—in terms of the radio power received at the earth— is known as Cygnus A. At a frequency of 100 MHz some 11,800 flux units are received from Cygnus A, although it is believed to be 6×10^8 light-years distant from us. This is approximately equal to the flux density received at the earth from the quiet sun, at 100 MHz.

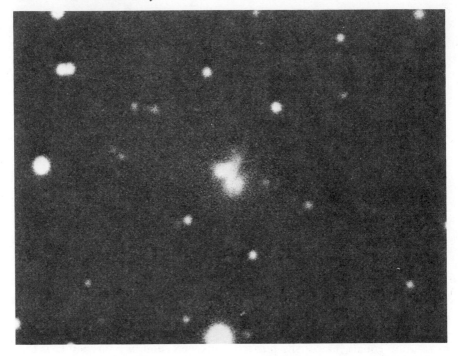

Figure 14.9   The Cygnus A radio source, seen through the 200-inch telescope. Some 7 × 10⁸ light-years distant, this is the most luminous discrete radio source in the sky. Once thought to be two galaxies in collision, it is now believed to be the result of one exploding galaxy. Courtesy Hale Observatories, California Institute of Technology.

Its peculiar appearance in the optical (see Figure 14.9) led people to believe that Cygnus A represents two galaxies in collision. In this model the stars in the two galaxies would probably miss each other, but the interstellar gas atoms would collide and heat up, giving rise to the radio output. Thus the radio energy would be supplied by the kinetic energy MV^2 of the two colliding galaxies, each of mass M, moving with a relative speed V.

However, the source is now thought to be much farther away than previously believed. Hence even the kinetic energy of two whole galaxies is inadequate to supply the observed radio luminosity, which is 10^{44} ergs per second. Cygnus A radiates as much in the radio alone as does our galaxy at *all* wavelengths.

Another strong source of radio energy, called Centaurus A, has also been identified optically (see Figure 14.10). One point that has been noted is that, as in Cygnus A, the radio source is *doubled*. That is, there are two main sources of radio energy, located on opposite sides of (and outside) the optical galaxy. Indeed, there are actually four approximately colinear radio sources in the case of Centaurus A; all four are spread over some 10° of the sky.

Figure 14.10 (a)   The Centaurus A radio source, as its central regions appear to the 200-inch telescope. The radio source covers 10° of sky and is composed of four colinear centers on either side of the optical object, which is centered at $\alpha = 13^h22^m31^s.6$, $\delta = -42°45'.4$ (1950.0) and is only about 10 arc-minutes in angular diameter. Courtesy Hale Observatories, California Institute of Technology.

We think that what we may be seeing is not two colliding galaxies but rather a single galaxy that is exploding. When a galaxy explodes, huge "blobs" of plasma are thrown out from the galactic center (in opposite directions, to conserve momentum). The ejected plasma then interacts with the intergalactic medium; a shock is set up along the outer surface of the blob, heating the plasma even further. In the case of the four sources, then, it is possible that we have evidence for *recurrent* galactic explosions, with the time between events given approximately by the ratio of the distance between outer and inner radio sources to the speed of expansion. All these extra-galactic radio sources are brightest at their outermost edges, just as if the ejected plasma were heated in the shock formed where it impinges on the intergalactic medium.

The spectrum of the radio energy has led most observers to conclude that the synchrotron mechanism is operative there. If the optical radiation is also due to the synchrotron mechanism (as seems likely, since weak polarization has been detected), we are dealing with huge clouds of 10^{11} eV electrons radiating in strong magnetic fields, x-ray reports imply 10^{13} eV.

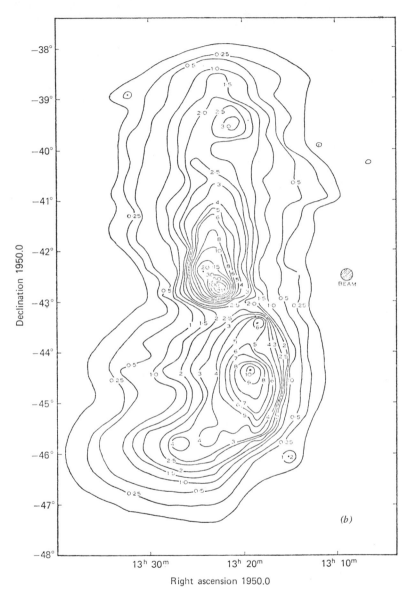

Figure 14.10*b*   Radio isophotes of Centaurus A. The numerals refer to the brightness temperature at 1410 Mhz (in °K), or a wavelength of 21 cm. The size of the beamwidth employed is shown. Taken from B. F. C. Cooper, R. M. Price, and D. J. Cole, *Australian Journal of Physics*, 1965.

The most documented case of a galactic explosion is the galaxy that is labeled M82. It is "only" some 10^7 light-years distant from us, so that the pictures of it taken with the Mount Palomar telescope may be expected to be quite revealing. Figure 14.11 shows the galaxy. Such photographs show that some 0.2% of the galactic mass is being thrown out of the center of the galaxy; a mass of 5 million suns may have been ejected.

Ultraviolet synchrotron emission may be responsible for the observed light. If so, the short wavelength of the emissions indicates that we are looking at a source of 10^{13} eV electrons. An alternate possibility is that the explosion's shock wave has heated the interstellar medium in M82. The observed intensity would indicate a temperature of $10^{8}°$K. Since a one-hundred-million-degree plasma presumably radiates in the x-ray range, it therefore seems possible that M82 may be an x-ray source as well as a radio source.

A very important parameter is the total energy E evolved in the explosion of a galaxy. We can measure—at least roughly—the present luminosity, in ergs per second, but we also need to estimate the time during which this energy has been liberated. One way of doing this is to find the distance S from the center of the galaxy to the cloud of plasma. The plasma cannot have been ejected with a speed in excess of c; we may assume $v \approx c/10$. The time τ is then $\tau = S/v$, and we may multiply the luminosity L by τ to find E.

We find $E \approx 10^{61}$ ergs for Cygnus A and perhaps 10^{59} ergs for Centaurus A. It is interesting that if our assumed value of v is too high, the total energies evolved are even higher. To put 10^{61} ergs in perspective, it corresponds to $10^{7} M_{\odot} c^2$; we would have to annihilate 10 million solar masses to produce this kind of energy release.

There are other types of radiogalaxies. One of the most famous is Virgo A. This has been determined to coincide with the strange optical galaxy known as M87. The most peculiar optical feature of M87 is a puzzling "jet" of intense blue light that extends out 20 arc-seconds (about 4500 light-years) from the galactic center (see Figure 14.12). A faint "counterjet" exists in the opposite direction.

The light from the jet is polarized and seems to emanate chiefly from "knots" within the jet. The jet is not trivial; it apparently contains a million solar masses. The radio energy also seems to come from the knots, but is mostly produced near the outermost edge of the jet. The radiation is now understood as synchrotron radiation. If the field strength in the knots is assumed to be 100 gammas, we have 10^{-8} electron cm^{-3} for electron energies between 10^{11} and 10^{12} eV.

14.15 Quasars

Quasistellar objects ("quasars" or QSOs) have recently been detected. The history of their detection is one of ever more precise radio-position measurements of discrete sources that had not been optically identified until it became

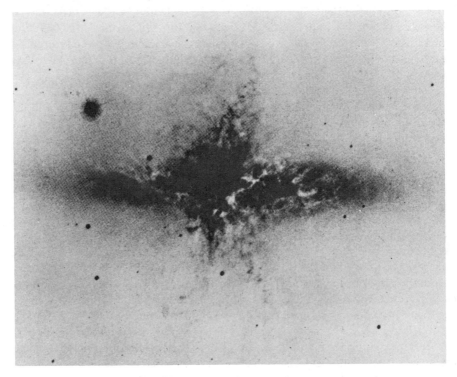

Figure 14.11   Photograph of M82 taken with the 200-inch Hale reflector in Hα light with an interference filter of total half-width 80 Å. From A. R. Sandage and W. C. Miller, *Science* **164,** 405–409 (April 1964). Copyright American Association for The Advancement of Science.

possible to pick them out of photographs of star fields. The discovery of these objects represents one of the most exciting scientific events of our times.

As their name suggests, quasistellar objects are starlike in appearance, although their only resemblance to stars as we know them is their small angular size. The astonishing fact is that each may liberate as much optical energy per unit time as 100 *galaxies* of normal stars would liberate, and it also generates some 10 times as much radio power as does a normal galaxy.

Perhaps the most remarkable feature of their spectra is the large value of the red shift that has been found. We define the red shift Z of a spectral line to be

$$Z = \frac{\lambda - \lambda_0}{\lambda_0}, \tag{14.31}$$

where λ_0 is the laboratory wavelength of a given line and λ is the observed wavelength. Values of Z up to 2.22 have been found for some of the emission lines in quasar spectra. Lyman alpha normally occurs in the far ultraviolet at

$\lambda_0 = 1216$ Å; if $Z = 2.22$, Lyman alpha is observable (at sea level) in the blue region of the *visible* spectrum at $\lambda = 3916$ Å.

One of the more amazing features of a quasar is the total energy released. Age estimates (which are most uncertain, and are based on the apparent length of what appear to be jets) are of the order of 10^6 years. If we accept the present luminosity of these objects as representative of the average luminosity, then $E \approx 10^{60 \pm 2}$ ergs for quasars. Objects of stellar dimensions may have liberated as much energy as would be produced by the annihilation of nearly 100 million suns!

When examined by even the largest optical telescopes, quasars are found to have diffraction-limited images. This type of image, the so-called "spurious disc," would be found for any distant object smaller than a few light-years in diameter. A better estimate of the linear dimensions may be made with radio telescopes, as the moon occults the source. Characteristic dimensions from the lunar occultation observations appear to be ~ 1 light-year, *if* the presumed vast distances to the objects are correct. Long-base radio interferometers yield angular dimensions $\leqslant 0.02$ second of arc.

The great optical and radio variability of the sources (see Figure 14.13) provides another way of estimating the size of QSOs. For example, the quasar 3C446 *doubled* in optical brightness in $\tau = 1$ *day* (during 1966). Old photographic plates reveal that the brightest quasar, 3C273, varies by about half a magnitude with a period of ~ 15 years. The average magnitude of 3C273 is about $+12$. If we assume that the whole QSO surface is pulsating, the diameter D can hardly be greater than $c\tau$; this implies values of D less than 1 light-year, which compounds the difficulty in understanding the objects.

The study of quasar spectra has revealed some startling oddities. It turns out that the red shift of the absorption lines in the spectra is different from the Z measured for the emission lines. Indeed (at least in some cases) the absorption line Z is about 1.95, *regardless* of what the emission line red shift is. Why this should be so can only be regarded as a mystery at the present time. Perhaps some light will be shed on this problem as more observations become available. We shall return to a discussion of the various possible interpretations of red shifts when we discuss cosmology.

The radio and optical spectra are consistent with synchrotron radiation. Quasars are bright blue in color; the radio spectra are synchrotronlike continua. We therefore expect that quasars are associated with intense relativistic plasmas.

An interesting and probably significant similarity of the shape of quasar spectra to the continuum spectra of the nuclei of some galaxies has been detected, particularly in the infrared. It may be that quasars are related to the nuclei of galaxies, perhaps in an evolutionary sense; a quasar could be the object around which galaxies subsequently form. Thus it is possible that a quasar or quasars may exist at the center of the Milky Way.

Figure 14.12   The galaxy M87 (NGC 4486): a short exposure with the 120-inch telescope showing the (blue) knotted
"jet" that emerges from the nucleus. "Counterjets" that are
much fainter also exist. This galaxy is the optical counterpart
of the radiogalaxy Virgo A. Lick Observatory photograph.

It is fair to say, however, that quasars are not understood. The major
problem is the nature of the energy source, if the luminosity estimates are
accepted. The luminosity figures depend on the distance values that have been
inferred from the measurements of Z (see Chapter 16). If it could be shown
that the QSOs were close by (say $\sim 10^7$ light-years) instead of 10^9 light-years
distant, the luminosity requirement would be greatly reduced, since the flux
of electromagnetic radiation varies as r^2. If the objects are relatively close in,
however, and if the red shift is due to a Doppler shift, tremendous energy is
required to accelerate them to the "observed" velocities; energies of the order
of $c^2 M$ are involved, where $M \sim 10^{11} M_\odot$. (In this "local" theory quasars
represent the ejecta of explosions in our own and nearby galaxies.)

We return to the subject of possible energy sources in the next chapter,
merely stating for now that many possibilities have been proposed, including

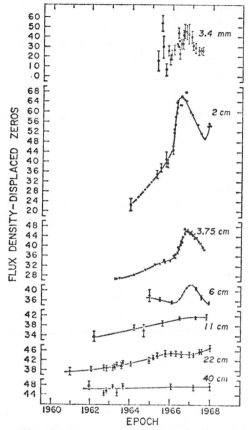

Figure 14.13 Radio variations of quasar 3C273 as observed at seven different wavelengths up to 1968. From K. I. Kellermann and I. I. K. Pauliny-Toth, *Annual Review of Astronomy and Astrophysics*, 1968.

gravitational collapse of supermassive objects and the annihilation of matter by antimatter. It is clear that the study of quasistellar objects will be one of the more active fields of space science for some time to come.

14.16 Pulsars

Early in 1968 four members of a new class of celestial objects were discovered by a group of British radio astronomers. The most extraordinary characteristic of these objects is that they periodically emit a brief radio pulse and the time between pulses is far more constant than any other known celestial object. They have been named "pulsars."

The first of the four to be found (in the constellation Vulpecula, at $\alpha = 19^h19^m23^s.4$, $\delta = +21° 46' 57''.4$ [1950.0]), called CP1919, has a repetition period of 1.3372795 $\pm$ 0.0000020 second. (The over-all pulse width is about 0.037 second.) The repetition periods of the other three differ only slightly from this, and the repetition rates of all four are sufficiently precise as to compare favorably with terrestrial time services such as radio station WWV.

The frequency spectrum of the radiation from the many ($\sim$50) pulsars found up to this writing is quite broad; the pulses have been detected at frequencies ranging from 40 to 611 MHz, although the signal-to-noise ratio at the earth appears to be a maximum near 100 MHz. The amplitudes of the pulses vary but are frequently as large as 200×10^{-26} watt meter^{-2} Hz^{-1} at 10^8 Hz. There is general agreement that at least some of the amplitude variation is caused by interplanetary scintillations.

Only one definite optical identification of a pulsar has been made. The south-preceding star of a pair of stars near the center of the Crab nebula has been found to vary its light output synchronously with a pulsar at about the same location that is known as NP0532 (the numbers in the pulsar designation refer to its right ascension).

Many explanations have been offered for the pulsars. The various possibilities include (a) signals from an intelligent civilization, (b) neutron star pulsations, (c) white dwarf pulsations, (d) neutron star rotation, (e) a close binary system with white dwarf or neutron star components, and (f) a gravitational lens effect.

Suggestion (a) cannot be ruled out, although the broad frequency spectrum and large power output seem uneconomical. Possibilities (b) and (c) have come into vogue because of the rapidity with which the pulses recur. However, theoretical models of neutron stars all lead to oscillation periods much shorter than 1.3 seconds, and the shortest period ever calculated for the oscillation period of a white dwarf is $\geqslant 3$ seconds; no pulsars have periods this long. Nothing is known about neutron star rotation, so that possibility (d) is difficult to evaluate; this model associates the precise short pulsation period with the fixed rotation period that might be expected of a normal star that has contracted to neutron star dimensions ($\sim$10 km) while conserving angular momentum. This model—a rapidly rotating neutron star—gained wide popularity after careful measurements of the period revealed that it is slowly lengthening, at least in some cases. In this model the rotational kinetic energy of the neutron star provides the energy source for the pulsar. It is therefore believed by some that the discovery of pulsars represents the first observational evidence of neutron stars.

Models (e) and (f) both suppose that "the" pulsar actually consists of two components. Thus (e) invokes two objects, each about one solar mass, that are close together (only about 10^3 km), so that the orbital period is only about 1 second. If they are white dwarfs or neutron stars, strong magnetic fields will exist around each; the action of one on the other will cause the magnetically opposite poles of the two components of the binary to face each other, creating a strong, substantially uniform field between them. Intense radiation is to be expected from particles injected into this strong-field region, and the radiation pattern will peak in a plane perpendicular to the line joining the two. This pattern then projects onto the sky as a rotating great circle; an observer sees two short pulses for each orbital revolution. It is not clear, however, how so small an orbit may be achieved. This same problem is associated with model (f), in which the pulsed radiation is thought to be the effect of gravitational focusing by members of a neutron star binary. As in (e) it must be explained how starlike objects can remain in orbits smaller than the earth. Both (e) and (f) may be tested experimentally, since velocity variations caused by deviations from linear motion will alter the pulse repetition frequency. This should tell us whether the pulsars are alone in space or are members of systems. The evidence to date indicates that they are not system members.

14.17   Primordial Fireball

Penzias and Wilson reported in 1965 the discovery of an omnidirectional flux of microwave energy. This discovery (at $\lambda = 7.35$ cm) has been followed by the detection of this radiation at other wavelengths in the centimeter range. The spectrum is shown in Figure 14.14. The brightness b is about 500 flux units at 7.35 cm, some two orders of magnitude more intense than the combined thermal and synchrotron radiation found at $\lambda = 1$ meter. These

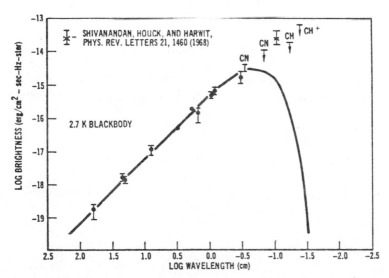

Figure 14.14   The frequency spectrum of the isotropic microwave radiation. The upper limits marked CN and CH are inferred from the relative populations of the rotational levels of CN, CH, and CH$^+$, as observed in interstellar absorption. From R. Webbink and W. Jeffers, *Space Science Reviews*, 1969.

latter radiations slowly decrease in brightness as the wavelength is decreased from 1 meter.

Most of the few published observations of the microwave flux are consistent with a brightness temperature T_b of 2.7 ± 1°K. The radiation has been interpreted as having originated at a time when the matter and radiation of the universe were in a very hot, contracted state—the primordial fireball. As the universe subsequently expanded, the cosmological red shift would have "cooled" this cosmic black body radiation from the original 10^{10}°K to the extent that the wavelength is now in the microwave region. In this view, therefore, the present-day "temperature of the universe" is 3°K.

If the "original temperature" of the universe was in fact 10^{10}°K, it would have provided sufficient thermal energy to "cook up" helium from hydrogen through nuclear reactions. As we have seen, the theory of stellar evolution requires that some 30% of the original mass be in the form of helium.

It will be most interesting to learn whether observations at shorter wavelengths confirm the black body hypothesis. For example, a 3°K black body spectrum should peak at $\lambda = 850\,\mu$, which is in the far infrared. If the spectrum should be found to depart from the black body law, some modification of the interpretation may be necessary.

It appears at this date that the isotropy of the radiation is complete. With a small improvement in precision, Oort has noted that we may be able to measure the effects of motion of the sun and of our galaxy relative to this universal radiation field; the motion of our galaxy is not presently known.

Also, as we shall see in Chapter 16, a measurement of the total energy density of this "background radiation" has cosmological implications. An integration under the smooth curve in Figure 14.14 results in an energy density of 1–10 eV cm^{-3}; there would appear to be at least as much energy in microwaves as there is in starlight ($\sim$ 1 eV cm^{-3}). This energy could, at the present writing, provide one source of the mass energy that cosmologists feel has not yet been detected in the universe.

14.18   Three-Halves Power Law

As ever more sensitive radio telescopes go into operation, it is of interest to inquire as to the number of radiogalaxies that we might expect to be observed. Suppose that there is an average number density $\bar{\rho}$ of radiogalaxies in space, and further suppose that this density is constant throughout space. The number N of radiogalaxies contained within a volume of space of radius r will then be just

$$N = \tfrac{4}{3}\pi\bar{\rho}r^3. \tag{14.32}$$

Now this value of N applies to all the radiogalaxies that are detectable above a given flux threshold, S_0. Let us assume that all radiogalaxies emit approximately the same radio power. The flux S from each of these depends on

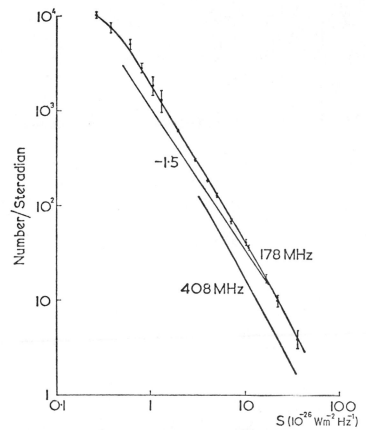

Figure 14.15   The $\log N - \log S$ relation presented by Gower in 1966 for $S_{178} \geqslant 0.27 \times 10^{-26}$ watt meter^{-2} Hz^{-1} and the 408-MHz relation from the Parkes (Australia) surveys. The straight line with a slope of -1.5 corresponds to a uniform population of sources in a static Euclidean universe. M. Ryle, *Annual Reviews of Astronomy and Astrophysics*, 1968.

distance as r^{-2}; the flux is given by $S = \text{constant} \times r^{-2}$. We may substitute this relation into (14.32). The result is that

$$N = \text{constant} \times S^{-3/2}. \tag{14.33}$$

Equation 14.33 is known as the "three-halves power law" of radio astronomy. It states that the number of radio sources should increase as the 1.5 power of the threshold flux; doubling the sensitivity of a telescope should result in detecting $(2)^{1.5} \approx 2.8$ times as many sources as previously observed.

It is well to state explicitly the assumptions that have gone into establishing this "law." We assumed that (a) all radiogalaxies were equally luminous, (b) there is a constant density of such sources throughout the universe, and (c) space is Euclidean. This last assumption was implicitly made when we wrote

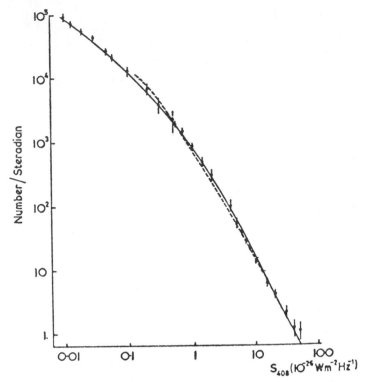

Figure 14.16   The log N — log S relation for $S_{408} \geqslant 0.01 \times 10^{-26}$ watt meter^{-2} Hz^{-1} (Pooley and Ryle, 1968). The dashed line corresponds to the 178–MHz source counts scaled by an amount corresponding to a mean spectral index $\alpha = 0.7$. From M. Ryle, *Annual Reviews of Astronomy and Astrophysics*, 1968.

relation (14.32). Thus it is possible to gain information on whether space is curved or not by examining the dependence of the number of radiogalaxies on the sensitivity of the radio telescopes.

The observational evidence is inconclusive. Figures 14.15 and 14.16 illustrate the 408-MHz data. The situation is perhaps best summarized by saying that N seems to be a constant $\times S^{-1.7 \pm 0.3}$. The spectral index itself may be dependent on S.

One of the reasons for the uncertainty is that it is clear that all radio-galaxies are *not* equally luminous; attention must be focused on particular galactic *types*, such as elliptical, irregular, or spiral. Furthermore, no allowance has been made in (14.33) for intergalactic absorption effects; various correction factors (which are at best poorly known) must be applied for telescopes looking in different directions in space.

Finally, nothing has been said concerning galactic *evolution* in time. When we look at extremely distant (i.e. faint) objects, we are looking at them as they were in the past. The results summarized in Figures 14.15 and 14.16 have led many investigators to suspect that when the universe was about one third as old as it is now believed to be, the population density of radio sources was some hundred times larger than at present. This factor is over and above the geometrical factor introduced by the presumed expansion of the universe. Many efforts are now being made to reduce the observational uncertainties through data selection and improvement of our knowledge of source obscuration, but it is clear that it will be some time yet before any firm conclusions may be reached.

Problems

14.1. Compute the Doppler shift of the 1420-MHz emission of a neutral hydrogen cloud moving at 30 km per second with respect to the earth. If the receiver bandwidth is 100 kHz, is the shift detectable?

14.2. How many times has the sun traveled around the galactic center during its life?

14.3. If pulsar CP1919 is about 300 light-years distant from the earth and if the frequency spectrum goes as $\nu^{-1.5}\,d\nu$, how much power is radiated by the object at radio wavelengths? Look up the annual rate of global electric power production and compare it with your answer.

14.4. At a wavelength of 3.5 meters it is found that the ratio of the brightness temperatures of the galactic center to that of the anticenter is 6.6. If the number of sources N in the galactic plane as a function of distance R from the center is given by $N = N_0 e^{-R/R_0}$, find R_0, assuming that $R_\odot = 8.2$ kiloparsecs.

14.5. The radio spectrum of NGC 4486 (the radio source otherwise known as Virgo A and identified with the optical galaxy M87) is proportional to $\nu^{-0.8}$. NGC 4486 is 10 megaparsecs distant from the earth, and the flux from it at 100 MHz is 1.2×10^{-22} erg cm^{-2} second^{-1} Hz^{-1}. Find the radio luminosity of Virgo A at frequencies between 20 and 1000 MHz.

14.6. Assume the moon to be a perfect isotropic reflector at the frequency used by a radar on the earth's surface. If the radar's receiver requires a signal of 1×10^{-13} watt for detectability, can it detect a reflected pulse from the moon, when the transmitter emits isotropic 1-megawatt (peak) pulses and the parabolic receiving antenna has an effective diameter of 26 meters?

14.7. (a) In the presence of a magnetic field the 21-cm line is split into its Zeeman components. In a 1-gamma magnetic field, compute the expected frequency splitting $\Delta\nu$.

(b) The radio spectrum of Cas A exhibits absorption lines in the 1420-MHz region. What is the width δv of these lines, if they are caused by clouds of interstellar gas whose internal velocity dispersion is only ~ 2 km per second?

(c) One may detect the splitting in (a) by switching periodically between the two differently polarized Zeeman components. If the brightness temperature of Cas A is T, the result is that the power delivered to the receiver at the *switching frequency* is ΔT, where $\Delta T = T(\Delta v/\delta v)$. Since present-day equipment requires $\Delta T \geqslant 1°K$, what value of T is required?

14.8. Assume that the inverse-square brightness law is valid over all distances in the universe and that all galaxies are the same size and are equally luminous. If the total number of galaxies detected by a radiotelescope having sensitivity S is given by

$$N_t(S) \times S^{-1.75},$$

how does the number density of galaxies depend on the distance from the earth?

14.9. (a) The radiogalaxy M87 is observed to have a radio flux that is nearly inversely proportional to the first power of the frequency. The flux at $\lambda = 3$ meters is 1×10^{-20} erg cm^{-2} second^{-1} Hz^{-1}. Taking the distance to M87 to be 3×10^7 light-years, what is the spectral index of the electron power law spectrum (assuming synchrotron radiation by electrons to be responsible)?

(b) What is the luminosity L_R of M87, integrated over the entire radio range, 1000 MHz? Compare L_R with the total luminosity of the sun.

14.10. Assume that the population of galaxies consists of two different types having the following properties:

Type I	Type II
Spherical, radius R_1	Spherical, radius R_2
Luminosity L_1	Luminosity L_2

Assume also that they are uniformly mixed over all space with relative number densities N_1 and N_2 everywhere (i.e., $N_1 = $ constant, $N_2 = $ constant, $N_1/N_2 = $ constant, $N_1 + N_2 = $ constant for any volume element in space). Obtain an expression for the total number of galaxies counted per unit solid angle as a function of detector flux sensitivity S, and compare the results with those obtained for only one type of galactic population of uniform density.

14.11. A large radio telescope having variable flux sensitivity S is employed in a survey of galactic density. It is found that the number of galaxies counted per unit solid angle increases with sensitivity according to the relation

$$N(S) \propto \frac{1}{S}.$$

Assume that all galaxies are spheres of radius A having surface flux intensity f_0, and that they are distributed according to a density function $N(r) = N_0(a/r)^n$, where r is the distance from us and a and n are constants.

(a) What is the value of n if an inverse square flux intensity law is valid [i.e., $f(r) = f_0(A/r)^2$]?

(b) What is the value of n if the flux intensity varies as $f(r) = f_0(A/r)^{2+\varepsilon}$, as we might expect from a Doppler effect or nonequilibrium absorption?

(c) Is there any value of n for an inverse square flux law for which there is no variation of $N(S)$ as a function of S?

Chapter 15

EXOTIC ASTRONOMY

Exotic astronomy—the study of the physical universe through electro-magnetic channels other than the optical and radio—has only become feasible with the recent rapid development of technology. Many developments, such as cryogenic devices, the modern scintillation counter, spark chambers, reliable sounding rockets with good attitude control, and large balloons that can lift heavy payloads, have all contributed to our ability to examine the universe in several new ways; "experimental astrophysics" has now become possible.

15.1   Absorption by the Terrestrial Atmosphere

The earth's atmosphere absorbs (with a few interesting exceptions) electro-magnetic radiation. The exceptions include wavelengths in the optical and radio regions and a few "windows" in the other spectral regions. Figure 15.1 illustrates the absorption by the atmosphere as a function of wavelength.

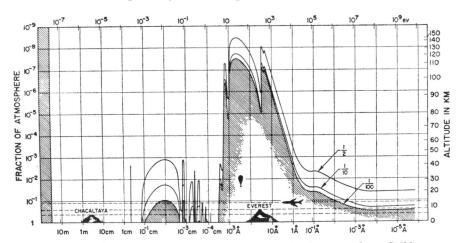

Figure 15.1   Attenuation of electromagnetic radiation in the atmosphere. Solid curves indicate altitude (and corresponding pressure expressed as a fraction of 1 atm) at which a given attenuation occurs for radiation of a given wavelength. From B. Rossi, *Space Research V*, North-Holland Publishing Company, 1965.

From the figure it is clear that x-ray and gamma ray observations must be conducted above, or at the top of, our atmosphere. The *optical depth* (the depth at which the intensity of the incident radiation is reduced to e^{-1} of its initial value) of 1-keV x-rays is only about 0.2 mg cm^{-2}. This corresponds to the pressure found at an altitude of about 100 km, so that 1-keV extra-terrestrial radiation must be studied with the aid of rockets or spacecraft.

The gases in interstellar space are also absorbers. Measurements of the interstellar Lyman alpha line width indicate densities of hydrogen of about 0.1 hydrogen atom per cm^3 within about a kiloparsec from the sun. Inter-stellar space is transparent at photon energies that vary from about 100 eV for the nearby stars to about 1 or 2 keV for the more distant galactic objects.

We also note that *neutrino* astronomy (nonelectromagnetic astronomy) requires the use of very massive shields. Some experimenters are burying their neutrino detectors in deep mines.

15.2   X-ray Observations

The first discovery of a discrete celestial source of x-radiation other than the sun was made in 1962 by Giacconi and his collaborators. About 30 such sources are now known to exist. We discuss only the observations (to date) in this section; the various production mechanisms are treated in Section 15.6.

It seems that stars like the sun (see Figure 15.2) will not produce detectable x-ray fluxes at the earth. The quiet solar corona produces less than 10^{-5} photon cm^{-2} second^{-1} keV^{-1} around 1 keV; even at only 1 parsec this is too

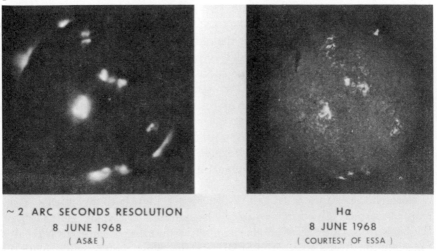

~ 2 ARC SECONDS RESOLUTION
8 JUNE 1968
(AS&E)

Hα
8 JUNE 1968
(COURTESY OF ESSA)

Figure 15.2   Photograph of the sun, taken with a soft x-ray camera. *Right:* photo of the sun, taken in Hα light about three hours before the rocket flight. Active regions show clearly in both photos. Courtesy *American Science and Engineering Corporation*, 1969.

small to be observable. Again extrapolating from the solar spectrum, radiation at energies below $\sim$0.1 keV may be detectable from normal stars within a few parsecs of the earth.

The brightest discrete source at wavelengths between 1 and 8 Å (and also the first to be found) lies in the constellation Scorpius; it has been named Sco XR-1. Some 5×10^{-10} watt per square meter is incident at the top of the earth's atmosphere from Sco XR-1. The low-energy spectrum of Sco XR-1 may be represented by an exponential, with an *e*-folding energy of $\sim$7 keV. If this energy is equated to kT, the "temperature" of the source is 50×10^{6}°K, over 10 times that of the solar corona (and probably greater than the temperature at the center of the sun).

The detection of an exponential, presumably thermal spectrum from Sco XR-1, together with a positional accuracy of less than one arc-minute, led to its tentative optical identification. An extrapolation of the thin plasma bremsstrahlung spectrum (see Section 15.6) down to the energies of optical photons led investigators to expect a blue object. There is one such object in the positional error-box; it is a peculiar, flickering object, with substantial light fluctuations that take place in minutes to hours.

While there is as yet no evidence for similar flickering in the soft, exponential x-ray spectrum, there have been some reports of an additional nonthermal hard component of the x-ray spectrum. It appears at this writing that much of the seeming contradictions in those reports may be eliminated if it is assumed that the hard component is both faint (compared with the exponential) in general and variable, as is the optical.

The identification of Sco XR-1 is not a settled matter. Radio emission has been detected from the object; the radio brightness seems too faint for an extrapolation of the bremsstrahlung spectrum, as are the optical emission lines. It is possible that the variability is responsible for many of the discrepancies.

Sco XR-1 is now believed to be $\sim$300 pc distant from the earth. No motion has been observed for the object, and the comparison of the optical and soft x-ray intensities permits an estimate of the interstellar absorption to be made. The results are consistent with a 10^9-cm diameter plasma cloud at a density of 10^{16} cm^{-3} and at the distance given above. An object earth-like in diameter that is at a plasma temperature of 10^7°K is consistent with a white dwarf.

Since, however, no other white dwarfs are known to be intense x-ray emitters, special circumstances would be required to explain Sco XR-1 on the basis of a white dwarf model. It may be that it is a white dwarf that is one component of a close-orbit binary system, which would help explain the fluctuations in brightness; the x-ray luminosity would be maintained by gravitational accretion of matter onto the white-dwarf surface from the other component of the binary.

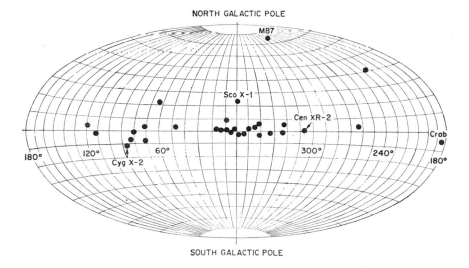

Figure 15.3   The distribution of discrete sources of x-radiation as it was known near the end of 1968. The coordinates are galactic longitude and latitude, l^{II} and b^{II}. From R. Giacconi, P. Gorenstein, H. Gursky, E. Kellogg, and H. Tananbaum, *American Science and Engineering Co.*, 1968.

The next strongest source is well known; it is Tau XR-1 and it is the Crab nebula. These x-rays are only about 15% as intense as those from Sco XR-1. The faintest sources detected to date produce about 3×10^{-13} watt per square meter at the top of our atmosphere.

The general distribution of sources shows a concentration toward the galactic plane, but the distribution in galactic longitude is highly irregular. Figure 15.3 shows the location of most of the sources.

Strong clustering is observed toward the galactic center, with some 15 sources detected in the 1–10-keV band in the Sagittarius-Scorpius region ($l^{II} = 315°$–$340°$). Table 15.1 lists the most recent coordinate determinations of those sources in the region of the galactic center. Other high concentrations are found in Cygnus and in Cepheus-Lacertus ($l^{II} = 60°$–$120°$). These groupings are suggestive of locations in the galactic spiral arms, but the positional errors are sufficiently large that the apparent association with the arms may be spurious. The only association that is clear at present is that with the galactic center.

If the various sources are in the galactic arms, we may, following H. Friedman, estimate their average distance from the earth to be ~2 kiloparsecs. In the 1–10-keV band the 25 sources within the Sgr-Sco and Cyg-Cas groups therefore occupy ~2% of the galactic disc. Using this, we may

Source Name	Coordinates (1950)				Error Limits[b]	Observed Intensity[c]
	α	δ	l^{II}	b^{II}		
	$16^h14^m.0$	$-39°\,31'$	$340°.40$	$+7°.80$	$20' \times 2°.2$	1.4
	$16^h45^m.2$	$-45°\,42'$	$339°.80$	$-0°.60$	$20' \times 5°.0$	1.5
	$17^h21^m.0$	$-46°\,39'$	$342°.80$	$-6°.10$	$17' \times 16°.0$	1.1
	$16^h40^m.2$	$-32°\,25'$	$349°.30$	$+8°.80$	$66' \times 30°.0$	0.6
GX349 + 2	$17^h04^m.0$	$-36°\,29'$	$349°.20$	$+2°.40$	$15' \times 1°.1$	1.2
Sco X-1	$16^h17^m.1$	$-15°\,31'$	$359°.10$	$+23°.78$		29.0
	$18^h21^m.1$	$-36°\,18'$	$357°.50$	$-10°.70$	$24' \times 30°.0$	0.6
GX3 + 1	$17^h43^m.6$	$-26°\,11'$	$2°.46$	$+1°.22$	$20'$	0.4
GX5 − 1	$17^h58^m.1$	$-25°\,07'$	$5°.04$	$-1°.05$	$10'$	2.2
GX9 + 9	$17^h28^m.4$	$-17°\,07'$	$8°.29$	$+9°.02$	$22'$	1.1
GX9 + 1	$18^h00^m.0$	$-20°\,35'$	$9°.20$	$+0°.82$	$20'$	1.2
GX13 + 1	$18^h10^m.3$	$-17°\,00'$	$13°.52$	$+0°.47$	$15'$	0.7
GX17 + 2	$18^h12^m.7$	$-13°\,48'$	$16°.60$	$+1°.50$	$15'$	1.8

[a] Data compiled by H. V. Bradt and his co-workers.

[b] Half-width of error limits (e.g., radius of error circle). The probability that a source lies within the limits is ~90%.

[c] Counts per square centimeter per second in the energy interval 1.5–8 keV. The error in intensity is ~±0.3 because of the uncertainty in the background level. An intensity of 1 cps cm^{-2} corresponds to an energy flux of 7×10^{-9} erg cm^{-2} second^{-1}.

estimate the x-ray luminosity of our galaxy, since the average luminosity is taken to be equal to that of the Crab nebula. For the entire galaxy, therefore, we estimate that there may be about 1250 sources of similar average intensity. Thus the 1–10-keV luminosity of our galaxy would be $\sim 7 \times 10^{41}$ ergs per second. The low energy x-ray luminosity is thus at least an order of magnitude greater than the radio luminosity.

It is not yet clear whether a similar situation exists with other galaxies, since x-radiation has thus far only been confirmed from one other galaxy, the peculiar galaxy M87. Radiation from M87 has been reported in the 1–10-keV and 40–100-keV bands; the x-rays appear, at this writing, to lie on an extrapolation of the power law radio spectrum of Virgo A (which is optically coincident with M87). The x-ray luminosity of M87 is ~ 70 times the radio luminosity of that "radiogalaxy"; perhaps it is better described as an x-ray galaxy. M87–Virgo A is the only extragalactic discrete source of x-radiation detected by several observers. Vir A is a strong radio source, but the brightest radiogalaxy Cyg A does not appear to emit detectable fluxes of x-rays; other types of galaxies may (see Section 14.14 and 15.7).

It is tempting to identify the "jet" that extends from M87 as the radio and x-ray source, but the x-ray data do not yet have nearly the required positional accuracy. The jet appears to be composed of a group of irregular concentrations of plasma, called "plasmons" by some. A bright blue continuum of polarized light, thought to be optical synchrotron radiation, has been detected from the plasmons (see Section 14.14).

X-radiation has also been detected and measured for a few discrete sources at energies higher than 10 keV. Data on the Crab nebula have been obtained by a number of observers; the spectrum has been measured out to 560 keV. Over the entire 1–560-keV range (see Figure 15.4) the energy flux differential spectrum may be fitted by a single power law, with a spectral index of -1.2 ± 0.2. Since the Crab exhibits a $\nu^{-0.27}$ spectrum at radio and infrared wavelengths, there appears to be a spectral break in the optical, where interstellar reddening corrections are difficult to determine. As Section 15.5 shows, a break like this is consistent with the synchrotron mechanism. The synchrotron explanation for 500-keV photons suggests that 2×10^{14}-eV electrons are present in the nebula, for a field of 2×10^{-4} gauss. This field is inferred from the value of the break frequency and from the age of the nebula.

The synchrotron mechanism may be used to explain how the hyperrelativistic electrons radiate their energy, but does not attempt to explain what the ultimate energy resource of the luminosity may be. About 15% of the total x-radiation of the Crab occurs in pulses; they are emitted in synchronism (~ 30 per second) with the optical pulses from pulsar NP0532, and have an energy spectrum that appears to be approximately parallel with that of the total nebula, at least for x-rays with energies between 1 and ~ 200 keV.

CRAB NEBULA AND NP0532

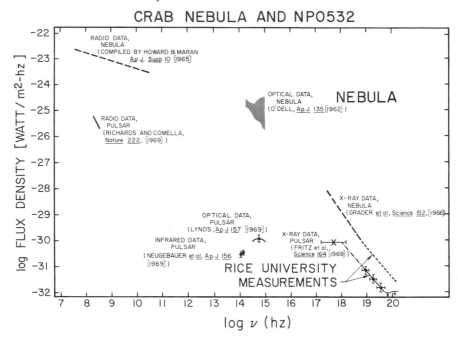

Figure 15.4 The spectrum, as measured by various observers, of the Crab nebula and of pulsar NP0532. The "Rice University Measurements" were performed by the author and his associates.

Therefore it may be that the energy of NP0532, assumed to be rotational energy, is the energy source of the nebula. If pulsars are in fact rapidly rotating neutron stars surrounded by dense plasma atmospheres and with intense magnetic fields near their surfaces, charged particles may well be accelerated to hyper-relativistic energies in their vicinity. The rapid rotation would take care of the lifetime problem associated with the synchrotron radiation of very energetic electrons; they radiate so copiously that they must be replenished frequently in order to maintain the luminosity of the nebula at a constant value.

At energies in excess of ~70 keV an object in Cygnus known as Cyg XR-1 is the brightest discrete source in the skies; it delivers 50% more power to the top of our atmosphere than the Crab nebula does, at energies above 30 keV. The spectrum of Cyg XR-1 exhibits a break at ~130 keV. The spectrum steepens from $E^{-1.8}$ by about one power of the energy in the break. There is strong optical obscuration in the direction of Cyg XR-1, and so this intense source has only been detected as x-ray energies; it has not been observed at radio or optical wavelengths.

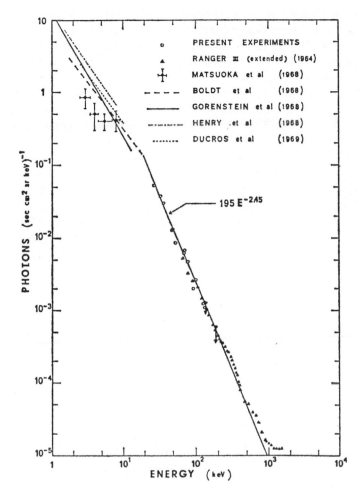

Figure 15.5 Energy spectra of the cosmic x-ray background as observed by various experimenters in the energy range 1 keV–1 MeV, as reported in 1970 by Bleeker and Deerenberg. *The Astrophysical Journal*, University of Chicago Press, 1970.

We should note at this point that the observations of the various sources are not free from discrepancy. It is possible that the x-ray emissions are time-dependent. For example, there have been reports of factor-of-2 fluctuations in the intensity of Cyg XR-1. It is entirely possible, though, that these "fluctuations" are in reality due merely to differences in the equipment used.

We have so far briefly discussed the various discrete sources that have been detected. A diffuse flux of x-rays has also been observed; it is as though the whole sky is glowing at x-ray wavelengths. This radiation forms a celestial background for observations of the discrete sources.

The data on the diffuse radiation are so far rather sketchy. Observations have been made at energies that range between 1 keV and about 6 MeV; an energy distribution that resembles a power law spectrum is present. The energy flux (intensity) spectrum may be approximated by $E^{-\beta}$, where $\beta \approx$ 1.4 $\pm$ 0.2, at energies between 1 keV and 1 MeV. Hence a *number* photon spectrum of the form $E^{-2.4\pm0.2}$ photons cm^{-2} second^{-1} keV^{-1} appears to fit the data in this energy range. Figure 15.5 shows the measurements reported through 1969. It is not clear at this writing whether the observed flux is truly diffuse or the result of a large number of unresolved discrete sources.

Recently a few observations have been conducted at energies below 1 keV. The first indications are that the diffuse flux at $\sim$0.25 keV is considerably in excess of that predicted by a downward extrapolation in energy of the power law. If it should be found to increase with galactic latitude, this will be evidence for an extragalactic origin. An alternative source for the low-energy photons is stellar coronae.

15.3   Gamma Ray Observations

There are three main energy regions that have been explored. One relies on electron-positron pair production in the detector; it is used at energies greater than about 30 MeV. Only one discrete source seems to have been detected in this high-energy region; G. Frye, V. Hopper, and their co-workers reported it. No optical identification has yet been made of this object, which lies off the galactic equator ($b^{\text{II}} = -20°$) at a galactic longitude $l^{\text{II}} = 3°$.

In addition, G. Clark, W. Kraushaar, and their associates have found evidence for some excess radiation from the galactic plane, about 1×10^{-4} photon cm^{-2} second^{-1} radian^{-1} having been detected from Sagittarius. This flux is for energies in excess of 100 MeV, and it is assumed that a line source such as the galaxy (viewed edge-on) is responsible for the radiation. It is not clear whether this represents the sum of many unresolved discrete sources or if the flux is truly diffuse.

At $E \geqslant 100$ MeV an upper limit to the diffuse photon flux out of the galactic plane of 10^{-4} photon cm^{-2} second^{-1} steradian^{-1} exists; this same upper limit appears to extend down to energies of the order of about 30 MeV. Upper limits to the fluxes from several possible discrete sources also have been measured by these and other workers, such as C. Fichtel and his collaborators. For example, the radio source Cygnus A produces less than 10^{-5} photon cm^{-2} second^{-1} at the top of the earth's atmosphere, for photon energies in excess of 30 MeV. With 95% confidence, an upper limit in the same energy region of 9×10^{-5} photon cm^{-2} second^{-1} has been set for Sco XR-1 and the

Kepler supernova. The quiet sun emits less than 4×10^{-5} photon cm^{-2} second^{-1} above 100 MeV.

Another gamma ray region of energies has already been referred to in connection with the results of x-ray astronomy. At photon energies between 0.1 and 10 MeV the photoelectric and Compton processes may be used for detection. Here there is some evidence for a diffuse flux of extraterrestrial photons. The Ranger 3 data, obtained in cislunar space with phoswiches, are shown in Fig. 15.5; the origin of this cosmic 70-keV–3-MeV flux is uncertain, as is the origin of the similar spectrum, measured in interplanetary space to extend beyond 6 MeV by J. Vette, L. Peterson, et al. in 1967.

The spectrum seems to be a prolongation of that given in the preceding section for the diffuse x-ray spectrum. A single power law may not provide an adequate fit over the entire energy range, however. As noted in the last section, there is a marked change below 1 keV, and there is evidence that the spectrum does more or less steadily steepen at energies higher than this. There may be a flattening of the spectrum in the 1–10 MeV range, but only a preliminary analysis of data from the ERS-18 satellite has been reported by Vette et al. at this writing.

Gamma ray investigations have also been conducted in a third, very-high-energy region. When quanta with energies greater than 10^{11} eV are incident upon the earth's atmosphere, large numbers of electrons are produced. These result from pair production and from the electron-photon cascade phenomenon (see Appendix C.2).

Charged particles in sufficiently large numbers may produce detectable flashes of visible light in the atmosphere, through the Cerenkov effect. Thus ground-based optical telescopes may, if suitably equipped for the detection of short flashes, be employed for research in very-high-energy gamma ray astronomy. As G. Fazio and his colleagues have shown, the fluxes of high-energy gamma ray photons from several candidates such as the Crab nebula are at most extremely small. If celestial sources of even higher-energy photons exist, they may be detected through studies of extensive air showers (see Section 15.10).

15.4 Production Mechanisms for X-rays

Only two mechanisms seem capable of explaining the x-ray continua in the discrete sources: synchrotron emission and thermal radiation from plasmas at temperatures of the order of 10^7–$10^{8\circ}$K. A third mechanism is described below in connection with the background radiation.

Line emission in the x-ray region (such as those from the K-shell) might also occur in the discrete sources; lines would follow ion excitation by inelastic electron collisions, and they would also follow when the doubly-excited states created by dielectronic recombinations decay. At higher energies the decay of any radioactive isotopes would also produce lines, giving rise to nuclear spectroscopy in astrophysical objects.

The origin of the background cosmic x-ray spectrum is difficult to explain. If it is not the result of a large number of unresolved discrete sources, the most likely mechanism would seem to be the *Inverse Compton* effect; relativistic electrons interact with microwave and optical photons in space, boosting the photon energy into the x-ray region at the expense of the electron's energy.

The power radiated through the Inverse Compton mechanism may be more or less readily estimated through the methods of classical electrodynamics, if the target photon energy is small compared with m_0c^2, where m_0 is the rest mass of the electron, although the details are beyond the scope of this book. The power radiated, P_{CT}, by an electron of energy E is

$$P_{CT} = -\frac{dE}{dt} = 2 \times 10^{-14}(1 + \beta^2/3)\left(\frac{E}{m_0c^2}\right)^2 \rho,$$

if the energy density of the target photons is ρ ergs cm^{-3} and where $\beta = v/c$.

15.5   Synchrotron Emission

As we noted above, it seems likely that the x-ray spectrum of at least some of the discrete sources is due to synchrotron radiation. The total power P_{ST} radiated by a single relativistic particle of charge e and rest mass m_0 in a uniform field B is

$$P_{ST} = -\frac{dE}{dt} = \frac{2e^4}{3m_0^2c^3}B_\perp^2\left(\frac{E}{m_0c^2}\right)^2\left[\frac{\text{erg}}{\text{second}}\right], \qquad (15.1)$$

where $B_\perp$ is $B\sin\theta$ and θ is the angle between the orbital plane of the electron and the line of force.

For electrons, in units more useful than the cgs units employed in (15.1), this expression becomes

$$-\frac{dE}{dt} = 3.79 \times 10^{-6}E^2B_\perp^2[\text{second}^{-1}]. \qquad (15.2)$$

This is the loss rate (for electrons) in terms of billion-electron volts, so that E is expressed in these units. $B_\perp$ is also expressed in gauss, in (15.2). It is interesting that the synchrotron process causes the electrons to lose energy at a rate that is proportional to the *square* of their energies; $dE/dt \propto E^2$, so that the electron spectrum steadily steepens.

Using (15.2) we may compute the *fractional* loss rate for the electrons; it is

$$-\frac{1}{E}\frac{dE}{dt} = 3.79 \times 10^{-6}EB_\perp^2. \qquad (15.3)$$

Relation 15.3 may be integrated. The result is

$$\frac{E}{E_0} = (3.79 \times 10^{-6}B^2)^{-1}\left(E_0 t + \frac{B_\perp^{\,2}}{3.79 \times 10^{-6}}\right)^{-1}, \qquad (15.4)$$

where E_0 is the electron energy at time $t = 0$. It is helpful to rewrite (15.4) in terms of $T_{1/2}$, the time by which the electron energy has dropped to $E_0/2$; the result is

$$\frac{E}{E_0} = \left(1 + \frac{t}{T_{1/2}}\right)^{-1}. \qquad (15.5)$$

Therefore we find that $T_{1/2}$ becomes

$$T_{1/2} = \frac{2.64 \times 10^5}{B_\perp^{\,2} E_0} \ \text{[seconds]} \qquad (15.6a)$$

for electrons. Since electron emission is so much more important than proton emission (as we saw in the previous chapters), we have so far ignored synchrotron emission by protons. However, for the sake of completeness, and because the emission by cosmic ray protons in planetary magnetic fields is of some interest, we include the proton expression that is analogous to (15.6a); it is

$$T_{1/2} = \frac{5.2 \times 10^7}{B_\perp^{\,2} E_0} \ \text{[years]} \qquad (15.6b)$$

for protons.

Now if we wish to explain 20-keV x-radiation by a discrete source as the result of the synchrotron process, we must assume that 10^{13}-eV electrons are present in the source (see Section 14.11). If the field strength $B \sim B_\perp \sim 10^{-4}$ gauss (like that in the Crab nebula), $T_{1/2}$ will be only about 30 years, which is, of course, much shorter than the age of the Crab. This result is customarily interpreted to mean that the 10^{13}-eV electrons in the Crab must be continuously replenished, or else the x-ray source would fade away in just a few decades. It is possible, however, to construct a whole variety of different synchrotron models, with different initial electron energy spectra and different injection times. Some of these models avoid the necessity for continuous replenishment.

The electron spectrum steadily steepens as a result of the radiative losses, but it does not cut off abruptly. Continuous injection can be avoided by postulating a flatter electron spectrum (than indicated by the radio observations) with a low-energy cut-off around 10^3 BeV. This seems rather artificial, but cannot be ruled out.

It is useful to measure the energy spectrum of the photons; from an analysis of this frequency spectrum, it is in principle possible to learn something about the production mechanism. For example, if the synchrotron process is operative and if relativistic electrons are injected at $t = 0$ with a power law energy spectrum, a power law photon spectrum results.

A possibility that has been suggested is that the x-radiation from the Crab is produced when electrons radiating at optical frequencies enter regions of high magnetic field. Tucker has shown how the resulting photon spectrum may be calculated.

If the electrons radiating at optical frequencies in the source with a weak field B_0 enter a field with higher strength B_1 (gauss), the resulting photon frequency ($\equiv F$) spectrum from the synchrotron process is of the form $F^{2-\gamma/2}$ if the initial electron spectrum is of the form $E^{-\gamma}$. This, however, is only true at frequencies up to a so-called "break frequency" F_b, where $F_b = (B_1/B_0)\, 1.1 \times 10^9 B_0 - 3_t - 2$; the break is determined by the age t (years). In this expression for F_b, F_b is in hertz. At photon frequencies in excess of the break frequency, the photon spectrum has a different slope; the slope at these higher energies is given by $t^{[(\gamma+2)3]} F^{2(1-\gamma)/3}$; it is a steeper slope (in this energy region) but still a power law under the assumptions of this example.

If one assumes a power law electron energy spectrum and postulates that the synchrotron process is responsible for the observed photon spectrum, a variety of photon spectra are possible; they all have power law energy dependencies with a break that is characteristic of the synchrotron mechanism. The time variation of the injection process, if any, determines the magnitude and frequency at which the break occurs.

Continuous injections would correspond to injection at times infinitesimally close together; the slope of the synchrotron frequency spectrum can be obtained in this case by smoothly joining the breaks. This spectrum is like that we discussed in Chapter 14, in our review of radio astronomy.

An interesting test of the synchrotron mechanism is to measure the *polarization* of the x-radiation (Section 14.11). Such experiments are now under way.

For a given electron spectrum, the synchrotron and Inverse Compton processes are competitive energy-loss mechanisms for the electrons. It is interesting to compare their relative importance in various regions of space.

Inverse Compton depends upon ρ, the energy density of the target electromagnetic photons, while the synchrotron process depends on B^2. Thus the relative importance of the two mechanisms will depend upon $\rho/(B^2/2\mu)$, the ratio of the two energy densities.

In interstellar space, this ratio $\geqslant 10$, since both optical and microwave photons with comparable energy densities are present. Compton is therefore expected to dominate the loss of energy by electrons between the stars. It also dominates in the intergalactic medium, even though little starlight is present there. The microwave background is present and the intergalactic magnetic field strength is probably orders of magnitude smaller than the 0.1–1 gamma field present in the spiral arms.

Synchrotron, however, is expected to dominate in galactic nebulae such as the Crab. It is true that the energy density of starlight may be 100 times the value at the earth, but the magnetic field strength within the Crab is typically milligauss; B^2 is up by four orders of magnitude.

15.6   Thermal Radiation

The most obvious type of thermal radiation is that given by Planck's law for a hot "black body" that is many wavelengths thick. It was thought at one time that a neutron or hyperon star would satisfy this requirement. It now appears, as discussed below, that the x-radiation from any such "stars" would be more likely to result from a hot, tenuous plasma atmosphere around them. The photon spectrum of a black body would be of the form

$$J(E)\, dE = \frac{\text{constant}}{E^3(e^{E/kT} - 1)} \text{ [keV cm}^{-2}\text{ second}^{-1}\text{ keV}^{-1}\text{]}. \tag{15.7}$$

The thermal x-radiation by a nearly transparent hot cloud of plasma, however, has a different photon spectrum that depends on the nature of the energy spectrum of the electrons. The fundamental process here is *bremsstrahlung*, or "deceleration radiation."

If we are dealing with hot electrons and protons, the photons emitted from electron-proton interactions of this kind may have any energy from zero up to the initial kinetic energy of the electron. Since the differential cross-section for the bremsstrahlung process varies as dE/E, the electron loses a large fraction of its initial energy in such an interaction.

If the electrons are in thermal equilibrium at an absolute temperature T (a Maxwellian energy distribution), an exponential photon spectrum will result from thermal bremsstrahlung. That is, a photon energy flux spectrum of the form

$$J(E)\, dE = \text{constant } e^{-E/kT}\, dE \text{ [keV cm}^{-2}\text{ second}^{-1}\text{keV}^{-1}\text{]} \tag{15-8}$$

will result. Of course, if the electrons have some other energy distribution, the photon spectrum given by (15.8) may not result from bremsstrahling.

There are at least three reasons why it now seems unlikely that the x-radiation observed from many of the sources comes directly from the surface of a neutron star. At high temperatures neutrino cooling becomes very important; one of the cooling reactions is $2n \rightarrow n + p + e + \bar{\nu}_e$. The emitted neutrinos probably bring down the surface temperature of a neutron star, which has been calculated to be $\sim 1\%$ of its central temperature, to less than $5 \times 10^{6\circ}$K in 1000 years or so; x-radiation from the Crab nebula therefore would not be expected, even on this basis alone, to be thermal radiation, since the expected surface temperature is too low. Second, many of the sources have power law spectra rather than exponentials.

Finally, a lunar occultation experiment conducted with a rocket in 1964 was the first to show that the x-ray source in the Crab was extended, not a point. A diameter of about 1 light-year was inferred from the results of that experiment and subsequent observations with collimators capable of resolutions down to 20 arc-seconds. This result is also inconsistent with direct emission by a neutron star, since such an object has a diameter of less than $\sim$20 km.

None of the observations conducted to date on the Crab exclude the possibility that the rotational energy of a rapidly rotating neutron star is the ultimate energy source for the present-day luminosity. This idea is currently popular because of the discovery of a pulsar within the nebula and the further discovery that the pulse period (approximately 33 milliseconds) is slowly lengthening (at the rate of $\sim$36.5 nanoseconds daily, or about 1 part in 2400 annually). An occasional "starquake," or irregular period change, may also take place. Pulsed emission has been detected from the Crab pulsar, NP 0532, at energies up to $\sim$200 keV. This pulsar, the only one known to radiate optical and hard x-ray energy as well as radio, is the pulsar with the highest known repetition rate.

One of the consequences of this suggestion is that extremely intense, possibly co-rotating, magnetic fields may exist near the surface of NP 0532; the plasma particles in the Crab may well be accelerated by this field. If the neutron star is the result of the collapse of a normal-size star by a factor of $\sim$5 $\times$ 10^4 in radius, one might expect the dipole surface field to increase by a factor of $(5 \times 10^4)^3$. If the sun were to suffer a similar collapse, surface fields of $\sim$10^{13} gauss would result, so that fields like these do not seem unreasonable for the surfaces of neutron stars.

15.7   Infrared Astronomy

The infrared is that region of the electromagnetic spectrum that lies between the radio and the optical. Hence we may define infrared wavelengths to range from about 1 μ to perhaps 1 mm, although the latter is within the scope of microwave techniques.

There are several atmospheric "windows" that permit limited ground level observations in the infrared. For example, in the "near infrared" region $(4\,\mu \geqslant \lambda \geqslant 1\,\mu)$ we find the 2-μ window. The "intermediate infrared" $(25\,\mu \geqslant \lambda \geqslant 4\,\mu)$ contains several windows such as those at 5 and 20 μ. Figure 15.6 indicates how atmospheric absorption depends on wavelength in the infrared. Atmospheric emission, principally by H_2O vapor, is a source of background that limits the sensitivity of infrared observations from within the atmosphere (see Figure 15.7). Indeed it may be that the most sensitive

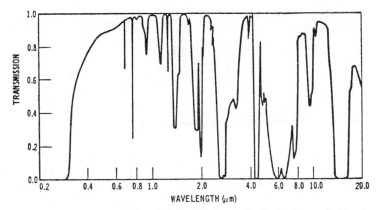

Figure 15.6   Atmospheric transmission (smoothed). At 2 km of altitude (zenith) there is assumed to be 0.25 cm of precipitable water vapor. From R. F. Webbink and W. Q. Jeffers, *Space Science Reviews*, 1969.

measurements of the distribution of water vapor in the stratosphere may be made with infrared techniques.

The terrestrial atmosphere is opaque in the far infrared; that is, at wavelengths between ~25 μ and 1 mm. It is to be expected that this region of the electromagnetic spectrum will be opened up when suitable detectors are lifted above the interfering constituents of the atmosphere (principally the triatomic molecules such as water vapor).

Infrared astronomy may be conducted with optical telescopes that focus infrared energy on special cooled detectors. Usually the beam is "chopped" at some frequency in order to reduce noise effects; the electronics are tuned

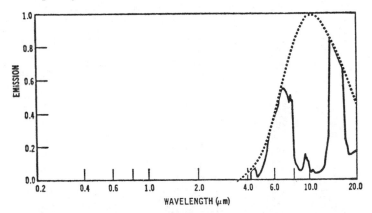

Figure 15.7   Atmospheric emission (zenith), relative to 275°K blackbody maximum. From R. F. Webbink and W. Q. Jeffers, *Space Science Reviews*, 1969.

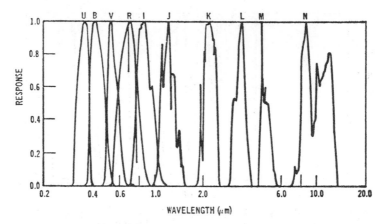

Figure 15.8   Response curves of photometer plus atmosphere. From R. F. Webbink and W. Q. Jeffers, *Space Science Reviews*, 1969.

to the chopping frequency. The resulting temperature rise in the detector then provides a measure of the amount of infrared energy that has been intercepted by the reflecting element of the telescope. (If a lens is used in a refracting system, it must be made of a substance, such as silicon, that will not absorb infrared. The detectors may be lead-sulfide photocells, germanium bolometers, or special photomultipliers; the wavelength used determines the detector employed.)

The UBV scheme of optical photometry has been extended to include infrared photometry. Figure 15.8 shows the UBV wavelengths, as well as the

TABLE 15.2. ABSOLUTE CALIBRATION OF INFRARED PHOTOMETRY

Filter Band	λ_0 (μm)	$\Delta\lambda$ (μm)	Absolute Flux (watts cm²-μm)	Magnitude 0.00 (watts m²-Hz)	m_{min}
U	0.36	0.066	4.35×10^{-12}	1.88×10^{-23}	23.0
B	0.44	0.098	7.20×10^{-12}	4.44×10^{-23}	24.0
V	0.55	0.087	3.92×10^{-12}	3.81×10^{-23}	23.5
R	0.70	0.207	1.76×10^{-12}	3.01×10^{-23}	21.0
I	0.90	0.231	8.30×10^{-13}	2.43×10^{-23}	18.0
J	1.25	0.297	3.40×10^{-13}	1.77×10^{-23}	15.5
K	2.20	0.578	3.90×10^{-14}	6.30×10^{-24}	10.0
L	3.40	0.701	8.10×10^{-15}	3.10×10^{-24}	8.0
M	5.00	1.128	2.20×10^{-15}	1.80×10^{-24}	5.0
N	10.20	4.330	1.23×10^{-16}	4.30×10^{-25}	2.0
Q	22.00	7.500	7.70×10^{-18}	1.02×10^{-25}	0.0

From R. F. Webbink and W. Q. Jeffers, *Space Science Reviews*, 1969.

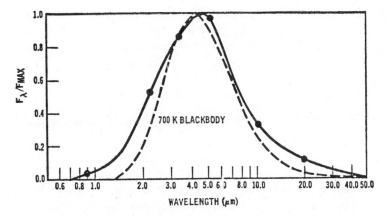

Figure 15.9 Spectral energy curve for NML Cygnus (solid line). From H. L. Johnson, E. E. Mendoza, and W. Z. Wisniewski, *The Astrophysical Journal*, University of Chicago Press, 1965.

R through N filter bands, which were designed to take advantage of the windows in the earth's atmospheric transmission. Table 15.2 gives the values of flux density in each band for an object of magnitude 0.00.

Discrete celestial sources of infrared radiation were first detected in 1937, but very little work was done in this region of wavelengths until the 1960s, when improved detectors became available. It is now clear that some enormous, cool objects do exist; some are so cool that they do not radiate detectably in the visible and are called "infrared stars." From the amount of infrared energy detected, one infrared star in Orion has a radius of 8 A.U. and may have a mass 6 times that of the sun, but the surface temperature is only a few hundred degrees. Another of these remarkable objects is NML Cygnus. Figure 15.9 illustrates the spectrum of that object.

Many of the objects so far observed are quite bright in the infrared; they radiate much more energy there than a black body at temperature T would radiate, according to the Planck law. Hence these objects exhibit an "infrared excess." The sun does not exhibit an infrared excess. At $\lambda \geqslant 1.6\,\mu$ the solar output is even less than would be predicted for a 5700°K black body, because of the opacity of the H^- ion.

Thus some of these objects are not sunlike. Various possibilities exist, and it is likely that the different objects have different natures.

One theory that explains some of the other infrared sources is that the central star is surrounded by a large envelope of dust. The circumstellar dust absorbs the output of the star and reradiates it in the infrared. The infrared spectrum of the interplanetary dust that causes the zodiacal light is remarkably like that of some of the infrared stars, in that the interplanetary dust spectrum shows an infrared excess. If this model is correct (which would mean that a sunlike central star exists), we are looking at the emission from dust that contains several planets' worth of mass. We may expect this dust to condense gravitationally into planets, so that we may be looking at proto-planetary systems.

A celebrated example found by Low and Smith is R Monocerotis, a 12th-magnitude T-Tauri variable. The dust could be associated with the mass loss detected for T-Tauri stars (Section 12.1). Figure 15.10 shows the circum-stellar spectrum of R Coronae Borealis, an object similar to R Mon.

Low has found that Jupiter's infrared temperature is about 27°K warmer than the 107°K that would be expected for a planet in radiative equilibrium with the sun. Since the energy radiated by a body varies at $T_e{}^4$, Jupiter radiates about 3 times as much energy as it receives from the sun. It therefore seems that Jupiter has an internal energy source. Gravitational contraction is one possible source that has been suggested.

To sum up, the main achievements of infrared astronomy to date include the discovery of infrared "stars" (such as the object in Orion) that may be *protostars*, contracting gas clouds that are dim or not yet glowing in the visible because thermonuclear reactions have yet to set in, but are

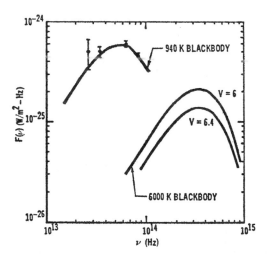

Figure 15.10 $F(\nu)$ versus frequency for R CrB. A black body of temperature 6000°K is shown normalized to the visual magnitude $V = 6.4$ of the object at the time of observation. From W. A. Stein, J. E. Gaustad, F. C. Gillett, and R. F. Knacke, *The Astrophysical Journal*, University of Chicago Press, 1969.

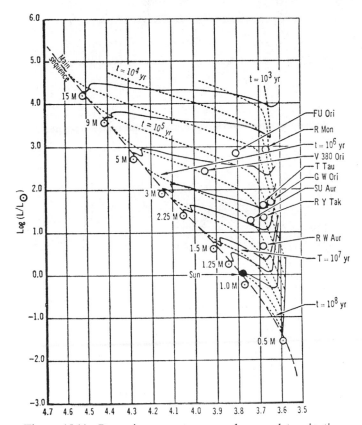

Figure 15.11   Pre-main sequence age and mass determination. After Iben and Mendoza. From R. F. Webbink and W. Q. Jeffers, *Space Science Reviews*, 1969.

extremely bright in the infrared. Figure 15.11 illustrates some theoretical estimates of the pre-main sequence evolution of various objects, including their predicted ages and masses.

Another very important infrared observation concerns quasars and galaxies. In the optical region quasars and many galaxies have common spectral features. Radio observations have revealed that both types of objects often have energetic central cores. At wavelengths intermediate between these two (the infrared), however, measurements made by Low and others have shown that many of both types are extremely bright; much of the energy in these central cores is emitted in the infrared (see Figure 15.12).

In both bright galaxies and quasars the infrared emission is several orders of magnitude greater than the energy radiated at all other measured wavelengths. Our galaxy is one of the many that has an "infrared core."

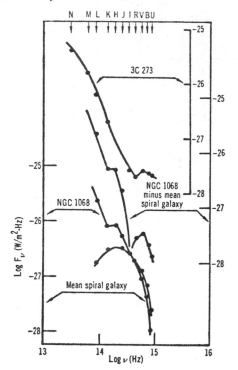

Figure 15.12 Comparison of fluxes of Seyfert galaxy NGC 1068, quasar 3C273, and a mean spiral galaxy. From R. F. Webbink and W. Q. Jeffers, *Space Science Reviews*, 1969.

It appears that there may be a connection between galaxies and quasars in spite of the tremendous disparity in size. This is because quasars and galactic cores, particularly the cores of those galaxies known as *Seyfert galaxies*, seem to have about the same size. Seyfert galaxies are spirals with small bright regions in their nuclei. About a dozen Seyferts are known; the optical spectra of the nuclei show broad, bright emission lines, presumably arising from hot gases there. Some Seyfert galaxies are strong radio emitters, and possibly also of x- and gamma radiation. The virial theorem implies that stars gravitationally bound in a small space, such as a Seyfert core, have high speeds, on the average. Hence many stellar collisions may occur in such cores, producing high-energy photons.

Thus it is possible that there is a quasar at the center of our galaxy, even though most detected quasars are most likely at cosmological distances. (This is inferred because at least one quasar, B264, shares the same red shift as that of four other objects known to be galaxies that appear close in angle to B264.)

There has even been speculation that quasars may be one of the stages in the evolution of galaxies. In any event, since so much of the radiation of quasars and galaxies is at infrared wavelengths, it is likely that infrared studies will reveal much about the physics of these objects.

The achievements of infrared astronomy also include the discovery of circumstellar envelopes. There are stars that are cool (and hence dim in the optical) from their black body spectra but exhibit large excess amounts of radiation at long wavelengths, say, 8–14 μ. That is, at infrared wavelengths, one sees much more energy than would be expected from Planck's law; planets may be forming.

The discovery by Low of the energy radiated by Jupiter is interesting. We apparently have close at hand a body almost massive enough to become a star.

Finally, a relationship between galaxies and quasars may have emerged from infrared studies.

15.8   Neutrino Observations

We have already seen (Section 11.20) how celestial electron-type neutrinos may be detected through the inverse beta decay process. These neutrinos are of interest in connection with stellar interiors.

We now inquire into the detection of muon-type neutrinos. These particles are produced whenever high-energy nucleons interact with the nuclei of gas atoms to make pions. The neutrinos come from the subsequent decay of the pions.

$$\pi^+ \rightarrow \mu^+ + \nu_\mu, \tag{15.9a}$$

$$\pi^- \rightarrow \mu^- + \bar{\nu}_\mu; \tag{15.9b}$$

$$\mu^+ \rightarrow e^+ + \nu_e + \nu_\mu, \tag{15.10a}$$

$$\mu^- \rightarrow e^- + \bar{\nu}_e + \nu_\mu. \tag{15.10b}$$

Thus the decay of one pair of charged pions produces twice as many mu-type neutrinos as electron-type neutrinos. Pion formation in space may be expected from the interaction of cosmic radiation with the interstellar and intergalactic matter.

When a large flux of high-energy ($E \geqslant 10^9$ eV) neutrinos is incident upon heavy nuclei, such as those found in the earth's crust, a few interactions may be expected to occur. If one uses as the neutrino target some thousands of tons of earth, several interactions per year will result from a 10^{10} cm^{-2} second^{-1} flux of neutrinos.

In particular, if we are dealing with muon-type neutrinos, μ-mesons will result from the interactions. μ-Mesons are charged particles; they may be detected through a variety of conventional techniques. The problem, however, lies in distinguishing these "daughter" muons from the cosmic ray muons produced in the earth's atmosphere. The latter flux (i.e., "noise") is perhaps 10^{10} times as intense as the "signal" at sea level.

The two fluxes may be distinguished by placing the muon detectors deep within the planet. Cosmic ray muons may have a zenith angle distribution that is sharply peaked toward the vertical, while the "daughter" muons are roughly isotropic because of the great penetrating power of the neutrinos.

As yet no definite results have emerged from neutrino astronomy. One only expects approximately 1 count per month from present-day detectors. Since these detectors only started operating in 1965, it seems that many years will be required before any conclusions may be drawn.

15.9   Gamma Ray Line Spectra

Considerable effort is now being expended in the search for lines in the gamma ray spectra of various candidate sources. The detection of optical radiation from an object tells us that atoms have been excited in it, and the detection of x-ray continua tells us that electrons have probably been accelerated to very high energies, but neither observation tells us *how* the excitation or acceleration took place. Gamma ray lines, however, are distinctive; they can only arise from nuclear processes.

One line that might be expected occurs at a photon energy of 0.511 MeV; it is due to the annihilation of positrons by electrons. Positrons may be produced by cosmic ray interactions as well as by the decay of some of the radioisotopes.

Another possible line occurs at 2.23 MeV. This line is caused by the capture of slow neutrons by hydrogen; deuterium results from this capture, and a 2.23-MeV photon is liberated. These energies (0.511 and 2.23 MeV) may be downshifted if the optical spectrum of a given source shows a red shift and if no dispersion in the shift as a function of frequency occurs.

The study of line emission in likely sites for nucleosynthesis, such as supernovae in both the x-ray and gamma ray regions of the electromagnetic spectrum, permits a check of astrophysical theories of element formation. Radioactivity may be found in supernovae and in supernova remnants, if heavy elements are synthesized in such explosions.

The study of line emission also relates to the possible energy source of the various large radio sources, including objects such as quasars and radiogalaxies. Various ideas have been proposed. Among them are the annihilation of antimatter and the multiple-supernova hypothesis, in addition to recent theories concerning the rotational angular velocity of neutron stars (pulsars).

In the next section we discuss antimatter annihilation in connection with the production of high-energy gamma radiation in quasars. The general idea (due to Alfvén) is that two dual universes may exist in Nature. One is made up of matter as we know it, and the other is composed of antimatter. This would satisfy the particle symmetry that has been observed in the laboratory. The two plasmas may "touch" each other in places; it could be that those places are the objects we call quasars. If this is so, a good deal of positron-electron annihilation radiation (red-shifted in accordance with the optical value of Z, if no dispersion in the shift occurs as a function of λ) should emanate from quasars. Thus a gamma ray line may be present at a (few) hundred kilo-electron volts if the colliding particles are at rest.

Radiogalaxy energy sources might be accounted for through the multiple-supernova hypothesis. The stars in the center of a galaxy are rather closely packed; if one explodes, it is at least conceivable that nearby stars might also be triggered by this explosion into similar explosions, giving rise to a truly titanic blast. This could be the energy source. If supernovae are in fact the sites of heavy-element synthesis, a characteristic line spectrum of radio-activity might be present in the gamma region of wavelengths emanating from radiogalaxies.

15.10   Production of High-Energy Gamma Radiation

High-energy nuclear interactions, such as proton-proton collisions, result in copious production of pions. Each neutral pion thus produced decays in 10^{-16} second into 2 photons of 70 MeV of energy in the rest frame of reference. Kaon decay is also an important source of photons.

Thus the interactions of cosmic radiation with the interstellar and inter-galactic media may be expected to produce high-energy photons. Since typical cosmic ray energies are $\sim 10^{10}$ eV, we may expect photons with energies of 500 MeV and more to result from bombardment by higher-energy cosmic rays. This may give rise to a diffuse, high-energy spectrum. Such a diffuse spectrum is not expected to be truly isotropic, for it should be governed by the structure of the galaxy, since the gas density depends on that structure.

Pions are produced through another mechanism. The annihilation of an antiproton by a proton produces $\pi^{\pm}$ and π^0 mesons; roughly one-third are neutral pions. If the antiproton is of low energy, photon energies of ~ 200 MeV result; higher-energy particles produce higher-energy photons.

In addition to pion decay, other production mechanisms for high-energy photons exist. One is the inverse Compton effect; another is bremsstrahlung by cosmic ray electrons in the various (discrete) possible sources.

It is of interest to attempt an estimate of the fluxes that are to be expected at the earth from these various mechanisms. For example, if one assumes that the locally observed cosmic ray density and interstellar matter density are the same throughout the galaxy (including the center), calculations indicate that $\sim 1 \times 10^{-5}$ cm^{-2} second^{-1} galactic center photons with energies in excess of

100 MeV should arrive at the earth from the first process (proton-proton collisions) noted in this section.

This estimate is arrived at by considering the product of four factors; this product gives the reaction rate. The factors are (a) the cosmic ray intensity, (b) the cross-section for P-P collisions, (c) the number density of target atoms, and (d) the path length L of the cosmic radiation in these "targets." (L is taken to be roughly two-thirds of the galactic radius for galactic center radiation.)

The other processes have been calculated to be less important sources. Thus bremsstrahlung is expected to produce only $\sim 0.1 \times 10^{-5}$ photon cm^{-2} second^{-1} at the earth, as is the inverse Compton effect when applied to starlight. (Inverse Compton, when applied to the fireball radiation phenomenon, may be responsible for only $\sim 0.05 \times 10^{-5}$ gamma photon cm^{-2} second^{-1}.)

We also see that discrete sources of high-energy photons may exist. Perhaps the quasars are among such sources, if matter-antimatter annihilation provides an energy source for quasars. It is hoped that interesting data will shortly become available, since experiments are now able to detect 10^{-5} photon cm^{-2} second^{-1} at $E \geqslant 100$ MeV.

All the above production mechanisms for high-energy gamma rays require the presence of matter. However, when cosmic ray protons with energies in excess of 10^{17} eV encounter starlight photons ($E \sim 1$ eV), gamma rays may be produced *without* nearby matter. Such a situation occurs in intergalactic space.

This "inverse pion" process occurs because at energies $\geqslant 10^{17}$ eV there is sufficient energy available for pion production in the rest frame. Near threshold the produced pion has almost the same velocity as the cosmic ray proton, so that the decay energy of the quanta in the laboratory reference frame is a characteristic 10^{16} eV. (There are exceedingly few cosmic ray particles with energies in excess of 10^{17} eV, so that there will not be many photons with energies greater than 10^{16} eV.)

Theoretical expectations are that the flux of 10^{16}-eV quanta is perhaps 0.001% of the proton flux at 10^{16} eV. Photons of these energies might be detected through a search for extensive air showers on the earth that are deficient in muons.

15.11   The Interstellar Medium and Exotic Astronomy

One of the very great, and potentially the greatest, reasons for interest in exotic astronomy lies in the tremendous penetrating power of radiation. We have already referred many times to the tiny cross-section ($\sim 10^{-43}$ cm^2) for neutrino absorption; it seems likely that the entire universe is rather transparent to neutrinos.

In addition, high-energy gamma radiation can penetrate a great deal of matter without serious attenuation, since a typical mean free path is $\sim$100 gm cm^{-2}. If the density of the intergalactic medium is 10^{-29}–10^{-30} gm cm^{-3}, it is clear that gamma ray observations may be conducted at distances far in excess of those useful in optical astronomy or even those explored by radio techniques.

Low-energy x-rays, such as 1-keV x-radiation, which has an optical depth of only $\sim$10^{-4} gm cm^{-2} in light elements, are of interest. At these low energies the spectra of discrete sources may be examined for the low-energy cut-offs that would be caused by interstellar absorption. This helps in establishing the distances to the sources. However, should the distance be independently known, the same data provide information, through the (measured) absorption coefficient, on the nature of the interstellar medium.

Problems

15.1. Estimate the number of head-on star collisions within the nucleus of M87, taking the stellar concentration there to be $\sim$5 $\times$ 10^4 as great as in the solar neighborhood and assuming a geometrical collision cross-section and a relative a collisional velocity of 10^3 km per second.

15.2. If our galaxy were to collide with and pass right through the Andromeda galaxy at a relative speed of 1000 km per second, how long would it be before our sun suffered its first direct collision with a star in Andromeda? (Assume that all the stars in Andromeda are about equal in size to the sun, the cross-section is geometrical, and the average density of stars in Andromeda is about 10 times the density in the neighborhood of the sun.)

15.3. (a) Suppose that all the discrete x-ray sources, of number N_x, were galactic and also supernova remnants. Determine the characteristic time τ for x-ray emissions (i.e., the time by which the source loses all energy available for radiation as x-rays) in terms of τ and the rate of supernova outbursts per galaxy (dN_s/dt).

   (b) Can this supernova process account for the x-ray luminosity of the Crab nebula?

15.4. Suppose that some fraction F of the energy released in a supernova is emitted as gamma rays in the million-electron volt range. Take F as the energy released in burning 0.01$M_\odot$ of hydrogen, a process that takes place in perhaps 10^3 seconds. Estimate the energy flux (ergs per square centimeter per second) at the earth, in the million-electron volt range, from an extragalactic supernova at a distance of 10 megaparsecs. Is this flux detectable?

15.5. Suppose that in a supernova part of the emerging hard flux is degraded to a few kilovolts and emitted as x-rays, and also that this x-ray number flux so produced has an intensity comparable for a few days with that of the flux at maximum light from a supernova. Taking the peak absolute magnitude of the supernova to be -18, compute the few-kilovolt x-ray flux to be expected at the earth from an extragalactic supernova at a distance of 10 megaparsecs. Is this flux comparable with that from Sco XR -1?

15.6. (a) Consider a hypothetical outburst in a galaxy at a distance d from the earth, involving the release of an amount E of energy, of which a fraction F_γ is emitted as high-energy gammas of mean energy $\bar{E}_\gamma$. If the outburst occurs during a time τ, what is the resulting high-energy flux J_γ at the earth?

(b) For $E = 10^{60}$ ergs, $d = 1000$ megaparsecs, $\bar{E}_\gamma = 100$ MeV (the mean photon energy resulting from π° decay), $\tau = 1000$ years, and $f_\gamma = 0.1$, estimate the flux to be expected, in photons per square centimeter per second.

15.7. A cosmological model was suggested in 1958 in which the intergalactic medium is at a very high temperature ($\sim 10^{9\circ}$K); this temperature was supposed to arise from the nearly 1-MeV electrons which would result after the decay of spontaneously created neutrons as envisioned by the "steady-state cosmology." The rate of production of bremsstrahlung photons ($1\,\text{Å} \leqslant \lambda \leqslant 10\,\text{Å}$) resulting from this model has been estimated to be 1.2×10^{-25} photon cm^{-3} second^{-1}. Take the cut-off radius for the universe to be $R = 5 \times 10^{27}$ cm, and compute the flux to be expected at the earth from this model. Does the flux estimate conflict with the observed diffuse flux?

15.8. (a) Suppose that the diffuse flux at x-ray energies is due to the contributions from many unresolved galaxies, each with an x-ray luminosity equal to that from our own galaxy. Taking the number density of galaxies to be $u = 3 \times 10^{-75}$ cm^{-3}, out to a cut-off distance for the universe of $R = 5 \times 10^{27}$ cm, calculate the isotropic background flux per steradian to be expected at the earth, in terms of $\langle L_x \rangle g$, the average x-ray luminosity for each galaxy.

(b) If the total x-ray flux received from all sources within our galaxy is ~ 50 photons cm^{-2} second^{-1} and if the galaxy is typical of all other galaxies, compute the isotropic background to be expected at the earth. How does this compare with observation?

15.9. Since 2 neutrinos are emitted in the conversion of 4 protons to a helium nucleus, estimate the total flux of solar neutrinos expected at the earth, using the solar constant and the energy released in the proton-proton chain of nuclear reactions (26.0 MeV). Compare this flux with that expected from all other nonterrestrial sources, assuming that they too emit because of the proton-proton chain.

15.10. (a) The electron-positron annihilation cross-section σ_a at nonrelativistic energies is $\sigma_a = \pi r_0^2 / \beta$, where r_0 is the classical radius of the electron and $\beta = v/c$. If α is the mean ratio of antimatter to matter in the universe, estimate the flux of 511-keV photons to be expected out to a distance of $R = 5 \times 10^{27}$ cm, assuming that the mean number density of electrons in the universe is $u = 10^{-5}$ cm^{-3}, so that the mean positron density is $\alpha \times 10^{-5}$ cm^{-3}.

(b) Present-day upper limits on the 511-keV flux in various directions average 1×10^{-3} photon cm^{-2} second^{-1}. Determine the upper limit on α from this. If there is appreciable antimatter in the universe, what can you say about the mixing with matter?

Chapter 16

COSMOLOGY

Cosmology is the study of the large-scale structure of the universe. The cosmologist attempts to synthesize an understanding of the distribution and behavior of matter in the universe on the basis of physical behavior observed both in the extraterrestrial universe and in the laboratory. Few facts that bear on the distribution of matter in the universe are known. Virtually all modern cosmology is based on two phenomena: one is that the night sky is dark; the other is known as the "red shift," interpretations of which have led to distance scales in the universe.

16.1 Distance Measurements

The central and most urgent problem of observational astronomy (at all wavelengths) is the accurate measurement of distances to the various celestial objects. Since cosmology rests on the results of these astronomical measurements, it is necessary for us to look at this problem.

There is trouble concerning the very notion of "distance" itself. The problem arises because we must know in advance the kind of geometry to employ before distances may be measured. For example, one might use the technique of *parallax* to determine the distance to a celestial object. In this method one uses two observatories that are separated on the earth's surface by a known distance. If the celestial object (a star, say) is sufficiently close, the two observatories will measure (slightly) different star-observer-baseline angles; from accurate measurements of the two angles and the length of the baseline, the "distance" to the star may be calculated. The baseline may be the diameter of the earth, the diameter of the earth's orbit, or some other "known" distance. This method is practical only for the closest stars; about 6000 stellar distances have been found in this manner. The distance obtained in this way is called the "local distance" by cosmologists.

Notice, however, that we *assume* space to be Euclidean in this calculation. That is, we assume implicitly that light travels in straight lines, so that plane geometry may be used. Of course, if space actually were curved, we could employ the appropriate geometry and then calculate the distance from the measurements. Thus, if space were spherical rather than Euclidean, we would merely employ spherical geometry rather than plane geometry; the calculation would still be straightforward. The catch, however, is that we do *not* know in advance what kind of geometry to employ.

We could, in principle, find out what kind of geometry to employ. "All" that we have to do is make some measurements of distance; the "radius of curvature" of space may be deduced from these. The reader will see, however, that we need to know distances in order to find the geometry so that we can measure distances; distance and geometry seem inseparable.

16.2 Local and Photometric Distances

As we shall see again in Section 16.8, we may write an expression for $(ds)^2$, the square of the separation between two events or the "line element," as

$$(ds)^2 = (dr)^2 - c^2(dt)^2,$$

where r is the radial "distance" as computed from Euclidean geometry and t is the time. This expression should be modified if space is curved rather than Euclidean.

One way of accomplishing the modification has been suggested by H. P. Robertson. He found a form for the line element that obeys the *cosmological principle*, which says that *aside from random fluctuations that may occur locally, the universe must appear the same for all observers*. The Robertson line element is given by

$$(ds)^2 = c^2(dt)^2 - R^2(t)(dr)^2, \tag{16.1}$$

where $R(t)$ represents a scale factor that is a function of time. Because $R(t)$ determines the scale of a spherical universe at any moment t, it is generally referred to as the "radius of a closed spherical universe," or "space dilation factor." The quantity r is, as before, the distance measured in the three-dimensional subsurface that represents space (see Figure 16.1).

Neglecting obscuration (which we may do, say, for radio or gamma radiation), it is possible to relate the apparent luminosity l of a distant galaxy to its absolute luminosity L (both of which may in turn be related to apparent and absolute magnitude, respectively). Robertson's relation is

$$l = \frac{L}{4\pi R_0^2 r^2(1 + Z)^2}, \tag{16.2}$$

where R_0 is the present value of the space dilation factor and Z is the red shift, $\Delta\lambda/\lambda$. In the case of the laboratory or even the local group of galaxies, where $Z = 0$, l is given by $L/4\pi r^2$, choosing $R_0 = 1$.

A photometric distance D can be defined by

$$D = R_0 r(1 + Z). \tag{16.3}$$

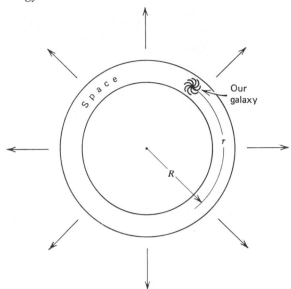

Figure 16.1   A two-dimensional illustration of the curvature of space. The radius of the (closed) space is given by R; r is measured within the hypersurface itself and is sometimes referred to as the co-moving radial co-ordinate in the Robertson line element. An expansion in the R direction (shown) causes a consequent increase in the "distance" r between spatial points.

This results in

$$l = \frac{L}{4\pi D^2},\tag{16.4}$$

so that it is still possible to speak meaningfully of distance, provided that we understand the word to refer to the luminosity distance D. The distances used herein are all photometric distances D, defined so that the intensity of a source of intrinsic power output P is proportional to P/D^2.

16.3   Cepheid Variables

We now discuss some of the techniques employed in celestial distance measurement. As we indicated previously, the most accurate method, parallax, can be used only for the closest stars. The vast majority of celestial objects do not display any detectable parallax. A variety of so-called "statistical" techniques have consequently been evolved in order to measure these distances. The methods are called "statistical" because they rely on the average properties of certain types of celestial objects. Perhaps the most important type includes those objects known as Cepheid variables.

Some stars pulsate regularly in apparent magnitude. One such group consists of yellow supergiants that are named cepheid variables (after the

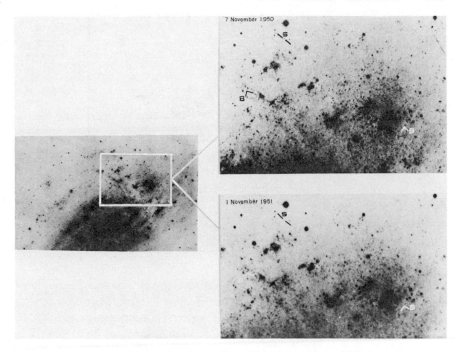

Figure 16.2 Three cepheids in NGC 2403 from blue plates taken 359 days apart with the 200-inch telescope. Periods are 46.460, 20.260, and 34.354 days for variables 5, 6, and 8. Magnitudes are 22.07 and 22.32 for variables 5 and 6 in the upper right and 22.02 for variable 8 in the lower right panel. From A. R. Sandage, in *Physics Today*, 1970.

first known, δ-Cephei, which was discovered to be of variable brightness in 1784 (see Figure 16.2). The light curve of δ-Cephei is shown in Figure 16.3. Some 600 cepheid variables have been detected in our galaxy alone (Polaris, presently the North Star, is a cepheid); their periods range from 1 to 50 days, with an amplitude that ranges from 0.1 to 2 magnitudes superimposed on a median absolute magnitude that ranges from −1.5 to −5. Theory explains the brightness variation as an actual radial pulsation of the star, because of the observed varying Doppler displacements of the spectral lines.

Henrietta Leavitt discovered in 1912 that a relation exists between the periods of pulsation and median luminosities, or average absolute magnitudes, of cepheids. This result has proved to be of extreme importance. She observed that the cepheids in the Small Magellanic Cloud show an approximately linear relationship between their *apparent* magnitudes and the logarithms of their periods of oscillation; the brighter the object, the larger the period of oscillation. Figure 16.4 shows the apparent magnitude versus period diagram for 48 cepheids in the Small Magellanic Cloud. It is clear that the statistics of 48 points do not permit a precise identification of the relation, but it is clear that

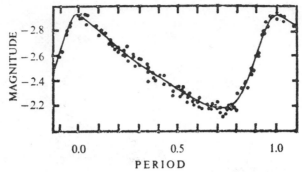

Figure 16.3   The light curve of δ-Cephei. From L. Goldberg and L. Aller, *Atoms, Stars and Nebulae*, The Blakiston Company, 1943.

the mean apparent magnitude depends on the period. If a straight-line relationship is found on a log-log plot like this, a power law dependence of brightness on period is indicated.

Since the stars in the Small Magellanic Cloud are all about the same local distance from the earth, it follows that the *absolute* magnitudes of the objects must be related to their periods. Hence, if we can independently measure the

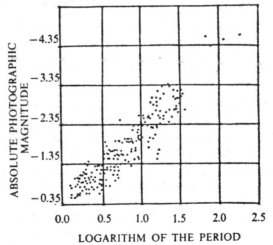

Figure 16.4   Absolute magnitudes of cepheids in the Small Magellanic Cloud, plotted versus their periods. Since the stars are all approximately the same distance from the earth, a plot of apparent magnitudes versus period would exhibit the same correlation. From L. Goldberg and L. Aller, *Atoms, Stars and Nebulae*, The Blakiston Company, 1943.

local distances of a few cepheids, we can "calibrate the scale" in absolute units, so that *we may determine the absolute magnitude of any cepheid merely by measuring its period*. This independent measurement has been indirectly made for a few cepheids within our galaxy, through a statistical method known as the technique of "secular parallax."

Secular parallax is a parallax method that uses baselines very long compared to 2 A.U., the diameter of the earth's orbit. It involves a determination of the velocity of the solar system in its motion toward the solar "apex" (which lies in the constellation Hercules). This velocity is measured with respect to the nearby stars, those within 60 light-years of the earth, and is an *average* speed. Once established, this speed permits the distance that the solar system has moved in some given long time (20 years, say) to be established. The velocity is such that our system moves ~80 A.U. in 20 years; secular parallax measurements of the apparent coordinates of distant stars made over this time scale are useful for determining stellar distances for faraway stars. In two decades it permits us to measure distances $80/2 = 40$ times as far as ordinary trigonometric parallax permits.

Assuming that all cepheids behave the same regardless of their location, we may use these results to find distances to other galaxies that contain detectable cepheids, for we can measure the mean apparent magnitudes with

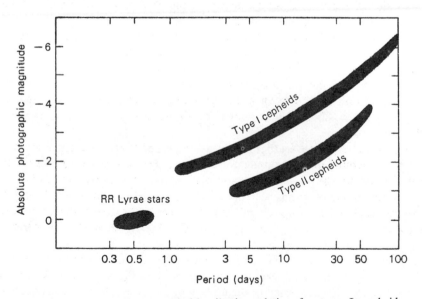

Figure 16.5 Approximate period-luminosity relation for type I cepheids (*upper curve*), type II cepheids (*lower curve*), and RR Lyrae stars, which are sometimes called "cluster-type cepheids," since they are found in globular clusters.

relative ease. This measurement, coupled with the information on the absolute magnitude obtained from the period measurement, enables us to deduce the distance to the galaxy, because of the $1/D^2$ dependence of the brightness.

There are catches to this, of course. First, as we have noted, the statistics on the Small Magellanic cepheids are not terribly good, and also the "known" distances to the cepheids within our galaxy are not precisely known. We have not included in this discussion the effects of light absorption by dust clouds between the unknown galaxy and the earth. These effects are not well understood, so that the apparent magnitude estimates are in some doubt. Nevertheless the study of cepheid variables is one of the most widely used ways of estimating distances in the cosmos.

There are two "populations" of cepheids; populations I and II are distinguished on the basis of their spectra. The *shape* of the curve relating the absolute magnitudes to the periods is the same for both populations, but population I cepheids are about 1.5 magnitudes brighter than those belonging to population II (see Figure 16.5).

16.4   Novae and Supernovae

We can measure angles to arbitrary accuracy, and the displacement of spectral lines may also be made with high accuracy. These facts may be used in studies of supernovae, and another means of distance measurement may be developed through these studies.

When a star explodes, luminous gases are expelled at high speeds into space. We can measure the angular "diameter" of the resulting nebula, and we can also measure the speed v of expansion, through observations of the Doppler displacement of the spectral lines emanating from the nebula. If we are fortunate enough to know *when* the supernova occurred, we can find the size of the nebula in absolute units, since the size is just $v\tau$, where τ is the age of the supernova.

This assumes, however, that the expansion is spherical. In at least one case, the Crab nebula, this is certainly not the case; the magnetic fields within the nebula restrain the plasma expansion differently, in different directions.

Another assumption has been made: that no acceleration occurs during the expansion phase. This too is highly questionable in the case of the Crab, since the interaction of the expelled matter with the interstellar medium apparently changes the speed of the ejecta.

With these qualifications in mind, we may proceed with the argument. Since we have measured the angular size of the object and now know the physical size, we can find its distance through Euclidean geometry. This method is in principle quite helpful, since supernovae flare up to approximately the brilliance of an entire galaxy and may therefore be readily detected. There are, however, practical difficulties.

We cannot, of course, measure supernova angular diameters in other galaxies; the angles are far too small. What we can do is attempt to learn the absolute brightness of supernovae at maximum brilliance, from observations of novae and supernovae that are at known distances. It seems at this writing that the peak absolute magnitude of a type I supernova is about -19 (only $\sim 1.5^m$ fainter than the absolute magnitude of our entire galaxy of 10^{11} sunlike stars). If we now assume that all type I supernovae achieve the same peak brightness, we can again measure local distances through the $1/r^2$ effect and measurements of the apparent magnitude, allowing somehow for intergalactic absorption.

16.5 Galactic Statistics

There is at least one other method of finding the local distance to a galaxy. If we can somehow determine the average luminosity of a galaxy, we can find its local distance by measuring its apparent magnitude and utilizing the r^{-2} effect. This method of distance determination may be calibrated by observations of cepheids in nearby galaxies. While it is widely used, this "galactic characteristics" technique is not very accurate. The average absolute magnitude of *all* galactic types is -15, but some spirals are much brighter; -19 is a more accurate figure for them. Hence at least a four-magnitude uncertainty exists. This uncertainty would give rise to an even greater error for a large spiral such as the Milky Way.

16.6 Olbers' Paradox

Olbers noted in 1803 that the fact that the night sky is dark is puzzling. He reasoned that if all the stars (he did not know of galaxies, but the argument applies to them too) are distributed uniformly throughout a Euclidean space that extends in all directions, *every bit of the night sky ought to be as bright as the sun*, if all stars are like the sun. He also assumed that all the radiation sources were at mutual rest, the universe of classical physics. This contradiction constitutes one of the most powerful reasons for discarding the classical model. The red shift, discussed below, provides a resolution of the paradox; the diminution of illumination that the red shift entails means that the total background radiation in a model universe is very small.

16.7 The Red Shift

The red shift is the single observational fact (other than the darkness of the night sky) on which most modern cosmology is based. Consequently the shift has received a great deal of attention, and the controversies that have strung up regarding its interpretation remain unsettled at present; they will most likely remain unsettled for a long time.

The observational fact is that in general, the larger the apparent magnitude of a remote galaxy, the more its spectral lines are displaced toward the longer-wavelength region of the spectrum, compared to a laboratory source of these spectral lines. Thus the fainter the galaxy, the redder it becomes.

It should be recognized that the red shift is a statistical effect. The statistics from which the functional form of the variation is found are constantly being re-examined. Also, it is not quite clear that no *dispersion* occurs, or, in other words, that the red shift, $Z \equiv \delta\lambda/\lambda$, does not depend on the wavelength λ. If dispersion does occur, so that $Z = Z(\lambda)$, a great many present interpretations will have to be revised. Exotic astronomy may help in resolving this question, since the wavelengths are considerably different from those measured in the optical spectrum.

Red shifts have been measured in the optical spectrum that range up to $Z = 0.4$ for clusters of galaxies. Emission-line red shifts for some of the quasars have been measured that are as large as 2.2; it is possible that even greater values of Z will be found. As an illustration of just how large this is, consider Lyman alpha radiation. First-excited-state hydrogen atoms normally emit at 1216 Å when they jump down to the ground state. Radiation of these wavelengths is in the hard vacuum ultraviolet; Lyman alpha from space cannot normally be observed at sea level or even from near the top of the earth's atmosphere. However, if we are looking at a source where $Z = 2.2$, Lyman alpha is observed at 3891 Å (in the blue!). This wavelength certainly can be detected at sea level.

Reference has already been made to the statistical nature of the red shift. Figure 16.6 shows work on 18 clusters of galaxies; the ordinate is proportional to the apparent magnitude of the cluster, and the abscissa is the measured

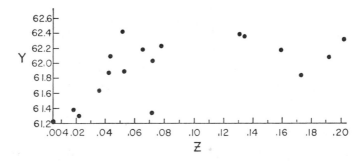

Figure 16.6   Red shift diagram for 18 clusters of galaxies. The parameter Y is defined to be the apparent magnitude minus five times the quantity cZ, where c is the speed of light and Z, the red shift, is $\Delta\lambda/\lambda$. From M. L. Humason, N. U. Mayall, and A. R. Sandage, *Astronomical Journal*, 1956.

red shift. In this diagram the so-called "*K*-correction" has been applied to the apparent magnitudes.

The *K*-correction attempts to correct for the spectral response of the detectors, such as photographic plates. A red shift may tend to remove the radiant energy from the wavelength "window" of the detector; the red shift makes us overestimate the apparent magnitude. The corrected apparent magnitude, m_c, is given by $m_c = m - K$.

Another correction, called the aperture effect correction, must be applied to the observed magnitudes. The total magnitude of a galaxy is difficult to measure because of the large angular size of the regions contributing to the total light. Because the observed magnitudes depend on the aperture used in the photometer, a systematic error in the magnitudes is introduced that depends on the galactic diameter itself. The correction factor is a function of the ratio of the aperture used relative to the angular diameter of the galaxy

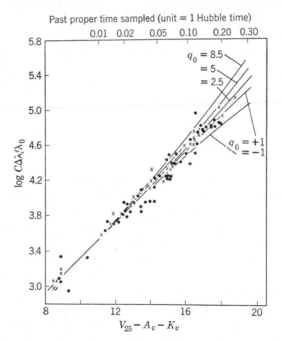

Figure 16.7 Combined data for radiogalaxies and first-ranked cluster members: x, brightest cluster galaxy; ●, radiogalaxies. Several theoretical relations for different values of q_0 are shown. From A. R. Sandage, in *High-Energy Astrophysics* (L. Gratton, Ed.) Copyright Academic Press, Inc., 1966.

or cluster; it has been chosen so as to correct the observed magnitudes to the light contained within a certain isophote. The standard isophote chosen is that point in the object that has a radial distance from the center or nucleus of 2.5 times the maximum radius visible on 48-inch Schmidt plates.

At any event, it is clear that the statistics are not very good. It is possible to find a correlation between the apparent magnitude and Z, but the form of the relationship is by no means precisely known.

Further work has combined optical and radio data. Figure 16.7 illustrates the observed red shifts. Plotted is the correlation between red shift and the apparent visual magnitude corrected for aperture effect to an isophote of ~25 magnitudes per square second of arc, visual absorption by material within our own galaxy, and the K-effect. Curves of the various values of the deceleration parameter q_0 (see Section 16.14) are shown for comparison.

16.8   Interpretations of the Red Shift

Many interpretations of the red shift are possible. To repeat, the observational facts are that the apparent magnitude of a galaxy is related to the red shift of its spectral lines, and the red shift also appears to be isotropic; red shifts of galaxies of the same apparent magnitude are the same, regardless of the directions to these galaxies.

One possibility is that the galaxies are at relative rest, but their light has become "tired" during its journey through space, so that the wavelength is increased—energy has been lost. Among the energy degradation mechanisms that have been mentioned in this connection are possible photon-neutrino interactions and photon-photon interactions. Neither interaction has ever been observed in the laboratory. Presumably, however, many such scatterings could occur over vast distances.

It is by no means obvious what the cross-sections for such reactions might be. It is doubtful that we could even see remote galaxies at all, much less the occurrence of dispersionless energy loss, should light interact with neutrinos of various energies. Alternatively, if photon-photon scattering is responsible for the red shift, one might expect to observe striking effects in eclipsing binary stars. These are not observed.

Another possibility is that the red shift is gravitational in origin. General relativity tells us, and experiment has confirmed, that masses can red-shift light; light has to do work in escaping from a gravitational "well" such as a star and is therefore reddened. Experimental confirmation of the gravitational red shift, has been provided by a Mossbauer measurement of the energy of gamma ray photons that are emitted and travel vertically upward through a distance of about 30 meters on the earth's surface, so that they must expend work against the planetary gravitational field. The energy of each such photon is found to be (minutely) less at the upper detector than it was at the source below.

If the cosmological red shift were to have a gravitational origin, however, the isotropic nature of the red shift would make it necessary to assume that our galaxy lies in a gravitational "hole" in the universe. Furthermore, we would be forced to conclude that as we receded from our galaxy, increasingly intense gravitational fields would be encountered. We find the belief that our position in the universe is at all unique to be rather repugnant. Also, the gravitational "explanation" would merely replace one mystery with another.

The most natural explanation of the red shift involves the ideas of relative velocity and the Doppler effect. In this interpretation the red shift is due to the galaxies receding from each other; the galaxies with the faintest apparent magnitudes are receding from us at the highest speeds.

We are thus led to the notion (through this one observational fact) of an expanding universe. There are no other observations that indicate relative motions of galaxies.

The nature of the dependence of the recessional velocity V on the distance r, however, is by no means clear, partially because of the poor statistics. The simplest relationship is a linear dependence of V on r. As we shall see, other possibilities exist.

16.9   Introduction to General Relativity

It is necessary to introduce the ideas of the theory of general relativity at this stage, as we attempt a more refined interpretation of the red shift. The reasons for the introduction of relativity are twofold.

First, if we assume a linear dependence of V on r, such that $V = Hr$, where H is a constant, we find that the inferred values of V can become quite large. If Z is about 0.5, for example, it turns out that V/c is also about $\frac{1}{2}$, and we know that classical mechanics breaks down when the speed of a body approaches that of light. Also, we know that intense gravitational fields are encountered in the universe and that classical mechanics has difficulties in such accelerated frames of reference; even special relativity must be discarded in favor of the general theory in such situations.

The absolute distance and absolute time of classical physics are discarded in relativity and replaced by a combination of these two factors. By absolute, we mean a quantity that has the same value in all reference frames. In relativity theory the "absolute" quantity is the separation of an event from any one of its neighbors; it is called the *interval ds* between them. The interval combines the space and time separations of the two events. Normally in classical physics we would write for the square of the separation between two points that

$$ds^2 = (dx^1)^2 + (dx^2)^2 + (dx^3)^2, \tag{16.5}$$

where in Cartesian coordinates we would identify x^1 with x, x^2 with y, and x^3 with z. But we also have a fourth dimension, time. In relativity we may define $x^4 \equiv ict$.

We must also allow for the fact (see Section 16.1) that we may have to make use of a geometry that is not Euclidean. Evidently, therefore, we cannot use in a general theory the simple extension of (16.5) that is suggested by our definition of x^4. Several complications are introduced by the possibility of various geometries.

Instead of simply summing the squares of the differentials, we must also consider the various products of the differentials. In other words, in addition to $(dx^1)^2$, $(dx^2)^2$, and so on, we must also consider $dx^1\, dx^2$, $dx^1\, dx^3$, and so on. Furthermore, in order to allow for non-Euclidean geometry, we have to introduce in front of each such product a function of the coordinates, $g_{\mu\nu}$, that specifies the geometry. This g-tensor is called the *metrical tensor*; the components $g\mu\nu$ (called the *coefficients of the metric*) are components of this tensor. For example, if space-time geometry really were Euclidean, we would have $g_{\mu\nu} = 1$ (when $\mu = \nu$) and $g_{\mu\nu} = 0 (\mu \neq \nu)$.

The product of the differentials $dx^1\, dx^3$ is multiplied by $2g_{13}$. The factor 2 arises because we assume that the coefficients of the metric are symmetric; we assume that $g_{\mu\nu} = g_{\nu\mu}$.

Our expression for the interval becomes

$$ds^2 = \sum_{\mu=1}^{4} \sum_{\nu=1}^{4} g_{\mu\nu}\, dx^\mu\, dx^\nu. \tag{16.6}$$

Some people call ds^2 the metric of space and time. In some other reference frame α, then, we may write the metric of space and time as

$$ds^2 = \sum_{\mu=1}^{4} \sum_{\nu=1}^{4} g_{\mu\nu}\, d\alpha^\mu\, d\alpha^\nu,$$

where the $g_{\mu\nu}$ are now functions of the new coordinates (α^1, α^2, α^3, α^4) but the numerical value of ds^2 is unchanged. To return to our Euclidean example, the expression (16.6) for the interval becomes

$$ds^2 = (dx^1)^2 + (dx^2)^2 + (dx^3)^2 - c^2 t^2. \tag{16.7}$$

This Euclidean expression for a line element is the one employed in the special theory of relativity.

The coefficients $g_{\mu\nu}$ are fundamental to relativity. As we shall see in the next section, it is possible to relate a tensor $T_{\mu\nu}$, called the *stress-energy tensor*, to $g_{\mu\nu}$. The tensor $T_{\mu\nu}$ describes such properties as the density, state of stress, and velocity of matter. Physical parameters are related to the (geometrical) metric. Thus, in a sense, the geometry of space-time arises from the distribution of matter in the universe.

16.10   The Einstein Tensor

It is possible to form a tensor, the so-called Ricci tensor $T_{\mu\nu}$, from $g_{\mu\nu}$, and the first and second derivatives of $g_{\mu\nu}$ with respect to the coordinates. The Ricci tensor is also symmetric; there are 10 functions $R_{\mu\nu}$ of the coordinates. We may also combine the metrical and Ricci tensors in a particular manner to form an invariant tensor. This invariant tensor is called the *scalar curvature R**.

It is useful next to consider the rate of change of these various tensors in space and time. It is possible to construct a combination of the tensors that has a zero rate of change; this is a property that is invariant, for it does not depend on the coordinate system. This invariant "combination tensor" is called the *Einstein tensor $E_{\mu\nu}$*, where

$$E_{\mu\nu} = R_{\mu\nu} - \tfrac{1}{2}g_{\mu\nu}(R^* - 2\lambda). \tag{16.8}$$

The Einstein tensor is again symmetric; $E_{\mu\nu} = E_{\nu\mu}$, so that there are 10 symmetric functions of the coordinates that comprise the Einstein tensor. In the case of Euclidean geometry (special relativity) it can be shown that both $R_{\mu\nu}$ and R^* are identically zero.

The vectorial divergence of $E_{\mu\nu}$ can be shown to be zero; this constitutes a mathematical proof that it is an invariant tensor. Let us merely say that the constant λ arises from the mathematical operations involved in forming the vectorial divergence. Hence λ rather resembles a constant of integration. The constant λ is usually called the *cosmical constant*. It may be positive, negative, or zero.

We now inquire whether other tensors have vectorial divergences equal to zero. If so, it may be that they are related to the Einstein tensor, which—so far—deals solely with geometry. There is such a tensor. It is the stress-energy tensor $T_{\mu\nu}$. As we mentioned in the last section, $T_{\mu\nu}$ describes the state of a distribution of matter in space.

We now assume that $E_{\mu\nu}$ and $T_{\mu\nu}$—both invariants and hence obeying "conservation laws"—are proportional to each other. The constant of proportionality is $-8\pi G$, where G is the "universal" gravitational constant.

Relating the two tensors in this way, we have what are known as Einstein's 10 field equations; they are

$$R_{\mu\nu} - \tfrac{1}{2}g_{\mu\nu}(R^* - 2\lambda) = -8\pi G T_{\mu\nu}. \tag{16.9}$$

These famous equations (16.9) imply that there is a connection between the metric of space and time and the distribution of matter.

It is interesting that the case of special relativity ($R_{\mu\nu} = R^* = 0$) indicates that the terms in λ and G must be negligibly small, since both the metrical and stress-energy tensors are non-zero. Hence special relativity will be useful only when both the "cosmic λ-force" and the gravitational force are negligible. This is indeed the case: special relativity has its most important applications is atomic physics, where electromagnetic effects are dominant.

16.11   Geodesics and the Bending of Light Rays

In plane geometry we can define a straight line as that distance between any two points which has a stationary value as compared with the distance measured along any nearby curve passing through the same two points. Similarly, we can define a *geodesic* by substituting the word "interval" for "distance" in the above sentence. The geodesic is a fundamental quantity. In special relativity the interval ds vanishes for a photon that moves through a distance $[(dx^1)^2 + (dx^2)^2 + (dx^3)^2]^{1/2}$ in time dx^4, because it moves with speed c. Since special relativity is only a "special case" of general relativity, the interval for a photon must also be zero in general relativity; it will be zero in the underlying space-time. A moving particle traces out a curve called the *world line* in space-time; it is the locus of the four-dimensional points successively occupied by the particle. Hence the world line of a photon is a *null geodesic*; the interval is zero.

The null geodesic equations involve the $g_{\mu\nu}$ coefficients. These, as we have seen from (16.9), depend on the gravitational and other forces taken into account in Einstein's equations. Thus we conclude that the path taken by light photons depends on the gravitational fields through which the light passes.

It is frequently said that relativity theory ascribes gravitational forces to geometry, not to a potential. Gravitational forces are due to the curvature of space, not to the derivative of a scalar potential field.

That the path taken by light depends on gravitational fields is a radically new conclusion of general relativity. In classical physics light moves in straight lines, not along curved paths. It is possible to test this conclusion observationally. The celestial coordinates of a star may be measured, and then measured again when the light from the star grazes the sun. Obviously this means that earthbound observers must wait until a visible solar eclipse occurs before the second measurement may be made. Accurate measurements may possibly be made in the future at radio wavelengths as the sun occults a spacecraft, but the effects of the corona on the transmission path will be difficult to sort out.

The optical measurements are very difficult, but they have been made. Apparent stellar deflections of 1.75 ± 0.19 seconds of arc have been measured; the error quoted is ± 1 standard deviation. The general theory predicts $(1.751/n)$ arc-seconds, where n is a number that ranges from 1 to 10, and $n = 1$ corresponds to light rays that just graze the sun. We therefore see that it is indeed possible that space-time is curved rather than Euclidean.

The curvature of space-time is perhaps most easily thought of with the aid of the "balloon analogy." Consider two-dimensional beings who live in two-dimensional galaxies in a universe that is curved in the third dimension. Beings such as these could only move from one place to another *in the surface* of their universe, which *we* can see to be balloon-shaped. To them, however, their universe is finite but unbounded. That is, they can move in it forever, but they cannot get infinitely far distant from their starting point.

Our universe may well be rather like this. We live in a four-dimensional hyperspace that, as we have just seen, may be curved in a fifth or higher dimension. Our universe too may be finite but unbounded; it may be possible to return to one's starting place by constantly moving forward. In principle, one could test this notion with a distinctive bright object such as a particular quasar. The radio and light from the quasar should come not only "directly" to the observer, but also to the observer the "long way around" through our universe. Thus the observer should be able to see the same quasar in front of him and also directly behind him, albeit considerably fainter in the latter case.

Another observational "proof" usually cited for the general theory concerns the advance of the perihelion point of Mercury's orbit. As we saw earlier, however, the recent observation that the sun may not be perfectly round means that the precession of Mercury's perihelion may be due to "causes" other than the curvature of space-time by a stellar mass. If the sun were to have a rapidly rotating core, as required by Dicke's explanation of Mercury's motion, the core would probably be flattened significantly. An oblate inner core would produce a slight oblateness in the photospheric layer. It is by no means completely proven, therefore, that general relativity is more than an interesting mathematical exercise.

There are cosmological theories, such as the one due to Brans and Dicke, that ascribe gravitation to scalar *and* tensor fields, unlike relativity (which ascribes it to geometry), or classical physics, which ascribes it to a scalar potential. One consequence of such theories is that the gravitational constant G may be changing with time. The expected rate of change is approximately 1 part in 10^{11} per year and is such that G was larger in the past than it is today.

If $\dot{G}/G \sim 10^{-11}$, several possibly detectable effects may result. As G decreases, the planets will steadily slow in their orbits about the sun. Planetary radar scanning may help to solve this problem by searching for such orbital changes. An upper limit of 5×10^{-10} has already been set by such radars.

Another effect is that the moon will recede from the earth in a predictable fashion. It is interesting in this connection to explore historical records of the time and locations of ancient eclipses. It may prove possible to measure the earth-moon distance to centimeter accuracy with lasers, and compare these results with the theory.

We noted in Chapter 12 that the sun is believed to have been less luminous in the past, as a consequence of theories of stellar evolution. However, the luminosity depends strongly on G; approximately a G^6 dependence is indicated. If G actually was greater in the remote past, the earth's temperature may well have been roughly constant in the past, rather than colder (as we indicated in Chapter 12), again assuming that other parameters, such as the greenhouse effect, have remained constant.

16.12   The λ-Force

Since general relativity theory has been confirmed at least partially by observation (through the bending of light rays by the sun), it is of interest to inquire as to other consequences of the theory. We have noted that the cosmic constant λ appears in (16.4). This constant arises during a proof that the vectorial divergence of $E_{\mu\nu}$ is zero. The constant λ has the dimensions of (a length)$^{-2}$.

We saw before that the value—indeed the sign—of λ remained open. If λ is negative, a repulsive force is present that drives all particles of matter away from one another. This deduction is arrived at from considering that a negative λ tends to reduce the effects of $R_{\mu\nu}$, which we know leads to gravitational attraction. If λ is positive, the force is attractive and acts in the same manner as, and in addition to, gravity. In any event, it must be so small that is is not detectable in the laboratory or even in the solar system; it can play a significant role only when the entire universe is considered.

It is clear that the λ-force will be important for cosmology, if it is indeed different from zero. We have mentioned the expanding universe idea previously. If the cosmic force is repulsive, remote galaxies will be subjected to a smaller gravitational pull than we would expect; their recessional speeds might in fact increase as $r \to \infty$. If λ is positive, the retarding force on galaxies that are flying away from us will be increased over that expected from gravity alone; the expansion might even come to a halt.

The nature of this force (if it exists) is presently a mystery. There have been many speculations about the cosmic force. For example, if the electric charge of the proton differs by only 1 part in 10^{19} from the charge of the electron (apart from the difference in sign), electrostatic repulsion will set in that is sufficiently great to provide the "observed" expansion of the universe. Laboratory measurements are now being attempted to measure, say, the charge on the neutron; is it identically zero?

16.13   The Expanding Universe and its Variants

The universe may be expanding at a uniform rate, or it may be accelerating or decelerating. For that matter, it may even be pulsating. All these ideas assume that it is indeed evolving in time; that is, the universe was different in the past from its present configuration. Evolutionary universes such as these are all possible and, in principle, accurate data on the red shift and distance could help in deciding among them. It is also possible that the universe is not evolving; this "steady-state" cosmology was quite popular until recently.

In the steady-state cosmology it is postulated that the universe is not only the same everywhere (an idea called the *cosmological principle*) but it is also the same at all times for all observers. One consequence of the steady-state idea is that the total amount of matter in the universe is not a constant but increases; creation takes place today such that one neutron per 1000 cm³ per 5×10^{11} years is created. The neutrons decay into "hot" (i.e., $E \lesssim 750$ keV) electrons.

It appears, however, that the observed intensity of the apparently diffuse (and therefore presumably background) cosmic x-ray spectrum may be incompatible with the steady-state theory. Very little $\geqslant 750$ keV radiation is found. The detection, however, of a break in spectral slope at ~ 1 MeV may have a bearing on this.

Let us assume for now that the universe is in fact expanding. The question we ask concerns the nature of the expansion: Is it accelerating, decelerating, or what?

The red shift, at least in principle, can supply the answer. We wish to construct a diagram where the red shifts are plotted as a function of the apparent magnitude, and see whether these points can be fitted best by a straight line or by a curved line that bends either up or down. We cannot quite do this directly, however, because of the effects of obscuration and the red shift. That is, the K-correction (Section 16.6) must first be applied to the apparent magnitudes.

Once the K-correction is made, we see that there is a relationship, at least statistically, but the functional dependence at which we arrive seems to depend greatly on who is selecting the data.

We here introduce one of the most famous relations of cosmology. This is known as "Hubble's law." Hubble, starting in 1929, examined the speeds of some nearby nebulae and arrived at the result that *the recessional speeds of the nebulae are strictly proportional* (i.e., linearly related) *to their distances*. In terms of observables, this law may be expressed as

$$Hr = cZ, \qquad (16.10)$$

where r is the local distance to the nebula, c is the speed of light, and H is a constant known as Hubble's constant. A currently accepted value for H is 100 km per second-megaparsec. It should be noted explicitly that we have assumed the red shift Z to be due to a Doppler shift (Section 16.7).

We see that large values of Z will result in very high recessional speeds. The speeds will approach that of light, and a relativistic correction to (16.10) will become necessary. If the correction is not made, values of $Z > 1$ imply $v > c$; and $Z = 2.2$ has been measured in at least one case. Relativity tells us that the wavelengths λ_0 are shifted at speeds v that approach the speed of light to a value λ, where

$$\lambda = \frac{\lambda_0(1 \pm v/c)}{(1 - v^2/c^2)^{1/2}}.$$
(16.11)

For recessional motion the plus sign is employed and the red shift $Z = [(\lambda - \lambda_0)/\lambda_0]$ becomes

$$Z = \frac{\Delta\lambda}{\lambda} = \frac{1 + v/c}{(1 - v^2/c^2)^{1/2}} - 1.$$
(16.12)

Optical and ultraviolet radiation can be shifted into the radio and infrared regions of the spectrum, and beyond. The relativistic form of Z (16.12) should be used in calculating the cosmological distances in the Hubble formula (16.10).

16.14 Models of the Evolving Universe

Hubble's constant H can be interpreted in several ways. If the universe initially started expanding from a point, H^{-1} represents the "age of the universe." If H has not changed from its present value H_0, this age is about 10^{10} years.

It is now necessary to pick a particular metric or, equivalently, a geometry. Thus we "build" a model universe. If we identify (t, r, θ, φ) with (x^4, x^1, x^2, x^3), the interval we use is given by

$$ds^2 = c^2\, dt^2 - R^2(t)\left\{\frac{dr^2 + r^2(d\theta^2 + \sin^2\theta\, d\varphi)}{1 - kr^2}\right\},$$
(16.13)

where the space-time curvature is determined by both $R(t)$ and the constant k, while the curvature of space alone is essentially given by just k. (Thus space could be Euclidean, but space-time could be curved.) The constant k may have any one of three values—$+1$, 0, or -1—depending on whether space is respectively closed, Euclidean and open, or hyperbolic and open.

One may now use the metric incorporated into (16.13) in the field equations (16.9). When this is done two differential equations result:

$$\frac{\dot{R}^2}{R^2} + \frac{2\ddot{R}}{R} + \frac{8\pi GP}{c^2} = -\frac{kc^2}{R^2} + \lambda c^2$$
(16.14)

and

$$\frac{\dot{R}^2}{R^2} - \frac{8\pi G\rho}{3} = -\frac{kc^2}{R^2} + \frac{\lambda c^2}{3}.$$
(16.15)

Equation 16.15 shows how the matter and energy density ρ of the matter in the universe depend on time, whereas (16.14) gives the time-dependence of the isotropic hydrodynamic pressure of matter, P. The solution of these two differential equations yields information on the geometry and history of the universe.

We may now introduce the present value of the Hubble constant, H_0, into these relations. We define H_0 through

$$H_0 \equiv \frac{\dot{R}_0}{R_0}. \tag{16.16}$$

Some people call R_0 the present-day "radius of the universe." It is also useful to define another quantity q; q is called the "acceleration (or deceleration) parameter" and provides a measure of how fast the speed of expansion is changing. The present value of q, q_0, is given by

$$q_0 = \frac{-\ddot{R}_0}{R_0 H_0{}^2}. \tag{16.17}$$

It is interesting to see how Hubble's relation (16.10) may be derived from considerations like this. The treatment that follows is due to Sandage. It has been shown that the photometric distance D, in all so-called Friedmann universes" (those where the cosmological constant is taken as zero), is given by

$$D = \frac{c}{H_0 q_0{}^2} \{q_0 Z + (q_0 - 1)[(1 + 2q_0 Z)^{\frac{1}{2}} - 1]\}. \tag{16.18}$$

[This result may be arrived at by considering (16.15), with $\lambda = 0$. We set $\rho = M/(4\pi R^3/3)$, where M is the mass of the universe, and find that $\dot{R}^2 = [2GM/R] - kc^2$. We desire to find r in Robertson's relation (16.2). The radial co-moving coordinate may be obtained from

$$r(R) = \int_R^{R_0} \frac{dR}{R\dot{R}}; \qquad r(T) = \int_{t_0-T}^{t_0} \frac{dt}{R(t)},$$

together with

$$Z = \frac{R(t_0)}{R(t_0 - T)} - 1 = \frac{R}{R_0} - 1.$$

We substitute the expression $(2GM/R - kc^2)^{\frac{1}{2}}$ for $\dot{R}$ and perform the indicated integrations, substituting $(Z + 1)$ for R/R_0. Equation 16.18 follows when the results of the integrations are simplified.]

In the (nonevolving) steady-state cosmology $q_0 = -1$; the photometric distance is given by

$$D = \frac{cZ(1 + Z)}{H_0}.$$

As we shall see later, however, $q_0 = +1$ is a good approximation to the available data. If $q_0 = +1$, (16.18) reduces to the usual form of Hubble's relation,

$$D = \frac{cZ}{H_0}$$

[see (16.10)].

The absolute luminosity L is given from (16.4) by

$$L = \frac{4\pi c^2 Z^2}{H_0^2} l \qquad (16.19)$$

when $q_0 = +1$. If the apparent luminosity of a distant galaxy is $l \times 10^{-16}$ watts meter^{-2} and if $H_0 = 100$ km per second-megaparsec,

$$L = 1.08 \times 10^{44} \, lZ^2 \text{[ergs per second]};$$

the largest Z-values to be found may depend on the power output of the sources.

If the present values of the density and pressure are ρ_0 and P_0, respectively, it is possible to rewrite (16.14) and (16.15) in terms of them as well as in terms of the observables H_0 and q_0. Let us assume that $\lambda = 0$ in the following. If we subtract (16.15) from (16.14), we find that

$$\frac{\ddot{R}_0}{R_0} + 4\pi G\left(\frac{\rho_0}{3} + \frac{P_0}{c^2}\right) = 0. \qquad (16.20)$$

Or, introducing q_0, (16.20) becomes

$$\rho_0 + \frac{3P_0}{c^2} = \frac{3H_0^2 q_0}{4\pi G}. \qquad (16.20a)$$

Inserting numerical values into (16.20a), we obtain

$$\rho_0 + \frac{3P_0}{c^2} = 3 \times 10^{-29} q_0 \left[\frac{\text{gm}}{\text{cm}^3}\right]. \qquad (16.20b)$$

In principle we can find a value of the parameter q_0 observationally from the slope of the apparent magnitude *vs* Z curve. If we are successful we can determine $\rho_0 + (3P_0/c^2)$.

We may rewrite (16.15), using the definition of H_0. This results in

$$\frac{kc^2}{R_0^2} = \frac{8\pi G\rho_0}{3} - H_0^2. \qquad (16.21)$$

Equation 16.20a may be combined with (16.21), yielding

$$\frac{kc^2}{R_0^2} = \frac{4\pi G\rho_0}{3q_0}\left[\rho_0(2q_0 - 1) - \frac{3P_0}{c^2}\right]. \qquad (16.22)$$

The spatial curvature of the universe, k/R_0^2, depends on P_0 and ρ_0. The geometry of space is determined by the energy content of the universe, as measured by the total density and pressure.

Now the total density ρ_0 is equal to the sum of the densities due to matter and to radiant energy. Thus

$$\rho_0 = (\rho_0)_{\text{matter}} + \frac{aT_0^4}{2}, \tag{16.23}$$

where a is Stefan's constant and T_0 is the present-day temperature of the universe.

Similarly, the hydrodynamic pressure P_0 depends on two factors. It is the sum of energy density and of the pressure due to the random radial motion (at a speed $\bar{v}$) of galaxies. We express P_0 as

$$P_0 = \frac{a}{3} T_0^4 + \rho_0 \bar{v}^2. \tag{16.24}$$

Equation 16.20a contains ρ_0 and P_0; rewritten, it becomes

$$(\rho_0)_m + \frac{2aT_0^4}{c^2} + \frac{3(\rho_0)_m \bar{v}_0^2}{c^2} = \frac{3H_0^2 q_0}{4\pi G}. \tag{16.25}$$

The present average density of the *visible* matter in the universe, $(\rho_0)_m$, is difficult to measure. One reason for the difficulty is the variation in mass-to-luminosity ratio, M/L, for different galactic types. (Expressed in solar units, the M/L ratio would be 1 for the Milky Way, if it consisted entirely of sunlike stars.) It turns out that $M/L \sim 100$ for ellipticals and is of the order of 20 for spiral galaxies, while it is near unity for irregulars.

As we look increasingly farther into space, we are also looking more at things as they *were*, not as they *are*, because of the fact that the speed of light is finite. Hence we are looking at the distribution of galaxies (and therefore of mass) as it was at different times in the past, not as it is now.

There is reason to believe that galaxies evolve in time, because stellar formation presumably takes place within them. Current astronomical opinion is that elliptical galaxies are older than spiral galaxies, because of the fact that the metal-deficient population II stars are found mainly in ellipticals (and in the nuclei of spirals), while the younger population I stars are found in the arms of spirals (see Figure 16.8) which also contain the dust and gases from which these sunlike stars may coalesce. Observations conducted in the ultraviolet region will help to check this idea; if ellipticals really are older they should be relatively faint at ultraviolet wavelengths compared to galaxies rich in young stars.

It is now thought that the evolutionary track for galaxies is such that ellipticals evolve from a globular, spherical shape toward a flattened,

Figure 16.8   The great galaxy in Andromeda is a spiral rather like our own Milky Way. The spiral arms are relatively rich in young blue population I stars; the nucleus contains population II stars. Courtesy Hale Observatories, California Institute of Technology.

elliptical one, since the latter look like the nuclei of spirals, and that spiral galaxies may then evolve from the ellipticals. If this analysis of galactic evolution is correct, determination of the mass density in the universe will be affected.

The determination relies on counts of galaxies and frequency of galactic types. In particular, as we have seen, ellipticals are underluminous; it is possible that the full number of ellipticals and hence the mass due to them has not been detected at great distances. The galactic counts indicate an average mass density of $\sim 1 \times 10^{-31}$ gm cm^{-3}, but if we take (M/L) into account it is believed that the average mass density due to *visible* matter (i.e., galaxies) is about 5×10^{-31} gm cm^{-3}.

Returning now to our equations, if we take $T_0 \sim 3°$K, we find $[aT_0^4/3]$ to be two orders of magnitude smaller than $(\rho_0)_m$. The random speed of galaxies appears to be less than 300 km per second; if so, the term $3(\rho_0)_m \bar{v}_0^2/c^2$ is also negligible compared with $(\rho_0)_m$. For these reasons we set

$$P_0 = 0.$$

Setting $P_0 = 0$ permits a considerable simplification of (16.20a), (16.20b), and (16.22). This tells us that

$$\rho_0 = \frac{3H_0^2 q_0}{4\pi G} = 3 \times 10^{-29} q_0 [\text{gm cm}^{-3}] \tag{16.26}$$

and

$$\frac{kc^2}{R_0^2} = H_0^2(2q_0 - 1). \tag{16.27}$$

Looking at (16.26), we see that if, as present observations suggest, $q_0 \sim +1$, the matter density in a closed universe, where radiation is negligible and the pressure of matter is zero, is nearly two orders of magnitude greater than observed (assuming, as before, that the cosmic constant $\lambda = 0$).

The value $q_0 = +0.5$ separates the Friedmann models of the universe into the oscillating and ever-expanding cases. Equation 16.27 shows that the three-dimensional subspace of the four-dimensional space-time continuum is Euclidean if $q_0 = +0.5$. Larger values of q_0 represent spaces of constant positive curvature, positive curvature of finite volume.

An interesting result emerges from the Friedmann modification of (16.15). If in this equation we multiply through by R^2 we obtain

$$\dot{R}^2 = \frac{\text{constant}}{R} - kc^2,$$

since the mean density ρ is of the order M/R^3, where M is the total mass of the universe.

The left-hand side is always positive; it is the square of a number. For $k = +1$ (a closed universe), however, the right-hand side can be positive only if its first term always remains larger than its second term. This means that R cannot exceed a certain limiting value, a value controlled by the amount of matter in the universe.

This means that a closed expanding universe must stop expanding after it has reached a critical size. Presumably it begins to contract again. It may continue to contract until it reaches a state of maximum density, after which it will once again expand. Thus a universe of this sort would be an oscillating or pulsating universe.

A further discussion of the consequences of the shape of space is outside of the scope of this book. Many theoretical universes exist, such as those where $\lambda \neq 0$ and where different metrics are chosen.

16.15 Tests of Relativistic Cosmological Models

Classically the most important test to be applied concerns the Hubble law. We need to refine the observational situation on the Hubble relation. The observational situation (as of 1967) has been summarized by Sandage, who treated the red shifts of quasars and of radiogalaxies separately.

The use of radio data has provided some new information on the expanding universe. It appears that the best value of the present-day deceleration parameter is

$$q_0 = +1.65 \pm 0.3.$$

The quantity q_0 is more than 9 times the probable error times the value of $q_0 = -1$ that would be required by the steady-state cosmology.

Our galaxy, however, is apparently one of a group of as many as 35 or so galaxies that are gravitationally linked to form a cluster known as the "local cluster" of galaxies. Figure 16.9 shows one cluster, in the constellation Hercules.

The precise number of galaxies in the local cluster is difficult to estimate because of optical obscuration by matter within our own galaxy. The red

Figure 16.9 Gravity operates over large distances, even as large as those between galaxies. Consequently clusters of galaxies form, such as this one (photographed with the 200-inch telescope) in the constellation Hercules. Courtesy Hale Observatories, California Institute of Technology.

shift would not be expected to operate at such "small" distances—that is, within the cluster—because of the linkage. The data used to estimate q_0 are based on other clusters that are bright enough to be seen.

It now seems that the local cluster may well be one member of a "local *supercluster*," the center of which is thought to be near the Virgo cluster. If such a supercluster exists, a correction must be made for that linkage; it is believed that such a correction would reduce q_0 from $+1.65$ but still may leave it positive.

As noted previously, a value of $q = +1.65$ would mean that space is positively curved with a finite volume. Equation 16.27 may be used to estimate the radius R_0 of such a universe. The relation (16.26) suggests that the density of matter and energy in such a universe is $\sim 5 \times 10^{-29}$ gm cm^{-3}. We saw in the last section that this is two orders of magnitude greater than the observed mass per unit volume.

It is interesting to speculate where the "missing" mass density might be found. One choice that many have eagerly seized upon is the energy of neutrinos. This has at least the virtue of being easy; neutrinos hide well and could be responsible for the mass energy thought to be missing.

Another possibility is fireball radiation. Does the energy density contained within it approach $c^2 \times 10^{-29}$ gm cm^{-3}? This possibility, too, is an attractive one for "closing the universe," since, as we have seen, matter (or energy) densities this high are required for finite spaces.

A very natural place to look for matter is intergalactic space. Although no obscuration is obvious, it may be that sufficient mass is present in this huge volume of space (and at kinetic temperatures of less than 10^{6}°K to satisfy the x-ray observations) to account for the missing mass.

If objects with very large mass-to-luminosity ratios exist in the universe, this matter may also be hidden. As we have seen (Section 16.7), gravity can red-shift light. Indeed, if an object of mass M is sufficiently small in size that its radius R is equal to $2GM/c^2$, the acceleration of gravity is so enormous that light cannot escape from it. At this "Schwarzchild limit," except for gravitational interactions, it would become "hidden mass." In the framework of cosmology, the local space closes in on itself, trapping the light within the volume and forming a "black hole" in space. An object that falls into a black hole (if they exist) will inevitably be crushed to indefinitely high density and lose its separate identity as indefinitely great tidal forces rip it "apart."

There are other data that bear on the evolution of the universe. Does G change with time? If so, and if G was larger in the past, gravitational collapse may have taken place more rapidly in objects at great distances; distant objects may be bright because $\dot{G} \neq 0$. In an expanding universe, does G depend on the distribution of matter? For that matter, does the value of e^2 depend on time, so that the charge on the electron or proton is not a constant?

We know that galaxies evolve; stars are born and die within them. Thus the stellar content of galaxies is variable in time; this will affect the luminosity data from which q_0 is derived. If q_0 is "actually" $+0.5$, a figure arrived at by allowing for galactic evolution, which the 1.65 figure did not, then it is possible than an oscillating universe exists, rather than an ever-expanding one. Hence more work is clearly called for on the evolution of galaxies.

16.16   Conclusion

We live in an exciting era. For the first time man can directly explore the universe around him. Experiments conducted from within the earth's atmosphere and in near-earth space have revealed much about the universe that surrounds us. It appears that the same physical laws appear to operate throughout space as are found to be valid on earth. Of course, we must always be prepared for surprises; the earth and its nearby space occupy only a minute fraction of the total volume.

One reason for our excitement is the rapid development of technology. Technology permits us to conduct many new observations from the earth, and spacecraft—at least those in earth orbit—are becoming commonplace. Craft have landed on the moon and also on Venus, and manned excursions to the planets may follow shortly. It is to be hoped that our understanding of the universe will increase at an ever greater rate.

Problems

16.1. Show that the energy of radiation E emitted by atoms in the gravitational field of a star of mass M and radius R is just

$$E = E_0 \left[1 - \frac{GM}{c^2 R} \right],$$

where E_0 is the energy carried by photons from unaccelerated atoms. (This problem is most simply solved by thinking about clock retardation on a test object freely falling toward the star and applying the formulae of special relativity.)

16.2. The red shift of a quasar is measured to be $Z = 2.1$. At what wavelength will Hα be observed? Is this wavelength detectable at sea level?

16.3. (a) Assume that a satellite is in a Keplerian orbit about a massive body with an average angular speed n radians per second. If the gravitational constant G is changing at a rate $\dot{G}$, what is the corresponding value of $\dot{n}$?

(b) If $\dot{G}/G = 3 \times 10^{-11}$, what would be the change in period of pulsar CP1919, if its radiation is indeed controlled by the orbital period of a binary system (or, for that matter, the rotation period of a single massive body)?

16.4. In the Einstein–de Sitter universe, $R = R_0(t/t_0)^{2/3}$ and $k = \lambda = 0$, where R_0 is the value of R at the present moment t_0. It also turns out that $q_0 = \frac{3}{2}$. What is the present-day matter density in this universe? How much of the material present in the universe has therefore been identified by astronomers?

16.5. Assume that all galaxies have equal luminosity Lg and that they are evenly distributed at rest in space with uniform density Ng. If $Lg = 10^{10} L_\odot$ and $Ng = 10^{-21}$ (light-year)$^{-3}$, what must the minimum radius of the universe be for the energy flux from other galaxies to equal the solar energy flux at the earth?

16.6. If the relativistic Hubble relation is

$$Hr = \frac{c + v}{(1 - v^2/c^2)^{1/2}} - c,$$

at what distance will a galaxy appear to rotate half as fast as it does in its own rest frame?

16.7. Suppose that you lived before the invention of the telescope, and that a well-known scientist of the time stated that the universe must be uniform, infinitely old, infinitely large, and completely static. Since you would not know about nucleosynthesis and stellar evolution, you could not obtain the age of the earth or the stars. How would you argue from simple visual observations that the scientist's hypothesis cannot be completely valid. (Support your argument mathematically!)

16.8. Assume that we live in a closed, finite, pulsating universe. From a thermo-dynamic point of view, the final state of one cycle must be the same as the initial state of the next in all respects. Yet we can see evolution during a cycle which does not seem to be symmetric about the time of maximum expansion (e.g., nucleosynthesis and stellar evolution, gravitational collapse). How would you explain the return to the initial state? That is, how do elemental abundances return to their initial values, and so on?

16.9. If stars are uniformly distributed in space with density N_0 (stars per A.U.^{3}) and have no relative motion, and if they all are equally luminous, with luminosity L_0 and radius R_0, how far out must we look (in astronomical units) to receive a uniform flux at the earth equal to the solar flux at the earth? In other words, how far out must we be able to see in order that the night sky is uniformly as bright as the surface of the sun? Assume perfect inverse square flux law and Euclidean space.

Appendix A

THE ORBITAL ELEMENTS

There are seven quantities that specify the position in space of an object moving under the influence of gravity. Should noncentral forces, such as atmosphere drag, be present, the orbital elements will change with time. It then is necessary to specify the rates of change of the elements and the time at which they were measured, in order to be able to predict the orbit at later times.

The seven orbital elements are the following (Figure A.1):

$$a = \text{semimajor axis,}$$
$$e = \text{eccentricity,}$$
$$i = \text{inclination,}$$
$$\Omega = \text{longitude of ascending node,}$$
$$\omega = \text{argument of perihelion,}$$
$$P = \text{sidereal period,}$$
$$T = \text{time of perihelion (or perigee) passage.}$$

The size of the orbit is given by a. When the orbit is not closed, the distance of closest approach q to the primary massive body (the sun, say) is used instead of a. Thus a body in a hyperbolic orbit around the sun will have the orbit described in terms of perihelion distance q.

The eccentricity defines the ellipticity of the orbit, as seen in Chapter 2. For example, it is zero for circular orbits and unity for parabolic orbits.

The inclination, i, is the angle between the orbital plane and a reference plane. In the description of objects within the solar system, an appropriate reference plane is the ecliptic. Since the ecliptic is also the orbital plane of the earth, the value of i for the earth's orbit is zero. For a satellite in orbit about the earth, an appropriate reference plane would be the equatorial plane of the earth.

The longitude of the ascending node, Ω, together with i, defines the position of the orbital plane in space. Ω is the angle measured eastward (in the reference plane) from the direction of the vernal equinox to the direction of the ascending node.

498

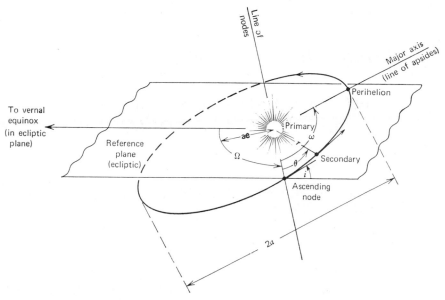

Figure A.1   Some of the orbital elements, as applied to interplanetary space. Two elements not shown in the figure are P, the sidereal period, and T, the time of perihelion passage. The angles θ and f, where $f = \theta - \omega$, are not orbital elements but are noted here because they are used in calculations of the type described in Appendix B.

We must still describe the *orientation* of the major axis *in* the orbital plane. This is accomplished through the use of the element ω. It is measured from the "line of nodes" (the line where the orbital and reference planes intersect) to perihelion (or perigee). More precisely, ω is measured from the ascending node to the major axis at point of closest approach.

It still remains for us to locate the object within its orbit. The remaining two orbital elements serve the purpose. The sidereal period, P, is the time required for the object to pass across a line from the massive primary (the sun, say) to a point in the "fixed" star field (the vernal equinox, for example). To locate how far around in the orbit the object has gone at a given time, we also specify T, the time of closest approach. It serves the purpose of a time reference.

Appendix B

ORBIT COMPUTATION AND COORDINATE TRANSFORMATIONS

The equation of an orbit of a celestial object or spacecraft was given in Chapter 2 as

$$r = \frac{a(1 - e)^2}{1 + e \cos(\theta - \omega)}.$$

(2.25)

Let us denote the angle $(\theta - \omega)$ by the letter f; this angle has come to be called the *true anomaly*. Another angle, E, is also useful in orbit problems; it is called the *eccentric anomaly*. Like f, it is measured from perihelion, but, unlike f, it is measured to a point on the circle whose radius is the semi-major axis of the elliptical orbit. Finally, a third angle is useful, the so-called *mean anomaly M*, where

$$M = \frac{2\pi}{P}(t - T).$$

We now desire to calculate r and f as functions of time. It is also desirable to convert from $r(t)$ and $f(t)$ to $\alpha(t)$ and $\delta(t)$ so that the object may be tracked by earth-based telescopes or antennae.

Once the eccentric anomaly is known as a function of time, f may be computed from it through the relation

$$\tan\left(\frac{f}{2}\right) = \left(\frac{1 + e}{1 - e}\right)^{1/2} \tan\left(\frac{E}{2}\right).$$

Now E may in turn be found, if M is known; the two are related through "Kepler's equation,"

$$M = E - e \sin E.$$

The mean anomaly itself is determined by Kepler's third law. That law yields

$$M = G(M_s + m)^{1/2} a^{-3/2}(t - T),$$

where the symbols have the usual meanings (see Appendix A) but the M_s on the right-hand side represents mass of the primary.

500

The eccentric anomaly, E, poses some difficulty for calculation, because of the nature of Kepler's equation. If the eccentricity, e, is small (less than 0.2, say), the following technique provides a satisfactory approximation.

(a) Compute

$$E_0 = M + e \sin M + \frac{e^2}{2} \sin 2M.$$

(b) Compute

$$M_0 = E_0 - e \sin E_0.$$

(c) Also compute

$$\Delta E_0 = \frac{M - M_0}{1 - e \cos E_0}.$$

Now, to a better approximation, (d) compute

$$E_1 = E_0 + \Delta E_0.$$

(e) Compute

$$M_1 = E_1 - e \sin E_1$$

and also (f)

$$\Delta E_1 = \frac{M - M_1}{1 - e \cos E_1}.$$

Keep continuing this iterative process until the value of E that is obtained satisfies Kepler's equation to the desired accuracy. Use this value of E to compute f, as shown previously, and also to compute r, from

$$r = a(1 - e \cos E).$$

Should the eccentricity of the orbit be near unity, special methods are required to compute the orbit. These methods are discussed in texts on celestial mechanics.

It now remains to convert from (r, f) to (α, δ) coordinates. It will also be necessary to know the orbital elements, for the position of the orbital plane in space will affect the values of the equatorial coordinates.

If the xy-plane is the reference plane (the plane of the ecliptic, say), we may denote the heliocentric coordinates of the object by l and b. We then have that

$$x = r \cos b \cos l,$$
$$y = r \cos b \sin l,$$
$$z = r \sin b.$$

In another, primed system, where the x'-axis lies along the line of nodes,

$$x' = r \cos b \cos (l - \Omega) = r \cos (f + \omega),$$
$$y' = r \cos b \sin (l - \Omega) = r \sin (f + \omega) \cos i,$$
$$z' = r \sin b = r \sin (f + \omega) \sin i.$$

A knowledge of the quantities f, ω, Ω, and i thus permits us to calculate the heliocentric longitude l and the heliocentric latitude b.

We may next transform from l and b to α_1 and δ_1, the equatorial coordinates of the spacecraft as seen from the sun. This may be accomplished through the use of some spherical trigonometry. The results of this transformation are that

$$\sin \delta_1 = \cos \epsilon \sin b + \sin \epsilon \cos b \cos l,$$

$$\sin \alpha_1 \cos \delta_1 = -\sin \epsilon \sin b + \cos \epsilon \cos b \sin l,$$

$$\cos \alpha_1 \cos \delta_1 = \cos b \cos l,$$

where ϵ, the obliquity of the ecliptic, is $23° 27'$.

Finally, for earthbound observers, it is necessary to convert from α_1 and δ_1 to α and δ, the coordinates as seen from the earth. For this transformation we need the geocentric rectangular coordinates of the sun (X, Y, Z), which are listed in the *American Ephemeris* for the time of observation. If ρ is the radius vector of the spacecraft from the earth,

$$\rho^2 = \xi^2 + \eta^2 + L^2,$$

while, as before, if r is the radius vector of the craft from the sun, $r^2 = x^2 + y^2 + z^2$. Here the $\xi\eta$ plane is the plane of the earth's equator; primed coordinates use the sun as the origin and unprimed coordinates refer to the earth as the origin. Relative to the *sun*,

$$\xi' = r \cos \delta_1 \cos \alpha_1,$$

$$\eta' = r \cos \delta_1 \sin \alpha_1,$$

$$L' = r \sin \delta_1.$$

But relative to the *earth*,

$$\xi = \xi' + x = \rho \cos \delta \cos \alpha,$$

$$\eta = \eta' + y = \rho \cos \delta \sin \alpha,$$

$$L = L' + z = \rho \sin \delta.$$

Hence

$$\tan \alpha = \frac{\eta}{\xi}$$

and

$$\sin \delta = L(\xi^2 + \eta^2 + L^2)^{-1/2},$$

the desired quantities. Of course, each tracking station must still convert from α to δ to local angles, such as elevation and azimuth, in order to track the spacecraft. This conversion requires knowledge of the station's longitude and latitude.

Appendix C

PARTICLE PHYSICS

Many elementary particles are now known to physics. The more important of these (for our purposes) are listed in Table C.1, along with a few of their properties. The electrical charge of each particle in the table is $\pm 1e$; each particle carries a charge of 1.6×10^{-19} coulomb. The mass is expressed in terms of the rest mass of the electron (9×10^{-28} gm) or, more conveniently, in terms of the rest mass energy E_0, where $E_0 = m_0 c^2$ and m_0 is the rest mass.

C.1 Particle Motion

We may describe the speed of a free particle in terms of β, where $\beta = v/c$; v is the velocity of the particle and c is the speed of light. Quantities related to β are the following:

(a) Momentum, pc:

$$pc = \frac{m_0 c^2 \beta}{\sqrt{1 - \beta^2}} .$$

(The momentum is a three-dimensional vector; its magnitude is pc.)
(b) Kinetic energy, T:

$$T = \frac{m_0 c^2}{\sqrt{1 - \beta^2}} - m_0 c^2.$$

(c) Total energy, E:

$$E = \sqrt{(pc)^2 + (m_0 c^2)^2} = \frac{m_0 c^2}{\sqrt{1 - \beta^2}} = T + E_0.$$

There are transformation rules for momentum and energy. Thus the quantity $(p_x c, p_y c, p_z c, E)$ transforms just as the four-dimensional space-time vector (x, y, x, ict) does, from one moving reference system to another.

That is, if there is relative motion βc along the x-axis of two systems that are in relative motion in *only* the x-direction (the primed system is taken to be

TABLE C.1. TABLE OF FUNDAMENTAL PARTICLES

Particle	Rest Mass	Mean Lifetime and Decay Modes	Interactions and Production
Neutrino, ν, $\bar{\nu}$	0	$\infty \cdots$	Inverse beta decay, $\sigma = 10^{-48}$ m^2 $= 10^3$ light-years of Fe
Photon, γ	$\leqslant 4 \times 10^{-51}$ kg	$\infty \cdots$	Electromagnetic
Electron, $e^{\pm}$	9.1×10^{-31} kg $= 0.511$ MeV	$e^+ + e^- \rightarrow 2h\nu$ in 6×10^{-9} sec	Electromagnetic, 1.6×10^{-19} coulomb
Muon, $\mu^{\pm}$	$206.8 m_e$ 105.7 MeV	$\mu^{\pm} \rightarrow e^{\pm} + \nu + \bar{\nu}$ in 2.22×10^{-6} sec	{Electromagnetic {Beta decay
Pion, $\pi^{\pm}$	$273.2 m_e$ 139.6 MeV	$\begin{cases} \pi^{\pm} \rightarrow \mu^{\pm} + \nu \text{ in } 2.54 \times 10^{-8} \text{ sec} \\ \rightarrow e^{\pm} + \nu (10^{-2}\%) \end{cases}$	Electromagnetic Nuclear (geometrical σ) $\sigma = 0.05\, A^{2/3} \times 10^{-24}$ cm^2 Beta decay
π^0	135.1 MeV	$\begin{cases} \pi^0 \rightarrow 2\gamma \text{ in } \sim 10^{-15} \text{ sec} \\ \rightarrow e^+ + e^- + \gamma \end{cases}$	Decays before interaction
Heavier mesons		Decay to π, μ, e, γ	Electromagnetic, nuclear
Proton, p^+	$1836.1 m_e$ 938.2 MeV	$\infty \cdots$	{Electromagnetic {Nuclear
Neutron, n^0	$1838.6 m_e$ 939.5 MeV	$n \rightarrow p + e^- + \nu + 0.783$ MeV in 1.1×10^3 sec	{Nuclear {Beta decay
Hyperons	$> M_{\text{nucl}}$	Decay to hyperons, n, p, mesons	{Electromagnetic {Nuclear

moving in the plus x-direction), we have that

$$p_x'c = \frac{1}{\sqrt{1-\beta^2}}\,(p_x c - \beta E),$$

$$p_y'c = p_y c,$$

$$p_z'c = p_z c,$$

$$E' = \frac{1}{\sqrt{1-\beta^2}}\,(E - \beta c p_x).$$

The quantity $[E^2 - (pc)^2]$ is an invariant; it is $m_0 c^2$.

It is also true that in any given coordinate system the vector sum of the momenta of several particles is constant for any interactions of the particles, as is the sum of the total energies.

We should also note that the motion of an unstable particle affects its apparent lifetime. If a particle with a mean life *at rest* ($\equiv \tau_0$) moves with a velocity βc, its observed mean lifetime will be

$$\tau = \frac{\tau_0}{\sqrt{1-\beta^2}}.$$

The mean distance l_D that this unstable particle will travel before decay is

$$l_D = \beta c \tau = \frac{\tau_0 p}{m_0}.$$

C.2   Electromagnetic Interactions

The different particles experience different forces. We say that they undergo different interactions. For example, pions and nucleons have *strong interactions*. These nuclear interactions are responsible for holding nucleons together in nuclei, and the quanta associated with the potentials from which these forces are derived are associated with the unstable particles.

Weak interactions, on the other hand, account for the beta decay of nuclei. They also account for the muon decay of π-mesons.

Gravitational forces probably occur between all particles, but are sufficiently weak that they may be ignored here. Finally, there are the electromagnetic interactions; these occur between all electrically charged particles and are perhaps the most familiar. We shall discuss separately those associated with charged particles and those associated with photons.

Particles

Classically the equation of motion in a magnetic field **B** and electric field **E** for a particle of charge q is

$$\frac{d}{dt}(pc) = q(c\mathbf{E} + \mathbf{v} \times \mathbf{B}).$$

If **B** is uniform, the particle moves in a circle perpendicular to **B**; the radius of curvature ρ of the circle is

$$\rho = \frac{pc}{zeB} \text{ [centimeters]}$$

when pc is expressed in ergs, the charge ze ($z \equiv$ atomic number) is expressed in electrostatic units, and B is in gauss. This is more conveniently rewritten as

$$\rho = \frac{pc}{300zeB} \text{ [centimeters]};$$

(pc/z) is now expressed in volts and B is still in terms of gauss. The quantity pc/ze is frequently denoted as the *rigidity* of the particle.

In passing through ordinary matter, charged particles lose energy to the atomic electrons through the so-called *Coulomb* interactions, which are just electrostatic interactions. This is the dominant source of continuous energy loss by particles heavier than electrons. It is dT/dx, where

$$\frac{dT}{dx} = \frac{4\pi e^4 z^2}{m_e c^2 \beta^2} N_0 \left(\frac{z_1}{A}\right) \left[\ln \frac{2m_e c^2 \beta^2}{I(1-\beta^2)} - \beta^2\right]\left[\frac{\text{eV}}{\text{gm/cm}^2}\right].$$

In this expression for dT/dx, N_0 is Avogadro's number ($N_0 = 6.03 \times 10^{23}$), z_1 is the atomic number of the target matter and A is its atomic weight, m_e is the rest mass of the electron, and $I = kz$ is the average ionization potential of the matter. A reasonable approximation to k is 11 eV. The quantity in brackets is called the *stopping power* per (atomic) electron of the matter.

The expression for the Coulomb loss is valid for all speeds βc that are greater than the (classical) orbital velocity of the target atomic electrons and for all speeds that are less than the extremely relativistic. In practice, electron capture and other effects modify the loss rate by heavy particles at low speeds.

Figure C.1 shows the proton energy loss in air. Electron capture by protons (forming neutral hydrogen atoms, which are not ionizing entities) becomes important at $T \lesssim 100$ keV. It is important for alpha particles at higher energies.

One may integrate the energy loss equation above. This leads to a definite *range R* as a function of initial energy, although experimental data are required for accurate work. Semiempirically the proton range R_P is found to be

$$R_P = 100\left(\frac{T}{9.3}\right)^{1.8} \text{ [cm of STP air]}$$

when T is expressed in million-electron volts.

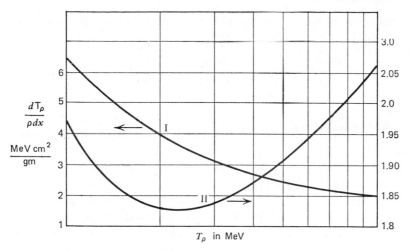

$\frac{dT_\rho}{\rho\,dx}$

$\frac{MeV\ cm^2}{gm}$

T_ρ in MeV

Scale I: 10^2 MeV $\leq T_\rho \leq 10^3$ MeV

Scale II: 10^3 MeV $\leq T_\rho \leq 10^4$ MeV

Figure C.1 Proton energy loss in air, taken from work by M. Rich and R. Madey, Report No. UCRL-2301, United States Atomic Energy Commission, 1954.

Notice that in general for any particle

$$R = \frac{m_0}{z^2}\ F(\text{initial velocity});$$

for a given initial velocity the range is inversely proportional to the square of the atomic number of the particle.

The passage of a charged particle leaves a track of excited atoms, molecules, ions, and electrons behind it in the medium. Also deposited in the medium are those faster electrons sometimes called *delta rays*.

It is an interesting and useful fact that in many cases the energy loss to form one ion pair does not depend on the nature of the incident particle. Some values are: air, 34 eV per ion pair; argon, 26 eV per ion pair; silicon, 3.5 eV per ion pair. It is roughly true that $\sim$30 eV are given up in most gases of practical interest in forming each ion pair (ion plus electron).

An interesting phenomenon occurs in transparent materials when $\beta c > c/n$, where n is the index of refraction of visible light in the medium. Part of the energy transferred to electrons of distant atoms appears in the special form of Cerenkov radiation instead of as ionization. This light is emitted in a cone of half-angle θ about the track of the particle, where

$$\cos \theta = \frac{1}{\beta n}$$

and where θ is measured from the forward direction of the particle (see Figure C.2). The energy lost as Cerenkov light is

$$\left(\frac{dT}{ds}\right)_{\text{Cerenkov}} = \frac{4\pi^2 e^2 z^2}{c^2} \int_{\nu_1}^{\nu_2} \left(1 - \frac{1}{\beta^2 n^2}\right) \nu \, d\nu,$$

where the integration is carried out over all frequencies ν of the emitted light for which $\beta c > c/n$. To cite a practical case, a relativistic particle in a material such as Lucite ($\beta n \sim 1.5$) produces ~ 200 quanta per cm with wavelengths between 4000 and 8000 Å. The particle loses about 1 keV per cm in the form of Cerenkov light. We should note that both heavy particles and electrons produce Cerenkov light.

The mechanisms by which electrons lose energy to matter are similar to those operative for heavier particles, but are modified by the incident electron's ability to lose *all* its energy to a single atomic electron. The result is

$$\frac{dT}{dx} = \frac{4\pi e^4}{m_e c^2 \beta^2} N_0 \left(\frac{z_1}{2A}\right) \left[\ln \frac{m_e c^2 \beta^2 T}{I(1 - \beta^2)} - \beta^2\right] \left[\frac{\text{eV}}{\text{gm/cm}^2}\right]$$

for an incident electron whose kinetic energy is T.

Unlike heavier particles, electrons are scattered and also lose energy by bremsstrahlung. Elastic multiple scattering reduces the macroscopic range by a factor that varies from 1.2 to 4, depending on the energy of the electron and on the atomic number z_1 of the target. We shall return later to a discussion of the scattering cross-section.

Bremsstrahlung (the German word for "deceleration radiation") is that radiation emitted by charged particles when colliding with other particles, so that inelastic scattering occurs. This mode of energy loss is proportional to the (deceleration)^2, which in turn is proportional to m_0^{-2}; only electrons therefore lose appreciable energy to bremsstrahlung.

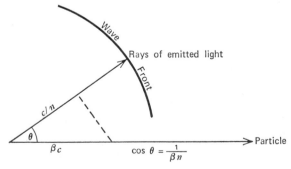

Figure C.2   Emission of Cerenkov light by charged particles.

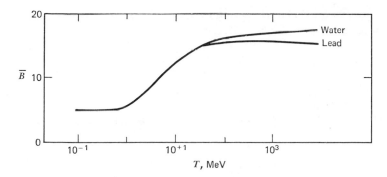

Figure C.3 The parameter $\bar{B}$ as a function of electron energy.

The average rate of energy loss to bremsstrahlung by an electron of energy T that collides with atoms of charge $z_1 e$ is

$$\left(\frac{dT}{ds}\right)_{\text{rad}} = N_0 \frac{z_1{}^2}{A} (T + m_e c^2) \sigma_0 \bar{B} \left[\frac{\text{ergs}}{\text{gm/cm}^2}\right],$$

where $\sigma_0 = (2\pi e^2/hc)(e^2/m_e c^2)^2 = 5.8 \times 10^{-28}$ cm^2 per nucleus and where $\bar{B}$ is a slowly varying function of the energy, as shown in Figure C.3.

Since $z/A \sim$ constant, $(dT/ds)_{\text{rad}}$ increases linearly with total energy and with the atomic number z_1 of the target. We may call the quantity $\sigma_{\text{rad}} \equiv \sigma_0 \bar{B} z_1{}^2$ the total radiative cross-section; at very high energies

$$\sigma_{\text{rad}} = 4 z_1{}^2 \sigma_0 \left[\ln \frac{183}{z_1{}^{1/3}} + \frac{1}{18}\right].$$

In this process photons are emitted with all energies up to T; the energy per unit frequency interval is approximately constant for $0 \leqslant h\nu \leqslant T$.

We now compare the cross-sections for the various processes. For *non-relativistic* electrons we have

$$\text{ionization cross-section} = \sigma_{\text{ion}} = \frac{2 z_1}{\beta^4} \ln \left(\frac{T\sqrt{2}}{I}\right)\left[\frac{\text{barns}}{\text{atom}}\right]$$

$$= \frac{1}{NT}\left(\frac{dT}{ds}\right)_{\text{ion}},$$

where N is the number of atoms per cubic centimeter and where 1 *barn* is defined to be 1×10^{-24} cm^2. Now the cross-section for elastic nuclear back-scattering (i.e., scattering by 180°) is σ_{nucl}, where

$$\sigma_{\text{nucl}} \simeq \frac{1}{4} \frac{z_1{}^2}{\beta^4}\left[\frac{\text{barns}}{\text{atom}}\right].$$

Elastic scattering (for angles greater than $\sim45°$) from atomic electrons has a cross-section σ_{elas}:

$$\sigma_{elas} \simeq \frac{2z_1}{\beta^4}\left[\frac{\text{barns}}{\text{atom}}\right].$$

Finally, the bremsstrahlung cross-section σ'_{rad}, expressed in terms of the fraction of kinetic energy, is

$$\sigma'_{rad} = \frac{T + m_e c^2}{T}\,\sigma_{rad} \simeq \frac{8}{3\pi}\cdot\frac{1}{137}\frac{z_1^2}{\beta^2}\left[\frac{\text{barns}}{\text{atom}}\right].$$

The general form of the ionization and bremsstrahlung cross-section (here not limited to nonrelativistic electrons) may be multiplied by N_0/A to give the energy loss per gram per square centimeter. The results are shown in Figure C.4 for energies up to ~5 MeV.

It is useful to calculate the ratio of radiation to ionization loss. For a particle of rest mass m_0 we find

$$\frac{(dT/ds)_{rad}}{(dT/ds)_{ion}} \simeq z_1\left(\frac{m_e}{m_0}\right)^2\frac{T}{1600 m_e c^2}.$$

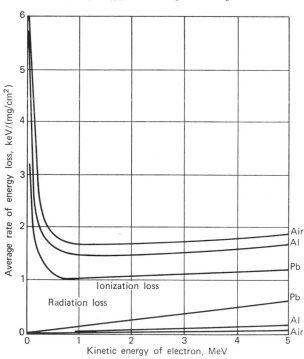

Figure C.4   Electron energy loss rate in various materials as a function of energy.

The ratio is unity at $T = 9$ MeV for electrons in lead, while it is unity for 100-MeV electrons in water. Ionization is more probable than radiation, at low energies.

The reader is cautioned that the energy loss rates in Fig. C.4 cannot be integrated to obtain electron range, because of scattering and because of fluctuations in bremsstrahlung loss. It is possible to calculate the bremsstrahlung emitted when electrons strike an absorber sufficiently thick to stop them completely. The photon energy spectrum from nonrelativistic electrons is

$$\frac{dI}{dv} = \text{constant} \times z_1(v_{\max} - v_{\min})\left[\frac{\text{ergs}}{\text{frequency interval}}\right],$$

where $hv_{\max} = T_{\text{initial}}$. This has been found to agree with experiment. The total energy emitted as bremsstrahlung is

$$I \simeq (0.7 \pm 0.2) \times 10^{-3} z_1 T^2 \left[\frac{\text{MeV}}{\text{incident electron}}\right],$$

where T, the initial electron energy, is expressed in million-electron volts. This result for I is valid when $0 \leqslant T \leqslant 2.5$.

There is a formula for the *practical range* of electrons, which has been found empirically. It is

$$R_{\text{PR}} = 530T - 106 [\text{mg/cm}^2]$$

when T (expressed in million-electron volts) is between 1 and 20. It is seen that the practical range in grams per square centimeter is roughly half the kinetic energy expressed in million-electron volts.

All the above applies equally well to positrons as well as electrons. Positrons, however, may also annihilate; $e^+ + e^- \rightarrow 2\gamma$, where each gamma photon (assuming $T_{e+} = T_{e-} = 0$) has $m_e c^2 = 0.51$ MeV of energy. The two photons move away from the annihilation event in opposite directions.

Photons

Photons passing through matter lose most of their energy in one of three processes, the dominant one being determined by the photon energy and the atomic number of the target material. These processes and the energy range over which they predominate are photoelectric absorption ($0 \leqslant hv \leqslant 0.1$ MeV), Compton scattering (~ 0.1 MeV $\leqslant hv \leqslant 3$ MeV), and pair production ($hv \geqslant 3$ MeV). In addition the processes of nuclear photodisintegration ($hv \geqslant 8$ MeV) and photoproduction of mesons ($hv \geqslant 150$ MeV) are sometimes observed.

Photoelectric absorption occurs when a photon interacts with an electron bound in an atom and loses all its energy to that atomic electron; the photon is completely absorbed in the one interaction.

The kinetic energy of the ejected electron is

$$T = h\nu - B_e,$$

where $h\nu$ is the energy of the incident photon and B_e is the binding energy of the electron in the atom; T is linearly related to $h\nu$.

Momentum considerations require that the atom carry away some momentum from the interaction. As we shall see later, a nearby nucleus is also required (for the same reasons) in the pair production process.

Photoelectric cross-sections σ_τ depend sharply on $h\nu$ and Z_1. When the photon energy $h\nu$ is not near any electron-binding energy level, it is found that

$$\sigma_\tau = \text{constant} \times \frac{Z_1{}^n}{(h\nu)^\gamma}.$$

In this expression n ranges from 4.0 to 4.6 (a rough value commonly used is $n = 4.5$) and n varies from 3 down to about unity, as $h\nu$ increases from 0.1 to 3 MeV. The cross-section σ_τ changes discontinuously as an absorption edge is crossed (in energy), as shown in Figure C.5.

In *Compton scattering* a photon interacts with either a bound or unbound electron, producing a scattered photon of lower energy and an increased kinetic energy for the electron. The presence of the atom is irrelevant, since both momentum and energy may be conserved with the two emerging "particles" (i.e., photon and electron).

The conservation laws require that in this scattering the wavelength change $(\lambda' - \lambda)$ be given by

$$\lambda' - \lambda = \frac{h}{m_e c}(1 - \cos\theta),$$

where θ is the scattering angle for the photon, measured from the forward direction. If $\alpha \equiv h\nu/m_e c^2$, so that α is the energy in units of $m_e c^2$ then of the scattered photon, $h\nu'$, is related to the energy of the incident photon, $h\gamma$, by

$$h\nu' = h\nu\,\frac{1}{1 + \alpha(1 - \cos\theta)},$$

and the kinetic energy T of the electron is $T = h\nu - h\nu'$.

Notice that the maximum energy that can be transferred to the electron by the Compton process, $T_{\max}$, is

$$T_{\max} = h\nu\left(\frac{2\alpha}{1 + 2\alpha}\right);$$

it occurs during *back-scatter*, where $\theta = 180°$.

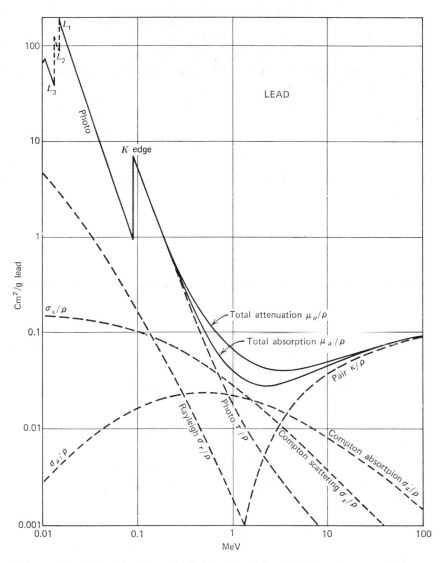

Figure C.5 Mass attenuation coefficients for photons in lead. The corresponding linear coefficients for lead may be obtained by using $\rho = 11.35$ gm/cm^3. Taken from R. D. Evans, *The Atomic Nucleus*, McGraw-Hill Book Co., New York, 1955.

The Compton cross-section, averaged over the polarization directions of the incoming photon, is

$$d\sigma_c = \frac{r_0^2}{2}\left(\frac{\nu'}{\nu}\right)^2\left[\frac{\nu}{\nu'} + \frac{\nu'}{\nu} - \sin^2\theta\right]d\Omega\left[\frac{cm^2}{electron}\right];$$

$d\sigma_c$ is the differential *collision* cross-section for scattering into a solid angle $d\Omega$ per target electron. Classically, r_0 is interpreted as the "radius of the electron"; $r_0 = e^2/m_e c^2 = 2.18 \times 10^{-13}$ cm. If I is the incident energy flux, in units of electron volts per square centimeter-second, the number of photons per second scattered into $d\Omega$ at an angle θ by 1 cm² of a target of thickness ΔS is just

$$\frac{I}{h\nu}(NZ, \Delta S)\, d\sigma_c.$$

We have assumed that ΔS is small, so that scattered photons do not suffer further interactions. The differential *scattering* cross-section, $d\sigma_s$, is defined by

$$I(NZ, \Delta S)\, d\sigma_s = \text{energy per second scattered into } d\Omega$$

$$= h\nu'\left[\frac{I}{h\nu}(NZ, \Delta S)\, d\sigma_c\right].$$

Hence

$$d\sigma_s = \frac{\nu'}{\nu}\, d\sigma_c.$$

These cross-sections may be integrated over $d\Omega$, taking account of the dependence of θ on ν', to yield the average *collision* cross-section, which for small α can be expressed as

$$\sigma_c \simeq \frac{8\pi}{3}r_0^2[1 - 2\alpha + 5.2\alpha^2 + \cdots]\,cm^2 \text{ per electron,}$$

$$\frac{1}{h\nu}(NZ, \Delta S)\sigma_c = \text{number of photons scattered per second}$$

$$= \text{number of electrons recoiling per second.}$$

The average scattering cross-section is similarly obtained by integration, and for small α equals

$$\sigma_s \simeq \frac{8\pi}{3}r_0^2[1 - 3\alpha + 9.4\alpha^2 + \cdots]\,cm^2 \text{ per electron.}$$

$I(NZ, \Delta S)\sigma_s = \text{energy in the form of photons scattered per second. Hence}$
the average energy of a scattered photon is

$$\overline{h\nu'} = \frac{\sigma_s}{\sigma_c}h\nu$$

and the average electron energy is

$$\bar{T} = h\nu - h\nu' = h\nu \left(\frac{1 - \sigma_s}{\sigma_c} \right).$$

Therefore $I(NZ, \Delta S)(\sigma_c - \sigma_s) = $ energy going to electrons per second, and it is convenient to define the absorption cross-section

$$\sigma_{\text{abs}} = \sigma_c - \sigma_s \quad \text{(small } \alpha\text{)}$$

$$\simeq \frac{8\pi}{3} r_0^2 [\alpha - 4.2\alpha^2 + \cdots] \text{ cm}^2 \text{ per electron.}$$

The scattered electrons and photons are generally projected forward, the direction being more strongly peaked as α increases. Notice that σ_{abs} is proportional to the energy deposited in a thin absorber, while σ_c is proportional to the number of photons removed from the incident beam. The actual reduction observed when the absorber is placed in the beam depends on the type of detector used.

When the energy of an incident photon $h\nu > 2m_ec^2$, it is possible for a photon to be absorbed in the Coulomb field of a nucleus and create an *electron-positron pair*. The nucleus recoils and absorbs some momentum, but essentially no energy, so that

$$T_{e^+} + T_{e^-} = h\nu - 2m_ec^2.$$

The distribution of energies of the e^+, say, is nearly flat from 0 to $h\nu - 2m_ec^2$, and the total cross section is

$$\kappa_a = \sigma_0 Z^2 \bar{P} \text{ cm}^2 \text{ per nucleus,}$$

where $\bar{P} \simeq 2\text{--}11$ for $h\nu = 5 - 1000$ MeV and drops to zero at lower energies. For very high energies

$$\bar{P} = \left[\frac{28}{9} \ln \left(\frac{183}{Z^{1/3}} \right) - \frac{2}{27} \right].$$

These cross-sections may be converted to attenuation lengths in grams per square centimeter as follows:

$$l = \frac{\rho}{N\tau + NZ\sigma_c + N\kappa} = \frac{A}{N_0 \tau + Z\sigma_c + \kappa} \frac{1}{} \text{ g per cm}^2.$$

The quantity $1/l$ in lead is plotted against $h\nu$ in Figure C.5.

One can deduce from the foregoing that when a high-energy ($\gg m_ec^2$) electron or photon falls on matter it will produce a cascade of $e^\pm$ and γ whose numbers multiply by alternate pair production and bremsstrahlung emission as the particle passes through the matter. Their number continues to increase with penetration, and the average energy per particle continues to decrease, until the electrons have so low an energy that they lose more energy by ionization than by radiation. Thereafter the cascade, or shower, is absorbed.

The multiplication occurs in a characteristic distance, which is

$$\sim \left(\frac{\sigma_0 Z^2 \bar{P} N_0}{A}\right)^{-1} \text{gm per cm}^2 \text{ for pair production,}$$

$$\sim \left(\frac{\sigma_0 Z^2 \bar{B} N_0}{A}\right)^{-1} \text{gm per cm}^2 \text{ for bremsstrahlung.}$$

In practice a single thickness or radiation length X_0 is defined by

$$\frac{1}{X_0} = 4\sigma_0 \left(\frac{N_0}{A}\right) Z(Z+1) \ln{(183 Z^{-\frac{1}{3}})} \text{ cm}^2 \text{ per gm,}$$

and is the distance in which the kinetic energy of an energetic electron or photon is reduced by the factor $1/e$. The energy below which multiplication is unimportant is called the *critical energy* T_c.

Material	X_0 (gm/cm^2)	T_c (MeV)
Air	37.7	84.2
Aluminum	24.5	48.8
Lead	6.5	7.8

If the initial particle has energy T_0, after traversing a thickness x gm/cm^2 there are $\sim 2^{x/(X_0 \ln 2)}$ particles ($\frac{1}{3}\gamma$, $\frac{2}{3}e^{\pm}$) of average energy $T_x = T_0/2^{x/(X_0 \ln 2)}$. The maximum number of particles occurs when $T_x = T$, and is equal to T_0/T_c. Detailed calculations have been made of both the longitudinal and lateral spread of these showers in the atmosphere.

Weak Interactions

The weak interaction forces couple the initial and final particles in beta decay, exemplified by pion, muon, and neutron decay and by the interaction of neutrinos with matter. These forces always involve some particles that do not experience nuclear interactions: electrons, muons, and neutrinos.

Neutrinos carry away, as kinetic energy, a significant amount of the energy released in beta decay. Once produced, they interact *very* weakly with matter. The interaction cross-section $\sigma_v = 10^{-44}$ cm^2 per nucleon, so that the mean free path of neutrinos in iron is 10^{22} gm per cm^2 or 1000 light-years of solid iron. Therefore the energy going into neutrinos is lost from the vicinity of the neutrinos' production.

Muon decay goes as
$$\mu^{\pm} \to e^{\pm} + \nu + \bar{\nu} + 105.2 \text{ MeV.}$$

The kinematics of three-body interaction allow electron energies of 0–53 MeV in the center of mass system. Most cosmic ray secondary muons decay in flight after traveling a considerable distance; for example, for a 10.3-BeV muon
$$\tau(\beta) = 2.16 \times 10^{-4} \text{ sec,}$$
$$l_d = 65.0 \text{ km,}$$

which is greater than the height of most of the atmosphere. Some muons are stopped by ionization loss and decay at rest. Negative muons are usually captured into Bohr orbits around a nucleus, and because of their large mass spend much time inside the nucleus. On the average a μ^- passes through 10^{17} gm per cm² of nuclear matter without interacting, but occasionally the *weak interaction*
$$\mu^- + p \to n + \nu$$

occurs. This is the inverse of neutron decay.

Charged pions decay according to the scheme
$$\pi^{\pm} \to \mu^{\pm} + \nu + 33.9 \text{ MeV,}$$

and the two-body decay fixes the muon energy at 4.1 MeV in the center of mass or zero-momentum (CM) system. For a 13-BeV pion
$$\tau(\beta) = 2.4 \times 10^{-6} \text{ sec,}$$
$$l_d(\beta) = 712 \text{ meters.}$$

Pions rarely stop before they decay. They suffer nuclear interaction with about the same probability as nucleons, and so an energetic charged pion may well have a nuclear interaction before decaying.

Neutral pions decay into two gamma rays so rapidly that nuclear interactions are unlikely.
$$\pi^{\circ} \to 2\gamma + 135.1 \text{ MeV.}$$
At 13 BeV a π^0 has
$$\tau(\beta) = 1.44 \times 10^{-13} \text{ sec,}$$
$$l_d(\beta) = 4.3 \times 10^{-3} \text{ cm.}$$

Strong Interactions

The pions and all heavier particles interact with strong (i.e., nuclear) forces that are essentially independent of electric charge except that at low energies Coulomb repulsion may prevent particles coming close enough to interact. At energies on the order of or less than the binding energy of nucleons in nuclei (8 MeV) nuclear reactions are extremely complex. At higher energies it is a pretty good approximation to assume that nucleons and nuclei interact if they come within a fixed distance, about equal to the sum of their classical radii ($R = 1.2 \times 10^{-13} A^{1/3}$) of each other, and otherwise not at all. Hence the cross-section for interaction between a pion or nucleon and a nucleus of weight A is

$$\sigma_{\text{nucl}} = 0.05A^{2/3} \text{ barns per nucleus,}$$

and the mean free path for nuclear interaction is

$$l_{\text{nucl}} \simeq 33A^{1/3} \text{ gm per cm}^2.$$

What happens depends on the kinetic energy of the collision. If the kinetic energy of the incident particle merely divides among the kinetic energies of the incident and target particles, the collision is elastic. In order for mesons to be created, the kinetic energy of the initial particles in the CM system must equal or exceed the rest mass energy of the mesons created. Thus the reaction

$$p + p \rightarrow p + n + \pi^+$$

requires $T_p \geqslant 293$ MeV in the laboratory system. If the target nucleon is bound in a nucleus, it has some kinetic energy of motion, and the required incident energy is somewhat less.

The probability of meson production increases slowly with increasing energy, but the inelasticity, defined as the fraction of kinetic energy in the CM system used for meson creation, does not equal $\frac{1}{2}$ until $T_p = 1.7$ BeV. With further increases of energy, the number of mesons increases slowly and the average energy per meson increases.

When an energetic particle strikes a C, N, or O nucleus of the atmosphere, or an Ag or Br nucleus in a photographic emulsion, particles emerge with various velocities, and these are categorized according to the density of the ionization tracks they make in the surrounding medium (grain density in an emulsion).

Particle	Ionization	Charge
Shower particles (thin tracks)	$<1.41 \times$ minimum	$E_p > 80$ MeV $E_\pi > 500$ MeV
Gray particles	1.4–$8 \times$ minimum	$80 > E_p > 25$ MeV 500 MeV $> E_\pi$
Black particles	$>8 \times$ minimum	25 MeV $> E_p$ 800 MeV $> E_\alpha$

It is found that about 80% of the shower particles are pions and the rest fast protons; the number of shower particles is denoted by n_s. The total number of

Figure C.6 Schematic appearance of a nuclear "star" in an emulsion viewed through a microscope. The moving particle approaches from the upper left and a shower of secondaries emerges from the interaction.

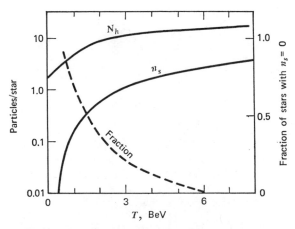

Figure C.7 Empirical nuclear—star parameters as a function of energy.

gray and black particles is denoted by N_h. The shower particles are projected forward in the direction of the incident particle, and are produced by multiple interactions of the incident particle and shower particles with nucleons inside the struck nucleus. The gray particles diverge more from the incident direction, while the black particles are emitted isotropically. The latter are "boiled off" from the struck nucleus as the energy left in the nucleus after the passage of the intranuclear cascade has been shared among the remaining nucleons. They are sometimes called evaporation particles. Figure C.6 shows the appearance of a nuclear star in an emulsion and Figure C.7 gives values of n_s and N_h produced by cosmic rays and machine-accelerated protons in an emulsion. Notice that π^+, π^-, and π^0 are produced in about equal numbers at the same energies as are p^+ and n^0, but that the neutral particles make no tracks.

Appendix D

ELECTRODYNAMIC ORBIT THEORY

The equations of motion in a dipolar magnetic field **B** are

$$\frac{d\mathbf{p}}{dt} = \frac{ze}{c} \mathbf{v} \times \mathbf{B} \tag{D.1}$$

and

$$\mathbf{B} = -M\nabla \frac{\cos \Theta}{r^2}, \tag{D.2}$$

where **p** is the momentum of a particle that has charge ze and velocity **v**. Θ is the colatitude; in the case of the near-earth environment, where the geomagnetic field is nearly dipolar, θ is measured from the north geomagnetic pole, and $M = 8.06 \times 10^{25}$ gauss per cm³ is the earth's magnetic moment. Since $|\mathbf{p}| = $ constant, the equation can be written

$$\frac{d}{dt}\frac{\mathbf{v}}{v} = \frac{Ze}{pc} M\mathbf{v} \times \mathbf{B}.$$

Let $ds = v\,dt$ be the element of arc length along the particle's trajectory, and define a unit of length C_{St}, called the *Störmer length*, by

$$C_{\text{St}}^2 = \frac{Ze}{pc} M \quad \text{or} \quad C_{\text{St}}^2 = \frac{300Z}{pc} M \tag{D.3}$$

when pc is expressed in electron volts. Then if all lengths are expressed in units of C_{St}, the equation of motion for any charged particle is

$$\frac{d^2}{ds^2}\mathbf{r} = \pm \left(\frac{d}{ds}\mathbf{r}\right) \times \frac{1}{r^3}(-\hat{\mathbf{r}}2\cos\theta - \hat{\boldsymbol{\theta}}\sin\theta), \tag{D.4}$$

where $\pm = $ sign of charge.

This cannot be integrated in terms of known functions, except in special cases such as motion in the equatorial plane, $\theta = \pi/2$. Planar motion can be expressed in terms of elliptic functions, and there is a circular orbit at $r = 1$

(Störmer). Energies of circular motion at two different distances from the earth's dipole are:

C_{St}	Pe/Z	T_{proton}
6378 km	59.6 BeV	59.6 BeV
10×6378	0.596 BeV	190 MeV

Positive particles move from east to west.

In general the equation of motion can be separated to describe motion in a meridian plane, and motion of the meridian plane about the dipole axis. Let w and z be rectangular coordinates in a moving meridian plane that contains the particle, and let α be the azimuthal angle of the meridian plane measured from some reference axis.

$$w = r \sin \theta = r \cos \lambda,$$

where $\lambda =$ latitude, and

$$z = r \cos \theta = r \sin \lambda.$$

Then

$$\frac{d}{ds} \alpha \equiv \alpha' = \frac{2\gamma}{w^2} + \frac{1}{r^3},$$

where γ is a second constant of the motion, related to the impact parameter. (The first constant is kinetic energy.)

Define

$$Q \equiv w'^2 + z'^2 = 1 - (w\alpha')^2 = 1 - \left(\frac{2\gamma}{w} + \frac{w}{r^3}\right)^2.$$

Then

$$z'' = \frac{1}{2}\frac{\partial}{\partial z} Q,$$

$$w'' = \frac{1}{2}\frac{\partial}{\partial w} Q.$$

The latter pair of equations will not be used here. However, a great deal of information can be extracted from the Q-equation by noting that if $\varphi =$ angle between the trajectory and a normal to the meridian plane ($\varphi = 0$ for W to E motion) then

$$Q = 1 - \cos^2 \varphi = \sin^2 \varphi.$$

Obviously regions in which $Q < 0$ are forbidden, and the boundaries of these regions, $Q = 0$, are turning points of the motion where $\varphi = 0$ or π. The boundaries of the totally forbidden regions, therefore, are obtained by setting

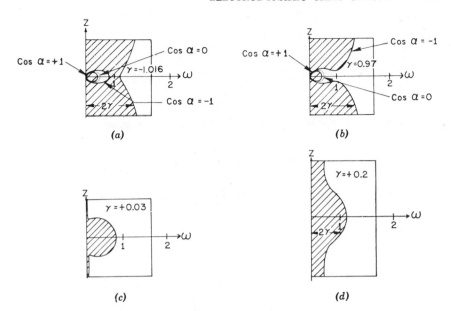

Figure D.1   Meridian plane, forbidden regions are shaded. Notice the "trapped particle" region in (a). Adapted from C. Stormer, *The Polar Aurora*, Oxford University Press, 1955.

$\cos \varphi = +1$ and -1 and solving the equation

$$\cos \varphi = \frac{2\gamma}{r \cos \lambda} + \frac{r \cos \lambda}{r^3} \qquad (D.5)$$

for r as a function of λ, with γ taking on various values. Figure D.1 shows the result, and in Figure D.2 are drawn some orbits in the equatorial plane for various values of γ. It is evident that $|2\gamma|$ equals the impact parameter; that is, the perpendicular distance between the asymptotic velocity and the dipole axis.

Now in the case of radiation that is isotropic and homogeneous at infinity, such as cosmic radiation, orbits with all values of γ occur. The problem of finding the cut-off rigidity at a given latitude and direction on the earth's surface is therefore reduced to finding the γ for which orbits reach a minimum distance from the dipole at the given λ and the value of φ corresponding to the given direction of approach. For all λ and φ this occurs when the "jaws" of the outer forbidden region have just opened and the inner forbidden region has a minimum extent.

This happens when $\gamma = -1 + \epsilon$, $1 \gg \epsilon > 0$. Hence the minimum value of r is found from (D.5) with $\gamma = -1$. The result is

$$\frac{1}{r_{\min}} = \frac{1}{\cos^2 \lambda} (1 \pm \sqrt{1 + \cos^3 \lambda \cos \alpha}), \qquad (D.6)$$

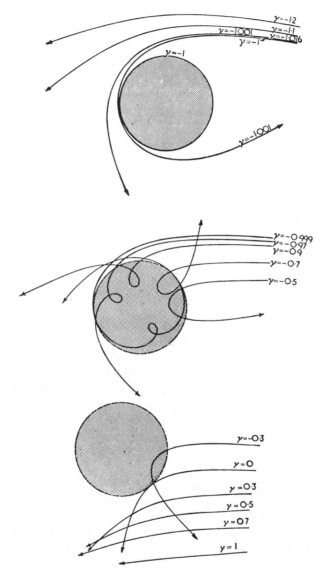

Figure D.2 Orbits in the equatorial plane of particles coming from infinity in a dipolar field. The shaded circle is the trajectory with radius unity. From C. Størmer, *The Polar Aurora*, Oxford University Press, 1955.

with the positive root required to make r a minimum. For the earth, $r_{min} C_{st} = r_e = 6378$ km. Substitution into (D.3) gives

$$\left(\frac{r_{min}}{r_e}\right)^2 300M = R_{min} = \frac{59.6 \cos^4 \lambda}{(1 + \sqrt{1 + \cos^3 \lambda \cos \alpha})^2} \text{ [BeV].} \qquad \text{(D.7)}$$

Thus the cutoff rigidity decreases to zero as $|\lambda| \to \pi/2$ and, at every λ, has a maximum for positive particles arriving from the east ($\varphi = \pi$) and a minimum for those coming from the west ($\varphi = 0$). The cut-off for vertical arrival is

$$p_m = \frac{59.6}{4} \cos^4 \lambda.$$

The cone of constant φ, corresponding to a particular p_{min} and λ, is called the Störmer cone. The theory of charged particle motion in a dipolar field was first developed by Störmer in 1934; Lemaitre and Vallarta went into much greater detail in their analysis 4 years later.

The Störmer cone may be represented by an orthogonal projection of its intersection with a unit sphere upon a horizontal plane (tangent to the Earth's surface). All directions east of the Störmer cone are strictly forbidden to positive particles with $R < R_{min}$, but not all directions to the west are necessarily allowed. This is the case because for every γ, such that $-1 < \gamma - 0.78856$, there are periodic orbits in the vicinity of the jaws. Consider a negative particle projected outward from the earth, which must follow the same orbit as a positive particle moving inward from infinity. For every λ there are directions asymptotic to periodic orbits (considering negative particles projected outward) without being re-entrant, and this set of directions is called the main cone. To the east of this are directions whose orbits (called re-entrant) approach the periodic orbit, then return to regions nearer the dipole, move outward again, and after repeating this one or more times are asymptotic to a periodic orbit. All such directions are forbidden to particles from infinity, but they form a set of measure zero; that is, they subtend an infinitely small solid angle. However, to the east of the cones of asymptotic orbits are directions whose orbits are re-entrant one or more times before escaping between the jaws to infinity. These directions are allowed, but in the presence of the solid earth one of the re-entrant orbits may intersect the earth so that the directions are shadowed. Such directions are called penumbral. In addition to these penumbral directions there are directions with small elevation above the horizon, for which an upward projected particle intersects the earth in its first loop about the earth's magnetic field, as shown in Figure D.3. These directions are called simple shadow cones. Figure D.4 shows the complete set of cones at a latitude of ~30°. Near the equator the penumbra is nearly opaque. At higher latitudes it becomes less

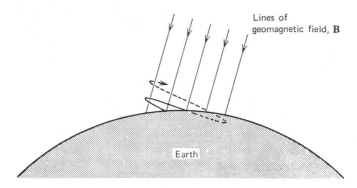

Figure D.3   The motion of a charged particle moving upward in the geomagnetic field. The altitude angle is so small that the trajectory intersects the earth in its first loop about **B**; these directions are called the simple shadow zones.

so. The simple shadow cones do not subtend much solid angle at any latitude.

Very exact calculation of Störmer orbits may not be very useful, because the geomagnetic field deviates from a dipole field. At geomagnetic latitudes $\lambda \leqslant 55°$ these deviations can be pretty well accounted for by using the eccentric dipole, or even better by calculating the Störmer cut-off that would be produced by a dipole located at the earth's center, or at the eccentric dipole location, but having the strength and direction required to produce the actual surface field at the point in question. By such approximations it is

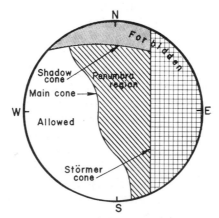

Figure D.4   Allowed cones at a magnetic latitude near 30°.

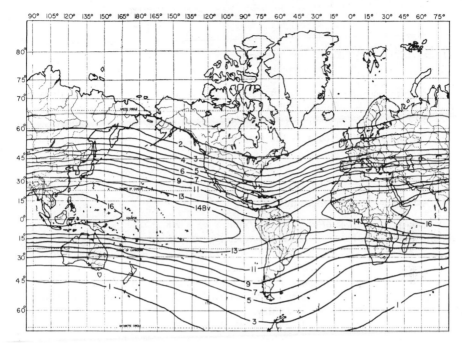

Figure D.5   Cosmic-ray cutoff rigidities at various positions on the earth. Adapted from M. Shea and D. Smart, *Journal of Geophysical Research*, 1965.

possible to fit the lines of constant cosmic ray flux at the earth's surface (or equivalently at the top of the atmosphere), the measured cut-off energies, and other data. For example, the line of minimum cosmic ray flux coincides more closely with the dip equator than with the geomagnetic equator. On this line the maximum cut-off occurs in the Indian Ocean. Tables of corrections to simple Störmer cut-offs have been published and the rigidities plotted on global maps (see Figure D.5).

At $\lambda \geqslant 55°$ the cut-offs are apparently considerably lower than are predicted by even modified Störmer theory, and may in fact be zero over a portion of the polar caps. This is thought to be due to the great distortion of the polar field lines, which are stretched some distance into the magnetospheric tail. The exact configuration of the polar lines, and the resulting cut-off rigidity, are not known at present.

BIBLIOGRAPHY

Books and journals that the author has found useful are listed below. A division into the areas of Astronomy and Astrophysics, Cosmology, Fields and Particles, and Planetary Atmospheres and Structure has been made. This distinction is not very meaningful in the case of the journals, however, since many of them carry articles in more than one area of space science.

BOOKS

Astronomy and Astrophysics

Abell, G. (1969) *Exploration of the universe*. New York: Holt.

Allen, C. W. (1963) *Astrophysical quantities*. London: University of London Press.

Aller, L. (1954) *Astrophysics*. (Vols. I and II.) New York: Ronald Press. *American Ephemeris and Nautical Almanac*. Washington, D. C.

Baker, R. H. (1964) *Astronomy*. (8th edition.) Princeton: Van Nostrand.

Blanco, V. M., and McCluskey, S. W. (1961) *Basic physics of the solar system*. Reading, Mass.: Addison-Wesley.

Brandt, J. C., and Hodge, P. W. (1964) *Solar system astrophysics*. New York: McGraw-Hill.

Chandrasekhar, S. (1957) *Stellar structure*. New York: Dover.

Doig, P. (1950) *A concise history of astronomy*. London: Chapman and Hall.

Gratton, L. (ed.) (1966) *High-energy astrophysics*. New York: Academic Press.

Hoyle, F. (1962) *Astronomy*. London: Crescent Books.

Hynek, J. A. (ed.) (1951) *Astrophysics*. New York: McGraw-Hill.

Motz, L., and Duveen, A. (1966) *Essentials of astronomy*. Belmont, Calif.: Wadsworth.

Moulton, F. R. (1928) *An introduction to celestial mechanics*. New York: Macmillan.

Norton, A. P. (1966) *A star atlas and reference handbook*. Cambridge, Mass.: Sky Publishing.

Schwarzschild, M. (1958) *Structure and evolution of stars*. Princeton: Princeton University Press.

Shklovsky, I. S. (1960) *Cosmic radio waves*. Cambridge, Mass.: Harvard University Press.

Shklovsky, I. S. (1968) *Supernovae*. New York: Wiley.

Steinberg, J. L., and Lequeux, J. (1963) *Radio astronomy*. New York: McGraw-Hill.

Cosmology

Bondi, H. (1960) *Cosmology*. Cambridge, England. Cambridge University Press.

McVittie, G. C. (1956) *General relativity and cosmology*. Urbana, Ill.: University of Illinois Press.

528

McVittie, G. C. (1961) *Fact and theory in cosmology.* New York: Macmillan.
Møller, C. (1952) *The theory of relativity.* Oxford, England: Clarendon Press.

Fields and Particles

Chamberlain, J. W. (1961) *Physics of the aurora and airglow.* New York: Academic Press.
Evans, R. D. (1955) *The atomic nucleus.* New York: McGraw-Hill.
Hess, W. N. (ed.) (1968) *Introduction to space science.* New York: Gordon Breach.
Hooper, J. E., and Scharff, M. (1958) *The cosmic radiation.* New York: Wiley.
Johnson, F. S. (ed.) (1965) *Satellite environment handbook.* Palo Alto: Stanford University Press.
LeGalley, D. P., and Rosen, A. (eds.) (1964) *Space physics.* New York: Wiley.
McCormack, B. M. (ed.) (1967) *Aurora and airglow: NATO Advanced Study Institute Proceedings.* New York: Reinhold.
Montgomery, D. J. X. (1949) *Cosmic ray physics.* Princeton: Princeton University Press.
Morrison, P. (1961) *Handbuch der Physik.* (Vol. LI.) Berlin: Springer.
Piddington, J. H. (1961) *Radio astronomy.* New York: Harper.
Richtmyer, F. K., Kennard, E. H. and Lauritsen, T. (1955) *Introduction to modern physics.* New York: McGraw-Hill.
Rossi, B. (1952) *High energy particles.* Englewood Cliffs, N.J.: Prentice-Hall.

Planetary Atmospheres and Structures

Bullen, K. E. (1953) *An introduction to the theory of seismology.* Cambridge, England: Cambridge University Press.
Eisberg, R. M. (1961) *Fundamentals of modern physics.* New York: Wiley.
Kuiper, G. P. (ed.) (1962) *The moon, meteorites and comets.* Chicago: University of Chicago Press.
Ratcliffe, J. A. (1960) *Physics of the upper atmosphere.* New York: Academic Press.

JOURNALS

Astronomy and Astrophysics

Astronomical Journal. New York: The American Institute of Physics.
Astronomy and Astrophysics. Berlin: Springer.
Astrophysical Journal (Parts I and II). Chicago: University of Chicago Press.
Monthly Notices. London: Royal Astronomical Society (Burlington House).
Publications of the Astronomical Society of the Pacific. San Francisco: California Academy of Science.
Sky and Telescope. Cambridge, Mass. Sky Publishing Corporation.
The Observatory. Herstmoneux Castle, Sussex, Great Britain: Royal Greenwich Observatory.

Fields and Particles

Journal of Geophysical Research. Washington, D.C.: Space Physics Section, American Geophysical Union.

Space Science Reviews. Dordrecht, Netherlands: D. Reidel Publishing Company.

Planetary Atmospheres and Structures

Journal of Geophysical Research. Washington, D.C.: Planetary Physics and Oceans and Atmospheres Sections, American Geophysical Union.

Icarus. New York: Academic Press.

Planetary and Space Science. Oxford, Great Britain: Pergamon Press.

GENERAL

The journals listed below do not emphasize research in any special area of space science.

Nature. London: Macmillan (Journals) Ltd.

Physical Review Letters. New York: American Institute of Physics.

Physics Today. New York: American Institute of Physics.

Science. Washington, D.C.: American Association for the Advancement of Science.

Index

Page numbers in boldface type indicate the most important entries; those in parentheses refer to illustrations.